Handbook of Microwave Technology

Handbook of Microwave Technology

Edited by **Hinata Kenshin**

NY Research Press

New York

Published by NY Research Press,
23 West, 55th Street, Suite 816,
New York, NY 10019, USA
www.nyresearchpress.com

Handbook of Microwave Technology
Edited by Hinata Kenshin

International Standard Book Number: 978-1-63238-256-6 (Hardback)

Printed in China.

Contents

Preface

Microwave science and technologies is one of the most important subjects of science. It has its roots in the extensively used technique of telecommunications. Microwaves, as the name clearly reveals is a wave with frequencies between 300 MHz and 300 GHz, including both UHF and EHF. The prefix of 'micro' essentially means that these waves are smaller, as compared to the waves which are used in typical radio broadcasting.

Microwave is considered important because of its narrower bandwidth of beams than radio waves. Microwaves are used frequently and effectively in spacecraft communication, TV among others. Even tele-communications are transmitted to longer distances using microwave between distant ground stations and satellites. Microwaves are also used in microwave ovens for heating purposes and in radar technologies by military operations and flight control. The term microwave has a technical definition when the term is used with respect to electromagnetism and circuit theories. Microwave science has grown into a vast area of scientific research due to its various uses in areas such as communication radar, radio astronomy navigation, heating and power application, spectroscopy etc. The current emphasis of engineers and scientists working in this field is on developing tools, which can further create effective applications using microwave science.

However, one must also keep in mind that microwaves can have a detrimental impact too, if excessive exposure happens. Unnatural microwave radiation interferes with all the bodily processes, which can prove to be hazardous for health. Thus, research is being conducted in this area as well.

This book is a combined effort of authors from all over the world, who're experts in this field. Without their hard work, support and timely submissions, this book would not have been possible. I also wish to extend my gratefulness to the team at the publishing house, for their constant support at each and every step. Lastly, I'd like to my family for their constant important support all through my life.

Editor

Fracture Toughness of Vinyl Ester Composites Reinforced with Sawdust and Postcured in Microwaves

H. Ku,[1,2] M. Prajapati,[1] and M. Trada[1]

[1] *Faculty of Engineering and Surveying, University of Southern Queensland, Toowoomba, QLD 4350, Australia*
[2] *Centre of Excellence in Engineered Fibre Composites, University of Southern Queensland Toowoomba, QLD 4350, Australia*

Correspondence should be addressed to H. Ku, ku@usq.edu.au

Academic Editor: Qing Quan Liang

A commercial vinyl ester resin supplied by Hetron Chemical Pty. was reinforced with varying percentages by weight of sawdust. The sawdust particles were sieved into 3 different sizes, which were <300 μm, 300–425 μm, and 425–1180 μm, respectively, with a view to increase its fracture toughness for civil and structural applications. The sawdust used varied from 0% w/t to 15% w/t in step of 5% w/t. For higher w/t% of sawdust, the mixture would be too sticky to be mixed and cast. The cast composites were cured in ambient conditions and then postcured in microwave irradiation. They were then tested for fracture toughness using short bar tests. The values of fracture toughness of the composites increased with increasing particulate size, and this is due to the size distribution of the filler. It was found that the optimum amount of sawdust (425–1180 μm) was 15% w/t, with which the increase in fracture toughness was 126% as compared to neat resin and the reduction in cost was 15%. Furthermore, the optimum amount of sawdust (300–425 μm) was also 15% w/t, with which the increase in fracture toughness was 28.3% as compared to neat resin and the reduction in cost was again 15%.

1. Introduction

The most commonly used thermosets as composite matrices are unsaturated polyesters, epoxies, and vinyl esters. Unsaturated polyesters dominate the market, whereas epoxies are preferred in high-performance applications. Unsaturated polyesters offer an attractive combination of low price, reasonably good properties, and simple processing. However, basic unsaturated polyester formulations have drawbacks in terms of poor temperature and ultraviolet tolerance. Additives may significantly reduce these disadvantages to suit most applications. Where mechanical properties and temperature tolerance of unsaturated polyesters no longer suffice, the epoxy is often used due to their significant superiority in these respects. These improved properties come at a higher price, and epoxies are most commonly used in areas where cost tolerance is the highest [1]. Epoxy vinyl ester range of resins (vinyl ester resins) was developed in the 1960s [2]. Vinyl esters (VE), as they are usually called, are closely chemically related to both unsaturated polyesters and epoxies and in most respects represent a compromise between the two. They were developed in an attempt to combine the fast and simple crosslinking of unsaturated polyesters with the mechanical and thermal properties of epoxies [1]. The pure vinyl ester resin is brittle and one approach to increase the performance and minimize the cost of the resin is to reinforce it with fillers. As the structural products are cast to shape, the best option to reinforce the vinyl ester resin is to mix it with particulate fillers.

A research centre in the University of Southern Queensland (USQ) manufactures a lot of composite structures for civil engineering applications for local governments and industries at a very competitive cost. A lot of researches have been done in the centre in finding out the most suitable combination of resin and reinforcer; up to date, the reinforcers used include cenospheres (ceramic hollow spheres or SLG), glass powder, calcium carbonate powder, and aluminium powder. This research project is aimed at, investigating the fracture toughness of a commercial resole vinyl ester resin reinforced with sawdust with a view to its applications in civil and structural engineering. The sample is simply made by casting the mixture of the resin, initiator, and the sawdust into the moulds.

2. Materials

The vinyl ester resin used is Hetron 922 PAS in summer and Hetron 922 PAW in winter. The vinyl ester is dissolved in 50% by weight of styrene. In this study, Hetron 922 PAW was used. The resin hardener (methyl ethyl ketone peroxide, MEKP) ratio used in this study was 98% resin by volume and 2% hardener by volume [3].

The sawdust used was pine waste from sawmills. It has a variety of practical uses, including fuel and manufacturing of the particle board. Until the advent of refrigeration, it was often used in icehouses to keep ice frozen during summer. In terms of hazards, it is flammable when in contact with fire [4]. It was sieved with three different sizes: <300 μm, 300–425 μm, and 425–1180 μm. The sawdust particles reduce the weight of the composites, lowering its cost and improving its performance.

The sawdust content in the resin was varied from 5% w/t to 15% w/t. Above this percentage, it became very difficult to mix. The sawdust was mixed with the resin without any treatment. MEKP (methyl ethyl ketone peroxide) was the hardener used for curing. Before subjecting the samples for short bar tests, they were cured under ambient conditions for 24 hours and then postcured in microwaves. The MTS 810 universal tensile testing machine was used for the testing.

3. Fracture Toughness

Fracture toughness is the maximum stress that a material can withstand if it contains flaws of a certain size and geometry; in other words, it measures the ability of a material containing a flaw to withstand an applied load. A typical fracture toughness test may be performed by applying a tensile stress to a specimen prepared with a flaw of known geometry and size as depicted in Figure 1. The stress applied to the material is intensified at the flaw [5]. The stress intensity factor is

$$K = f\sigma\sqrt{\pi a}, \quad (1)$$

where f is a geometry factor for the specimen and flaw σ is the applied stress, and a is the flaw size.

If the specimen is assumed to have an "infinite" width then $f \cong 1.0$; for "semi-infinite" width, $f \cong 1.1$ [6, 7].

The critical stress intensity factor is defined as fracture toughness, K_c which is the K required for a crack to propagate and

$$K_c = f\sigma c\sqrt{\pi a}. \quad (2)$$

K_c is a property that measures a material's resistance to brittle fracture when a crack is present and its unit is MPa$\sqrt{m}$. The value K_c for this thick-specimen situation is known as the plane strain fracture toughness K_{Ic}; furthermore, it is also defined by Munz [8]:

$$K_{Ic} = f\sigma\sqrt{\pi a}. \quad (3)$$

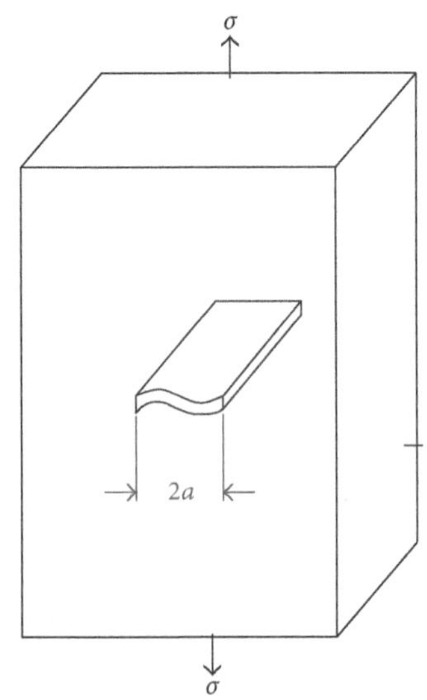

FIGURE 1: Schematic drawing of fracture toughness specimens with edge and internal flaws.

4. Short Bar Test and the Samples

Baker described the background, selection criteria, and specimen geometry options for short rod and short bar methods [9]. The background, selection criteria, and specimen geometry options for the samples were clearly explained, which made the manufacture of the samples easier. Figure 2 shows the short rod and short bar specimens with straight chevron slots. The load line is the line along which the opening load is applied in the mouth of the specimen. The specimen parameter, B, is the specimen diameter (for short rod) or breath (for short bar). The other parameter, W, is the length of the specimen. The relationship between the peak load (to fracture the sample), the breath, and the length of the specimen was given in (4). The equation for fracture toughness in a short bar test can be derived from basic fracture mechanics using the assumptions of linear elastic fracture mechanics (LEFM). The requirements for LEFM were used in this study [9]. The equation for the material plane strain critical stress intensity factor, K_{ICSR} [8] is

$$K_{ICSR} = \frac{(F_{max} Y_m^*)}{B\sqrt{W}}, \quad (4)$$

where F_{max} = Peak load; B is the breath of the sample; W, is the length of the sample; Y_m^* is the compliance calibration according to ASMT E-399-78 and $Y_m = 15.7025$, $\omega = W/H = 1.90$, $\alpha_0 = \alpha_0/W = 0.31$, $\alpha_1 = \alpha_1/W = 0.97$.

All parameters like a_0 a_1, W, and H are shown in Figures 2 and 3, respectively. Also, $B = 50$ (by design) and $F_{max} = 903N$ (average peak load of six samples of 10% w/t of sawdust (300–425 μm) reinforced vinyl ester composites).

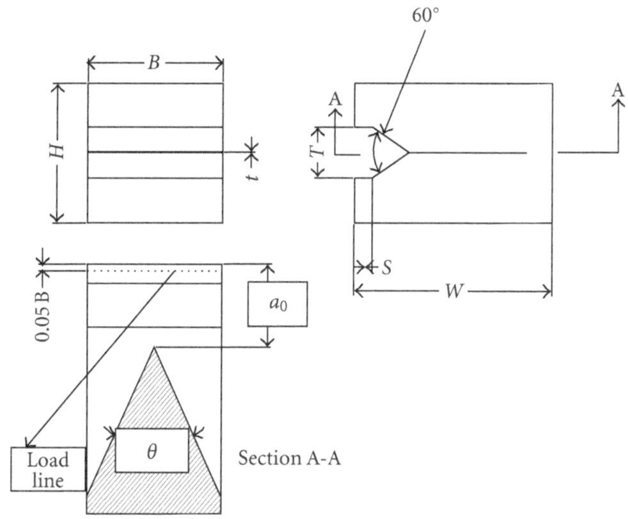

Symbol	Definition	Value	Tolerance
B	Breadth	B	
W	Length	$1.5B$	$\pm .010B$
H	Height	$.870B$	$\pm .005B$
a_0	Initial crack length	$.513B$	$\pm .005B$
θ	Slot angle	55.2	$\pm 1/2$
t	Slot thickness	See Table 3 (of Barker, 1981)	
S	Grip groove depth	$.130B$	$\pm .010B$
T	Grip groove width	$.313B$	$\pm .005B$
R	Radius of slot cut	See Figure 4 (of Barker, 1981)	$\pm 2.5B$

Figure 2: Short bar specimen with Straight chevron slots. The load line is the line along which the opening load is applied in the mouth of the specimen.

The fracture toughness for 10% w/t of sawdust (300–425 μm) was calculated as

$$K_{\text{ICSR}} = \frac{(F_{\max} Y_m{}^*)}{B\sqrt{W}} = \frac{(903x15.7025)}{50\sqrt{75}} = 32.75 \text{ MPa}\sqrt{m}. \quad (5)$$

Vinyl ester resin was reinforced with sawdust, and the percentages by weight of sawdust, were varied from 0 wt% to 15 wt% in step of 5 wt% in the cured vinyl ester composite, that is, VE/Sawdust (X%), where X is %w/t of the filler; the 20 wt% of sawdust was tried but it was too viscous for mixing. The resin was first mixed with the catalyst, after this, the sawdust was added to the mixture, and they were then mixed to give the uncured composite. As the raw materials of the composites were liquid and sawdust particles, the short bar specimens were cast to shape. Table 1 shows the mass in grams of resin, catalyst, and sawdust required, respectively, to make 1000 grams of uncured composite with 15 wt% of sawdust (<300 μm). The mould was made from PVC (polyvinyl chloride) sheets with six pieces of short bar specimens each mould. This is depicted in Figure 4. The slots were made by inserting plastic sheets of suitable thickness. After preliminary curing, the samples were taken out of the mould and postcured in a modified Sanyo microwave oven (Figure 5). The samples were cured at 480 W for 10 minutes to bring the temperature to 100°C and then cooled in the cavity to room temperature. One hundred °C was the highest temperature of heating when the samples were cured conventionally in an electric oven in air [10]. The temperature was measured using an infrared thermometer. These specimens were then subjected to short bar tests. Figure 6 shows one of the VE/Sawdust (X%) short bar specimens after test.

The number of samples for each percentage by weight of sawdust of particular size range is six. An MTS 810 Material Testing System was used for the test. The rate of extension was 1 mm per minute. The short bar tests involved an opening load being applied near the mouth of the specimen, causing a crack to initiate at the point of the chevron slot. Ideally, the opening load should be less than the load required to further advance the crack. A continually increasing load must be applied until the crack length reaches

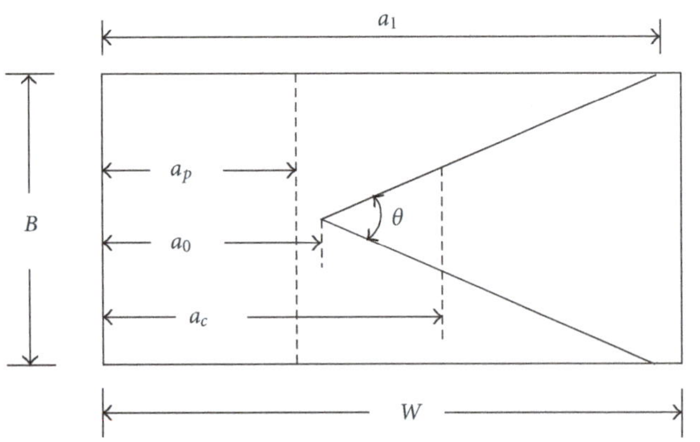

FIGURE 3: Cross-section dimensions of short bar specimen showing a_1.

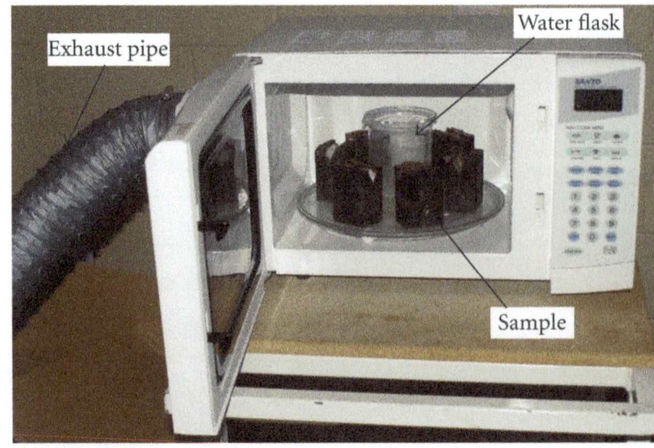

FIGURE 5: Modified SANYO microwave oven with samples.

FIGURE 4: The PVC short bar test mould.

FIGURE 6: The short bar specimens after testing.

the critical crack length, a_c. Beyond a_c, the load should decrease, as shown in Figure 7.

5. Results and Discussion

Figure 8 shows the fracture toughness of the composite specimens filled with varying percentages by weight of sawdust with diameters of <300 μm. It was found that the fracture toughness of neat resin was 27.15 MPa$\sqrt{m}$; the value of fracture toughness increased to 33.31 MPa$\sqrt{m}$ when the sawdust was 5% w/t, which was the highest fracture toughness value. The value then decreased again with higher particulate loading. The fracture toughness values for 10% w/t (26.82 MPa$\sqrt{m}$) and 15% w/t (27.32 MPa$\sqrt{m}$) of sawdust were more or less the same as that of neat resin (27.15 MPa$\sqrt{m}$). Table 2 depicts the fracture toughness of VE/Sawdust (<300 μm) with varying percentage by weight of filler with standard deviation in bracket. As the values of the standard deviations were low, it can be argued that the values of fracture toughness were reliable.

The value of fracture toughness of the same filler reinforced phenolic resin (Hexion Cellobond J2027L) was the lowest with the neat resin and was 8.43 MPa$\sqrt{m}$. Its value then increased slowly from 11.39 MPa$\sqrt{m}$ to 11.50 MPa$\sqrt{m}$ as the percentage by weight of sawdust increased from 5% w/t to 10% w/t. After this, it increased rapidly to 19.16 MPa$\sqrt{m}$ as the sawdust increased from 10% w/t to 20% w/t, which was 127.3% higher than that of neat resin [4].

The fracture toughness of neat vinyl ester resin (27.15 MPa$\sqrt{m}$) is expected to be much higher than its counterpart (8.43 MPa$\sqrt{m}$) because the former is a higher class resin with higher cost. However, the trends of the curves were not the same; the fracture toughness values of vinyl ester composites peaked at 5% w/t of sawdust and its fracture toughness was 22.7% higher than that of neat resin (Figure 8). For other percentages by weight of sawdust, the fracture toughness values were within 5 percent markers of that of the neat resin as depicted in Figure 8. On the other hand, the fracture toughness values of sawdust reinforced phenolic composites increased with increasing particulate loading. The fracture toughness of composite with 15% w/t of sawdust (<300 μm) was 127% higher than that of neat (phenolic) resin [4].

Figure 9 illustrates the fracture toughness of the specimens filled with varying percentage by weight of sawdust with diameters of 300–425 μm. The values and their standard deviations of the fracture toughness were also given in Table 2. As the values of the standard deviations were low, it can be argued that the values of fracture toughness were reliable. It was found that the values of fracture toughness of 5% w/t and 10% w/t by weight of sawdust were 31.14 MPa$\sqrt{m}$ and 32.75 MPa$\sqrt{m}$, respectively;

TABLE 1: Weight of materials required to make 1000 g of VE/Sawdust (15%).

Parameters	Materials	Resin (R)	Hardener (H)	R + H	Sawdust	Composite
Percentage by weight		40	1	—	—	—
Percentage by weight		—	—	17	3	—
Weight of materials in 1000 g of PF/sawdust (15%)		809.5 (g)	40.5 (g)	850 (g)	150 (g)	1000 (g)

TABLE 2: Fracture toughness of VE/sawdust with varying particle sizes and percentage by weight.

	w/t%			
	0%	5%	10%	15%
Fracture toughness (MPa $\sqrt{m}$), sawdust (<300 microns)	27.15 (1.53)[#]	33.31 (4.10)	26.82 (2.36)	27.32 (3.67)
Fracture toughness (MPa $\sqrt{m}$), sawdust (300–425 microns)	27.15 (1.53)	31.14 (2.75)	29.86 (3.99)	34.83 (3.96)
Fracture toughness (MPa $\sqrt{m}$), sawdust (425–1180 microns)	27.15 (1.53)	38.45 (4.12)	50.65 (3.30)	61.79 (1.90)

[#] Standard deviation.

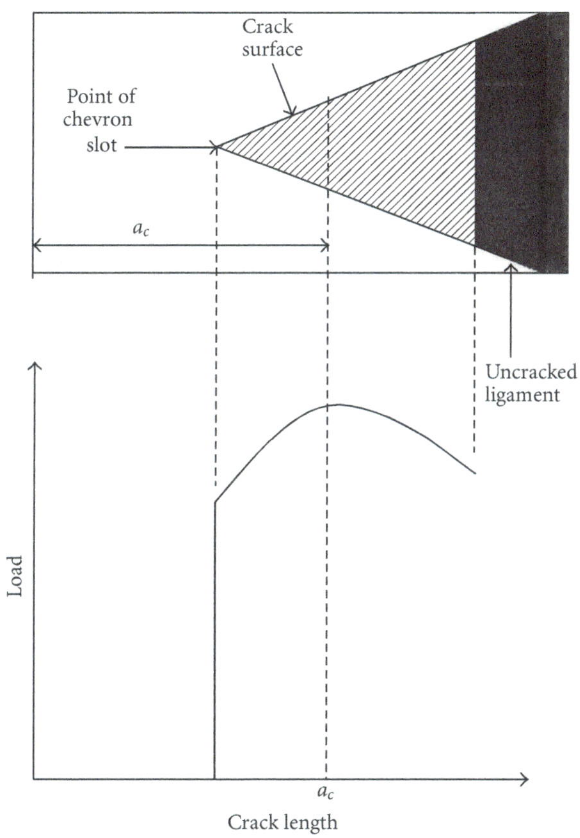

FIGURE 7: Variation of load versus crack length [8].

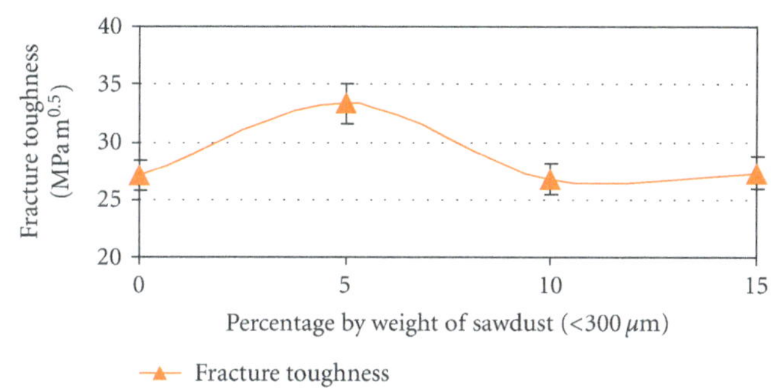

FIGURE 8: Fracture toughness of vinyl ester composites reinforced with sawdust with diameters of <300 μm.

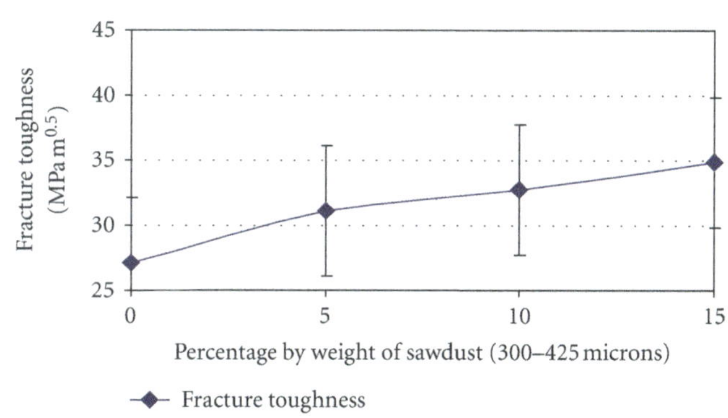

FIGURE 9: Fracture toughness of vinyl ester composites reinforced with sawdust with diameters of 300–425 μm.

both, values were within 5 percent markers of that of neat resin (27.15 MPa$\sqrt{m}$). The fracture toughness of 15% w/t of sawdust was 34.83 MPa$\sqrt{m}$ which was 28.3% higher than that of neat resin. The fracture toughness of the (300–425 μm) composites increased with increasing particulate loading.

On the other hand, the values of fracture toughness of the sawdust reinforced phenolic resin increased slowly to 14.46 MPa$\sqrt{m}$ when the sawdust was 10% w/t, and to 16.90 MPa$\sqrt{m}$ when the sawdust was 20% w/t, which was 100.5% higher than that of the neat resin. The trend of the curve was increasing with increasing particulate loading as shown in Figure 10. With sawdust particles of 300–425 μm in diameter, the situations were the same as in the case of sawdust with 300 μm in diameter; the fracture toughness of phenolic resin increased drastically with increasing reinforcement, while that of vinyl ester resin increased moderately with particulate loading rather than remained stable as in the previous case (Figure 8).

Figure 11 shows the fracture toughness of the specimens reinforced with varying percentage by weight of sawdust of

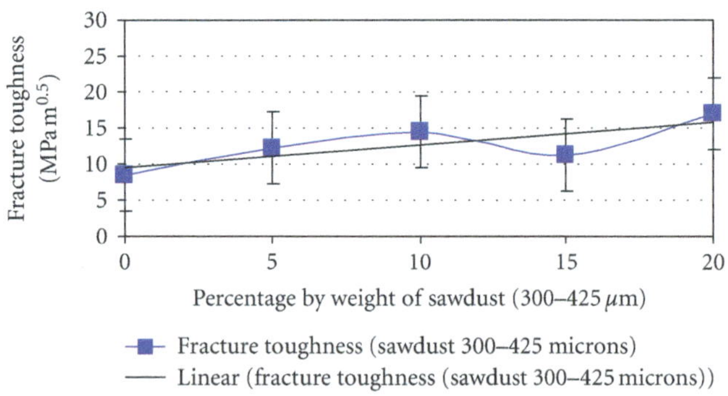

Figure 10: Fracture toughness of phenolic composites reinforced with sawdust with diameters of 300–425 μm.

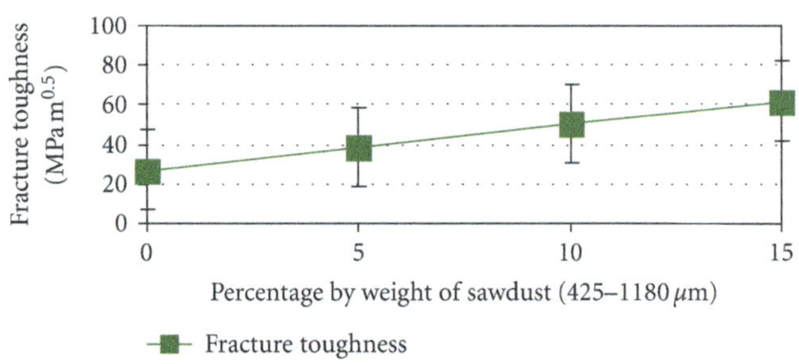

Figure 11: Fracture toughness of vinyl ester composites reinforced with sawdust with diameters of 425–1180 μm.

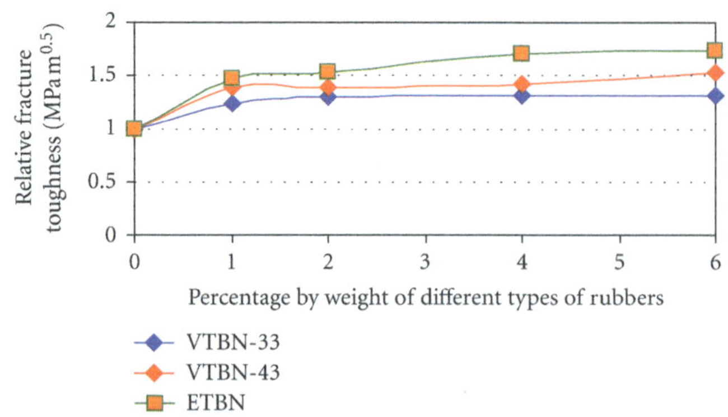

Figure 12: The effect of rubber addition on the fracture toughness of vinyl ester specimens, {adapted from [11]}.

diameters of 425–1180 μm. The values and their standard deviations of the fracture toughness were also given in Table 2. As the values of the standard deviations were low, it can be argued that the values of fracture toughness were reliable. The values of fracture toughness increased steadily from 38.45 MPa$\sqrt{m}$ at 5% w/t of reinforcer to 61.79 MPa$\sqrt{m}$ at 15% w/t of filler.

The fracture toughness of phenolic resin reinforced with sawdust of diameters of 425–1180 μm was 13.98 MPa$\sqrt{m}$ at 5% w/t of filler; the value then dropped to 11.03 MPa$\sqrt{m}$ at 10% w/t of sawdust loading; it then flattened around this value until the sawdust loading was 20% w/t. If trend lines were added, it can be found that the fracture toughness increased slightly with increasing particulate loading.

With reference to Figures 8, 9, and 11, the fracture toughness of the composites increased with increasing diameters of the sawdust; for example, at 15% w/t of sawdust, the values of fracture toughness of the composites increased from 27.32 MPa$\sqrt{m}$ (<300 μm) to 34.83 MPa$\sqrt{m}$ (300–425 μm) and to 61.79 MPa$\sqrt{m}$ (425–1180 μm) with increasing particulate size. From the above discussion, it can be argued that size distribution of the reinforcement plays an important part in increasing the fracture toughness of the composites. In general, the higher the size distribution of the reinforcement, the higher the fracture toughness values. Hameed et al. carried out a study on the effect of size distribution of the fibre on the flexural properties of metallic fibre reinforced concrete (MFRC) and found that the level of improvement in flexural properties of MFRC varied with the size distribution of the fibres; fibres with larger size distribution showed better efficiency in improving the flexural properties of MFRC [12]. Though the mechanical properties studied in the two cases were different, it can be argued that the higher the size distribution of the reinforcement, the higher the mechanical properties, flexural properties, and fracture toughness.

In another study, a model was developed by Kim to investigate the effects of size distribution of the fibre in short fibre reinforced discontinuous composite materials. The model was based on the theoretical development of conventional shear lag theory developed by Cox. He found that the effect of size distribution of the fibre is significant to composite strengthening through load transfer from the matrix to the fibre and the higher the size distribution of the fibre, the stronger the composites [13].

Dreerman et al. [11] toughened regular vinyl ester and endcapped vinyl ester resins with three types of rubbers: VTBN-33, VTBN-43, and ETBN. VTBN-33 is vinyl terminated rubber with acrylonitrile content of 17%; VTBN-43 is vinyl terminated rubber with acrylonitrile content of 16%; ETBN is epoxy terminated rubber, containing 50% w/t styrene. Figure 12 illustrates that the fracture toughness of regular vinyl ester specimens toughened by three types of rubber increased with rubber addition. The ETBN was found as the most effective toughening agent: 5% w/t caused about 70% increase in fracture toughness, whereas the addition of the same content of VTBN-33 or VTBN-43 resulted in 30% and 50% increases, respectively. Most of the fracture toughness improvement was made by the addition of 2% w/t rubber. The general trend of the addition of rubber to the endcapped vinyl ester was the same as the regular vinyl ester: fracture toughness increased with rubber loading; however, gradual increase was found in this case. The ETBN was again the most effective toughening agent, but it was not as effective as in the case of regular vinyl ester: an addition of 10% w/t caused just a 50% increase in fracture toughness [11]. The general trend of the fracture toughness of this study is the same as that of Dreerman et al.'s study, that is, fracture toughness increases with increasing particulate loading; this is particularly true with sawdust of larger diameters [11]. Salemane et al. [14] discovered that the

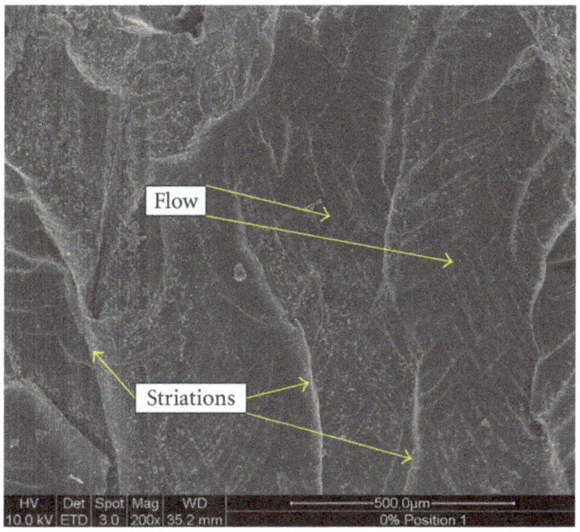

Figure 13: Fractured surface of neat vinyl ester resin, 200x.

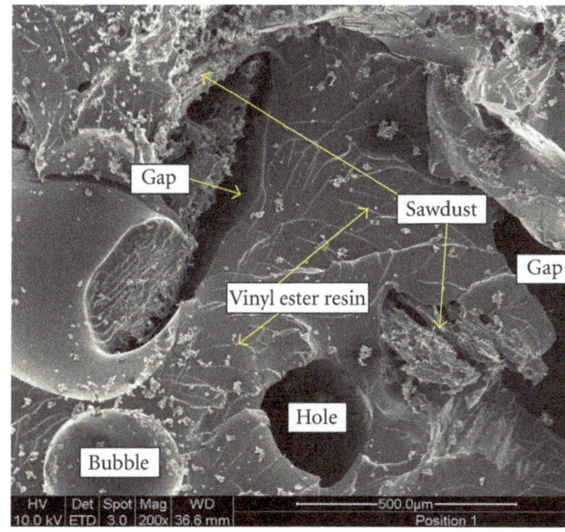

Figure 14: Fractured surface of 5 w/t% sawdust (300–425 μm) reinforced vinyl ester composite, 200x.

presence of MAPP (maleic anhydride grafted polypropylene) generally improved the tensile properties of the wood powder reinforced polypropylene composites. Larger wood powder particles gave rise to better tensile properties. This is again true in this study except that fracture toughness replaces tensile properties. However, some other researchers claimed that fine sawdust particles were difficult to disperse, and they agglomerated and behaved as large single particles. Aggregation of particles, especially finer particles, is another factor that can influence the final properties of the composites. Fine or small filler particles were found to improve the mechanical properties, for example, the stiffness of polymer composites more than the large ones. The only drawback to the use of finer particles was their tendency to agglomerate [15–17].

In Dreerman et al.'s study, the values of fracture toughness of neat regular and endcapped vinyl ester resins were 1 MPa$\sqrt{m}$ and 1.55 MPa$\sqrt{m}$, respectively, which were much lower than that (27.15 MPa$\sqrt{m}$) of the neat vinyl ester used in this study. The vinyl ester neat resin used by Dreerman et al. was Derakane 411-45 based on methacrylated diglycidyl ether of bisphenol A, containing 45 w/t% styrene [11]. Dreerman et al.'s work was 10 years ago, and it can be argued that improvement in vinyl ester manufacturing technology has been achieved in that period of time and the fracture toughness of current neat vinyl ester resin would be higher than that used 10 years ago.

5.1. Cost Analysis. The cost of 1 kg of vinyl ester used was \$10, and the sawdust was free or nearly free as it is disposed as waste in sawmill. The fracture toughness of the 15 w/t% sawdust (300–425 μm) reinforced vinyl ester composite was 34.83 MPa$\sqrt{m}$, while that of neat resin was 27.15 MPa$\sqrt{m}$. The increase in fracture toughness was 34.83 MPa$\sqrt{m}$ − 27.15 MPa$\sqrt{m}$/27.15 MPa$\sqrt{m}$ × 100% = 28.3%, and at the same time the reduction in cost was $(1 - ((0.85x\$10 + 0.15x\$0)/1x\$10)) \times 100\% = 15\%$. The fracture toughness of the 15 w/t% sawdust (425–1180 μm) reinforced vinyl ester composite was 61.79 MPa$\sqrt{m}$, while that of neat resin was 27.15 MPa$\sqrt{m}$. The increase in fracture toughness was (61.79 MPa$\sqrt{m}$ − 27.15 MPa$\sqrt{m}$)/27.15 MPa$\sqrt{m}$ × 100% = 126%, and at the same time the reduction in cost was $(1 - ((0.85x\$10 + 0.15x\$0)/1x\$10)) \times 100\% = 15\%$.

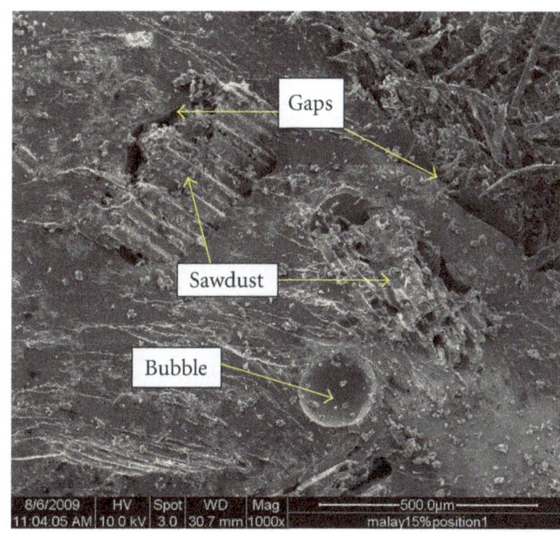

Figure 15: Fractured surface of 15% w/t sawdust (300–425 μm) reinforced vinyl ester composite, 1000x.

5.2. Fracture Morphology. Figure 13, magnified at 200 times, illustrates the fractured surface of neat vinyl ester resin, illustrating faint striations followed by a turbulent flow pattern of the fractured zone. The fractured surface is relatively smooth, typical of brittle materials, and it is similar to the fractured surface of neat vinyl ester by other studies [18, 19].

Figure 14 shows the fractured surface of 5% w/t sawdust (300–425 μm) reinforced vinyl ester composite, 200X. The fracture toughness of this composite was stronger than that of the neat resin because some of the stress was successfully transferred to the reinforcer. However, the fracture toughness of this composite was weaker than that of composite with 15 w/t% of sawdust (Figure 15) because gaps, bubble, and hole were found. The bubble was due to the evolution of gaseous styrene moieties at the degradation temperature of vinyl ester resin. Bubble and gaps were also found in 15% w/t of sawdust reinforced vinyl ester composite, but they were smaller as the magnification was 1000x [20]. With more sawdust and less resin in 15% w/t of sawdust composite, there would be less evolution of gaseous styrene moieties and hence less bubbles. The larger sawdust particles were likely to mingle better with each other and the resin and hence less gaps.

6. Conclusion

The values of fracture toughness of the composites increased with increasing particulate size, and this is due to the size distribution of the filler. It was found that the optimum amount of sawdust (425–1180 μm) was 15% w/t, with which the increase in fracture toughness was 126% as compared to neat resin and the reduction in cost was 15%. Furthermore, the optimum amount of sawdust (300–425 μm) was also 15% w/t, with which the increase in fracture toughness was 28.3% as compared to neat resin and the reduction in cost was again 15%.

It can be argued that if the composites had not been cured in microwaves but conventionally, the values of the fracture toughness were likely to be lower as in other cases of microwave curing of resins [10, 21]. The fracture toughness of sawdust reinforced vinyl ester resin composites cured in microwaves had been measured and it was found that, in general, the larger the particle size of the reinforcer, and the higher the particulate loadings, the higher the values of fracture toughness. This is due to the fact that sawdust particles with larger size distribution had a better efficiency in improving the fracture toughness of the composites.

References

[1] L. M. Baker, "Development of the short rod method of fracture toughness measurement," in *Proceedings of the Conference on Wear and Fracture Prevention*, pp. 163–180, ASM, Metals Park, Ohio, USA, May 1980.

[2] G. Pritchard, Ed., *Reinforced Plastics Durability*, Woodhead Publishing, UK, 1999.

[3] S. W. Davey, T. Heldt, G. Van Erp, and S. R. Ayers, "Vinylester/cenosphere composite materials for civil and structural engineering," *FRP International*, vol. 2, no. 3, pp. 2–5, 2005.

[4] H. Ku, M. Prajapati, and F. Cardona, "Thermal properties of sawdust reinforced vinyl ester composites post-cured in microwaves: a pilot study," *Composites Part B*, vol. 42, no. 4, pp. 898–906, 2011.

[5] H. S. Ku, S. C. Fok, and E. Siores, "Contrasts on fracture toughness and flexural strength of varying percentages of SLG-reinforced phenolic composites," *Journal of Composite Materials*, vol. 43, no. 8, pp. 885–895, 2009.

[6] W. D. Callister, *Materials Science and Engineering: An Introduction*, John Wiley and Sons, New York, NY, USA, 7th edition, 2006.

[7] H. Wang, *Engineering Materials—Study Book 1*, University of Southern Queensland, 2009.

[8] D. Munz, "Determination of fracture toughness of high strength aluminum alloys with chevron notched short rod and short bar specimens," *Engineering Fracture Mechanics*, vol. 15, no. 1-2, pp. 231–236, 1981.

[9] L. M. Barker, *Mechanics Applied to Brittle Materials*, STP 678, ASTM, 1979.

[10] H. Ku, M. Trada, F. Cardona, D. Rogers, and W. Jacobsen, "Tensile tests of phenol formaldehyde SLG reinforced composites post-cured in microwaves: preliminary results," *Journal of Composite Materials*, vol. 42, no. 26, pp. 2783–2795, 2008.

[11] E. Dreerman, M. Narkis, A. Siegmann, R. Joseph, H. Dodiuk, and A. T. Dibenedetto, "Mechanical behavior and structure of rubber modified vinyl ester resins," *Journal of Applied Polymer Science*, vol. 72, no. 5, pp. 647–657, 1999.

[12] R. Hameed, A. Turatsinze, F. Duprat, and A. Sellier, "Metallic fiber reinforced concrete: effect of fiber aspect ratio on the flexural properties," *Journal of Engineering and Applied Sciences*, vol. 4, no. 5, pp. 67–72, 2009.

[13] H. G. Kim, "Effects of fiber aspect ratio evaluated by elastic analysis in discontinuous composites," *Journal of Mechanical Science and Technology*, vol. 22, no. 3, pp. 411–419, 2008.

[14] M. G. Salemane and A. S. Luyt, "Thermal and mechanical properties of polypropylene-wood powder composites," *Journal of Applied Polymer Science*, vol. 100, no. 5, pp. 4173–4180, 2006.

[15] B. Tang, "Fibre reinforced polymer composites applications in USA," in *Proceedings of the 1st Korea USA Road Workshop*, FHWA, January 1997.

[16] K. Oksman and C. Clemons, "Mechanical properties and morphology of impact modified polypropylene-wood flour composites," *Journal of Applied Polymer Science*, vol. 67, no. 9, pp. 1503–1513, 1998.

[17] K. L. Pickering, A. Abdalla, C. Ji, A. G. McDonald, and R. A. Franich, "The effect of silane coupling agents on radiata pine fibre for use in thermoplastic matrix composites," *Composites Part A*, vol. 34, no. 10, pp. 915–926, 2003.

[18] D. Ray, D. Bhattacharya, A. K. Mohanty, L. T. Drzal, and M. Mishra, "Static and dynamic mechanical properties of vinylester resin matrix composites filled with fly ash," *Macromolecular Materials and Engineering*, vol. 291, no. 7, pp. 784–792, 2006.

[19] H. Ku, M. Jamal-Eddine, and M. Trada, "Fracture toughness of calcium carbonate powder reinforced vinyl ester composites: pilot study," *Journal of Reinforced Plastics and Composites*. In press.

[20] F. Cardona, D. Rogers, S. Davey, and G. Van Erp, "Investigation of the effect of styrene content on the ultimate curing of vinylester resins by TGA-FTIR," *Journal of Composite Materials*, vol. 41, no. 2, pp. 137–152, 2007.

[21] F. Cardona, H. Ku, N. Pattarachaiyakoop, D. Rogers, and M. Trada, "Fracture toughness of phenol formaldehyde composites post-cured in microwaves," *Journal of Electromagnetic Waves and Applications*, vol. 21, no. 14, pp. 2137–2146, 2007.

2

A Dual-Band SiGe HBT Frequency-Tunable and Phase-Shifting Differential Amplifier Employing Varactor-Loaded, Stacked LC Resonators

Kazuyoshi Sakamoto and Yasushi Itoh

Graduate School of Electrical and Information Engineering, Shonan Institute of Technology, 1-1-25 Tsujido-Nishikaigan, Kanagawa, Fujisawa 251-8511, Japan

Correspondence should be addressed to Yasushi Itoh, itoh@elec.shonan-it.ac.jp

Academic Editor: Juan Carlos Bohórquez Reyes

A dual-band SiGe HBT frequency-tunable and phase-shifting differential amplifier has been developed for the future active phased array antennas with a multiband, multibeam, and multitarget tracking operation. The amplifier uses varactor-loaded, stacked LC resonators in the design of the output circuit in order to provide frequency-tunable and phase-shifting capabilities for dual frequencies. By utilizing the varactor-loaded LC resonator, which has a variable resonant frequency and a large insertion phase variation, frequency-tunable and phase-shifting performances become available. Moreover, by using the stacked configuration, the frequency and insertion phase can be varied independently for dual frequencies. A dual-band SiGe HBT differential amplifier has achieved a lower-frequency tuning range of 0.56 to 0.7 GHz for a higher fixed frequency of 0.97 GHz as well as a higher-frequency tuning range of 0.92 to 1.01 GHz for a lower fixed frequency of 0.63 GHz. A lower-frequency phase variation of 99° and a higher-frequency phase variation of 90.3° have been accomplished at 0.63 and 0.97 GHz, respectively. This is the first report on the dual-band differential amplifier with frequency-tunable and phase-shifting capabilities.

1. Introduction

Recent and future wireless systems require a wide range of data rates over several frequency bands, and thus adaptive and reconfigurable transceivers become necessary to support high performance and flexibility [1]. Meanwhile, microwave and millimeter-wave sensors and radar systems require multiband, multibeam scanning phased array antennas as well as a multi-target tracking capability [2]. To address these requirements, multifunction capabilities are strongly required. Especially multiband amplification, frequency tunability, gain and phase control are crucial for realizing adaptive and reconfigurable phased array systems. Most of the traditional multiband amplifiers, however, provide a single function [3]. Meanwhile, the authors have presented multiband amplifiers with frequency-tunable as well as gain control capabilities for bandpass and bandstop types [4, 5]. To enhance the operational capability, a novel design approach is proposed in this paper for the multiband low-noise differential amplifier with frequency-tunable and phase-shifting capabilities. In order to realize both frequency-tunable and phase-shifting performances, varactor-loaded LC resonators are used in the design of the output circuit of the differential amplifier. LC resonators are widely used in the reflection type phase shifter because they provide a full 360° phase shift [6–8]. With the use of the parallel LC resonator in the output shunt circuit, frequency-tunable and phase-shifting performances with high gain can be achieved at around resonant frequencies. In addition, by cascading the parallel LC resonators in a stacked form, frequency and insertion phases can be varied independently for multiple frequencies. As a multiband phase shifter, the resonator-switching or diplexer method has been proposed for the loaded-line type [9], reflection type [10], and LPF/HPF type [11]. These phase shifters, however, produce additional losses due to the switch or diplexer.

Since the stacked LC resonator does not require additional switches or diplexers, a low insertion loss and a large phase variation can be expected. To make sure the usefulness of the novel design approach proposed in this paper, the dual-band SiGe HBT differential amplifier with frequency-tunable and phase-shifting capabilities has been actually designed, fabricated, and tested.

2. Circuit Design

A schematic diagram of the dual-band frequency-tunable and phase-shifting differential amplifier is shown in Figure 1. It employs varactor-loaded, stacked LC resonators in the design of the output circuit to achieve frequency-tunable and phase-shifting capabilities for dual frequencies. L_1, L_2, C_1, and C_2 are inductors and capacitors consisting of LC resonators. C_{V1} and C_{V2} are a variable capacitance of the varactor diode which is controlled by V_{C1} and V_{C2}. C_C is a decoupling capacitor having a large capacitance value. Resonant frequencies f_1 and f_2 are given as follows:

$$f_1 = \frac{1}{2\pi\sqrt{L_1(C_1 + C_{V1})}}, \tag{1}$$

$$f_2 = \frac{1}{2\pi\sqrt{L_2(C_2 + C_{V2})}}. \tag{2}$$

When two parallel LC resonators are cascaded in a stacked form, the following bandstop frequency f_3 is generated. Consider

$$f_3 = \frac{1}{2\pi}\sqrt{\frac{1/L_1 + 1/L_2}{C_1 + C_{V1} + C_2 + C_{V2}}}. \tag{3}$$

Now $f_1 < f_2$ ($L_1C_1 > L_2C_2$) then f_3 is necessarily sandwiched between f_1 and f_2 as

$$f_1 < f_3 < f_2. \tag{4}$$

C_{V1} and C_{V2} are carefully designed in accordance with the target frequency range and phase-shifting amount. In order to realize C_{V1} and C_{V2}, several varactor diodes with a fixed capacitance ratio are connected in series or parallel.

A locus of the reflection coefficient Γ_L of the varactor-loaded, stacked LC resonator is graphically shown in Figure 2. As the frequency increases from zero to f_1, Γ_L moves from the short to the open-circuited point **P**. Since the LC resonator is actually lossy, the point **P** has finite impedance. Then as the frequency increases up to f_3, Γ_L moves to the series resonant point **q** where the impedance is not ideally short due to the lossy circuit. As the frequency increases from f_3 to f_2, Γ_L comes back to the point **P** again. Finally Γ_L goes to short. At around the resonant frequencies f_1 and f_2, the insertion phase drastically varies from +180° to −180°. Therefore, with the precise control of V_{C1} and V_{C2}, the insertion phase can be tuned. In addition, at the resonant frequencies f_1 and f_2, a peak gain of the amplifier can be obtained. The mechanism is depicted graphically in Figure 3 by using a frequency response of the amplifier.

It is clearly demonstrated that the frequency can be varied with the coarse control of V_{C1} and V_{C2} and that the insertion

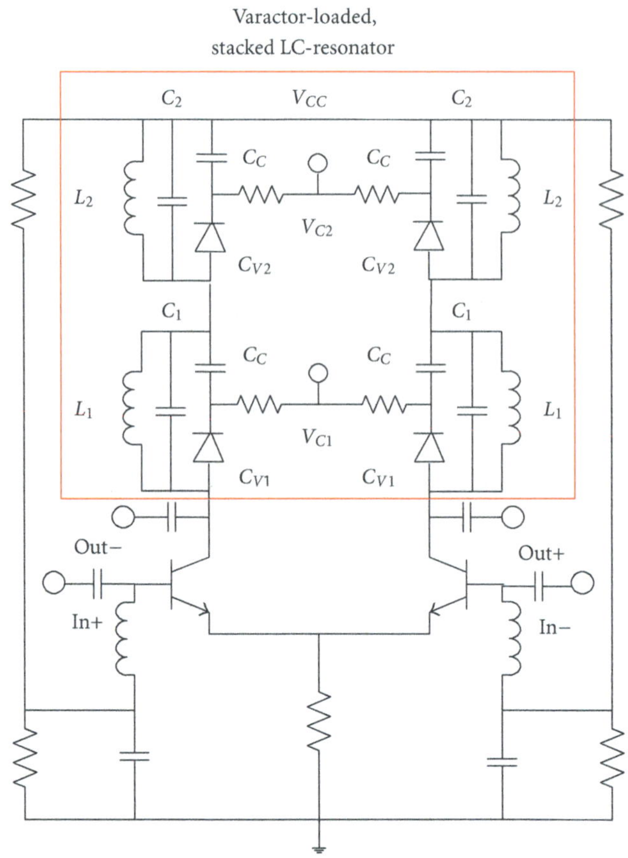

FIGURE 1: Schematic diagram of the dual-band frequency-tunable and phase-shifting differential amplifier.

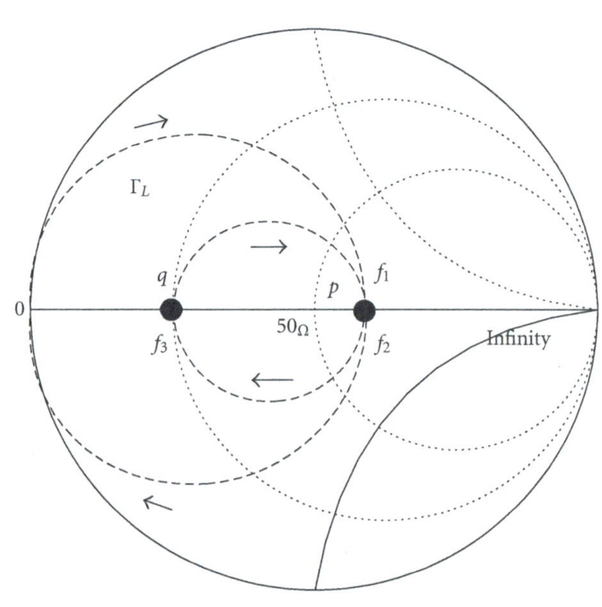

FIGURE 2: Locus of the reflection coefficient Γ_L of the varactor-loaded, stacked LC resonator.

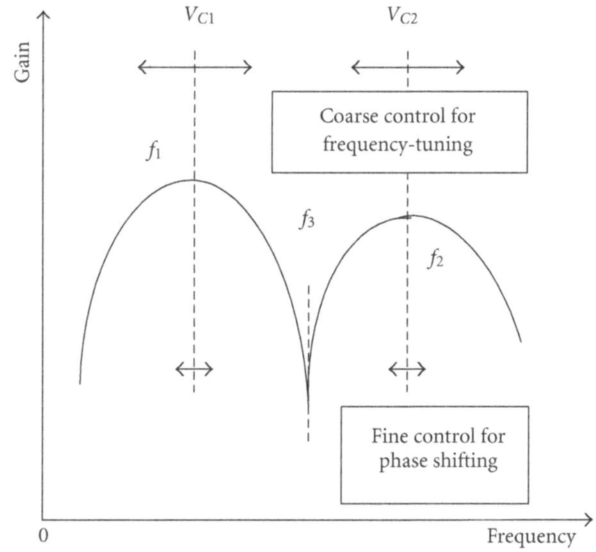

FIGURE 3: Frequency response of the dual-band frequency-tunable and phase-shifting differential amplifier.

phase can be also varied with the fine control of V_{C1} and V_{C2}. f_1, f_2, and f_3 are calculated by using from (1) to (3) for the circuit element values listed in Table 1. The circuit element values were chosen to meet the condition of (4). When C_{V1} varies from 6 to 15 pF, C_{V2} has a fixed value of 6 pF. Meanwhile, when C_{V2} varies from 6 to 15 pF, C_{V1} has a fixed value of 15 pF. The calculated results are shown in Figures 4 and 5, respectively. It is noted that f_1 and f_2 can be tuned independently and that f_3 is necessarily sandwiched between f_1 and f_2.

TABLE 1: Circuit element values of Figure 1.

Element	Value
L_1 [nH]	1
C_1 [pF]	10
C_{V1} [pF]	6–15
L_2 [nH]	1
C_2 [pF]	3
C_{V2} [pF]	6–15
C_C [pF]	1000

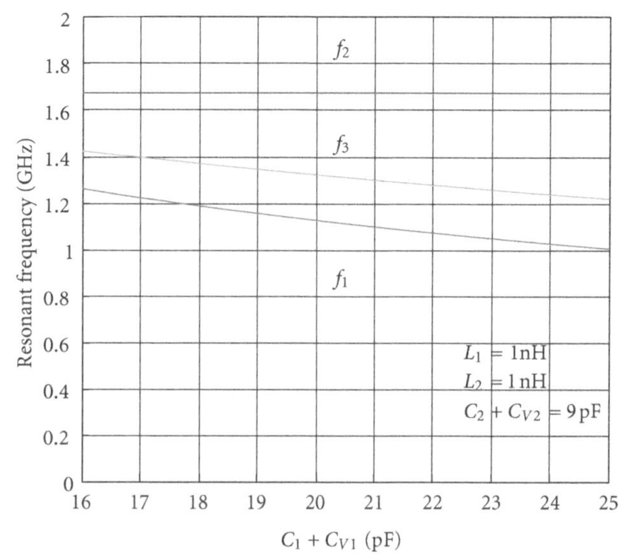

FIGURE 4: Calculated f_1, f_2, and f_3 (C_{V1} varies from 6 to 15 pF. C_{V2} has a fixed value of 6 pF).

3. Circuit Fabrication

A photograph of the dual-band frequency-tunable and phase-shifting differential amplifier is shown in Figure 6.

The differential amplifier was fabricated on the FR-4 substrate with a dielectric constant of 4.5. 0.35 μm SiGe HBT with an f_t of 25 GHz (Toshiba MT4S102T), Si varactor diode with a capacitance ratio of 2.5 : 1 (Toshiba 1SV279), 1005-type chip resistors, and inductors and capacitors are mounted on the substrate by soldering. The circuit size is $16 \times 16 \times 1.2\,\text{mm}^3$.

4. Circuit Performance

4.1. Frequency-Tunable Performances. Measured gain, input and output return losses for a fixed $V_{C1}(C_{V1})$ and a variable $V_{C2}(C_{V2})$ are shown in Figures 7, 8, and 9, respectively. As a coarse tuning, $V_{C2}(C_{V2})$ was varied from 6 to 18 V with 1 V step since the anode voltage is equal to V_{CC} of 6 V. The maximal reverse voltage is 15 V. f_2 moves from 0.73 to 1.31 GHz for a fixed f_1 of 0.56 GHz. The gain varied from 0 to 7.2 dB for f_2 and keeps a constant value of 8.9 dB for f_1. V_{CC} and V_{C1} were 6 V. Since the LC resonators are used in the output circuit, the output return loss is varied with the resonant frequency, but the input return loss keeps constant. Since the amplifier employed a reflective match approach for low-noise figure, the input and output return losses were less than 10 dB, which can be greatly improved by employing the lossy match approach using resistors [4]. In [4], the input and output return losses can be improved up to greater than 10 dB across the same frequency band by using the lossy match approach. This approach, however, provides a low gain and a poor noise figure. Thus, the reflective match approach was chosen. Meanwhile, the measured gain, input and output return losses for a variable $V_{C1}(C_{V1})$ and a fixed $V_{C2}(C_{V2})$ are shown in Figures 10, 11, and 12, respectively. In a similar way, $V_{C1}(C_{V1})$ was varied from 6 to 18 V with 1 V step. f_1 moves from 0.57 to 0.97 GHz and f_2 from 1.31 to 1.34 GHz. The gain varied from 8.7 to 13.1 dB for f_1 and 1.3 to 6.5 dB for f_2. V_{CC} and V_{C2} were 6 V. If the varactor-loaded LC resonator shows an ideally open-circuited performance, f_1 and f_2 can be controlled independently. But actually the gain and frequency shift at f_2 was affected to a certain extent by controlling $V_{C1}(C_{V1})$. This is because the parasitic resistance and low-Q factor of the varactor diode (C_{V1}) have made a serious effect on the gain and frequency at f_2. Therefore, $V_{C2}(C_{V2})$ in addition to $V_{C1}(C_{V1})$ must be adjusted simultaneously for keeping f_2 constant. Moreover,

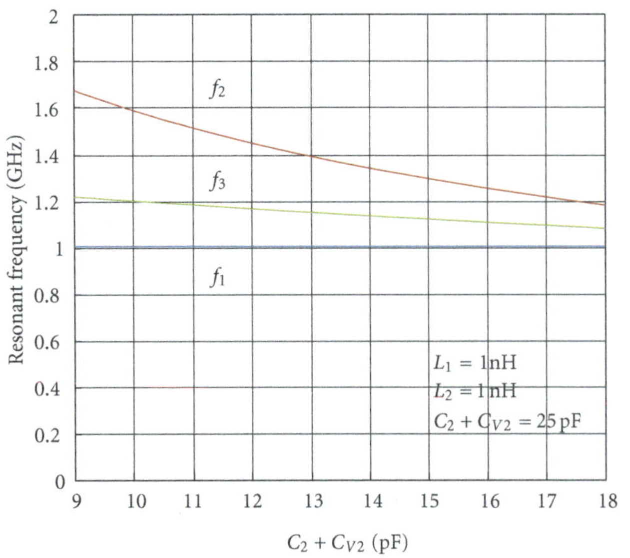

FIGURE 5: Calculated f_1, f_2, and f_3 (C_{V2} varies from 6 to 15 pF. C_{V1} has a fixed value of 15 pF).

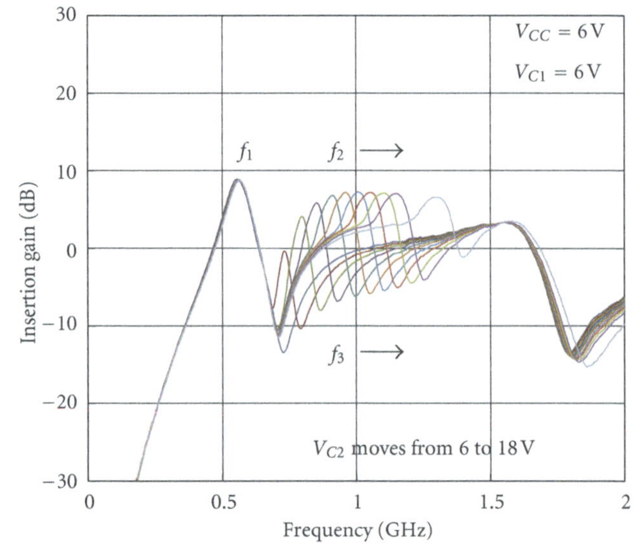

FIGURE 7: Measured gain for a fixed $V_{C1}(C_{V1})$ and a variable $V_{C2}(C_{V2})$.

FIGURE 6: Photograph of the dual-band frequency-tunable and phase-shifting differential amplifier: 16 × 16 × 1.2 mm^3.

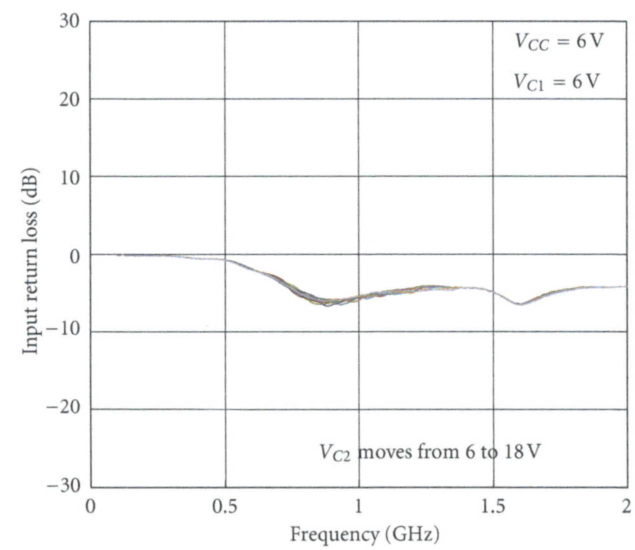

FIGURE 8: Measured input return loss for a fixed $V_{C1}(C_{V1})$ and a variable $V_{C2}(C_{V2})$.

as f_1 comes close to f_2, the bandwidth at f_1 becomes wider, and the gain decreases in Figure 10. This is considered to be the same reason of the parasitic resistance and low-Q factor of the varactor diode. In Figures 7 and 10, the 3rd gain peaking can be seen around 1.6 GHz. This is mainly due to the fact that the LC resonator in Figure 1 shows a dual-band resonation due to the lead inductance of the varactor diode.

4.2. Phase-Shifting Performances. Measured phase variations for a fixed $V_{C1}(C_{V1})$ and a variable $V_{C2}(C_{V2})$ are shown in Figure 13. As a fine tuning, $V_{C2}(C_{V2})$ was varied from 10 to 12 V with 0.2 V step. f_2 moves from 0.92 to 1.01 GHz for a fixed f_1 of 0.63 GHz. The phase variations at 0.97 GHz and 0.63 GHz were 90.3° and 2.5°, respectively. V_{CC} and V_{C1} were 6 V and 7 V, respectively. The gain variation, however, was 7 dB at 0.97 GHz. The gain variation can be improved by using a lossy match approach, which in turn degrades a phase-shifting amount; that is, there exists a design tradeoff between phase and gain variations. As one approach to address this problem, an additional variable gain amplifier (VGA) with a constant phase is utilized to compensate for the gain variation. Meanwhile, the measured phase variations for a fixed $V_{C2}(C_{V2})$ and a variable $V_{C1}(C_{V1})$ are shown in Figure 14. In a similar way, $V_{C1}(C_{V1})$ was varied from 6 to 8 V with 0.2 V step. f_1 moves from 0.56 to 0.7 GHz for a fixed

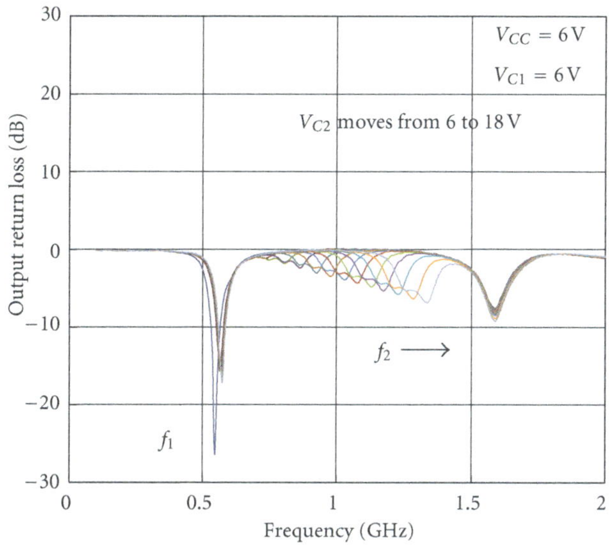

FIGURE 9: Measured output return loss for a fixed $V_{C1}(C_{V1})$ and a variable $V_{C2}(C_{V2})$.

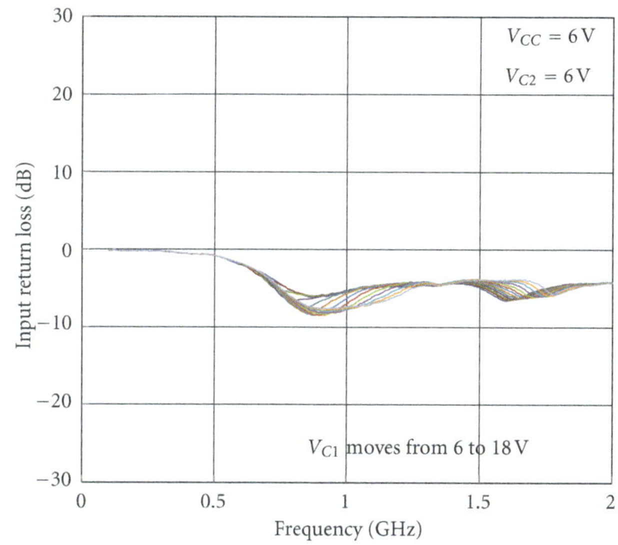

FIGURE 11: Measured input return loss for a fixed $V_{C2}(C_{V2})$ and a variable $V_{C1}(C_{V1})$.

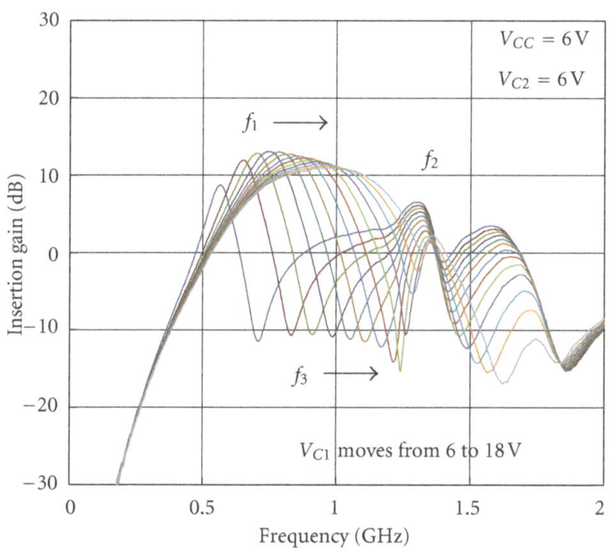

FIGURE 10: Measured gain for a fixed $V_{C2}(C_{V2})$ and a variable $V_{C1}(C_{V1})$.

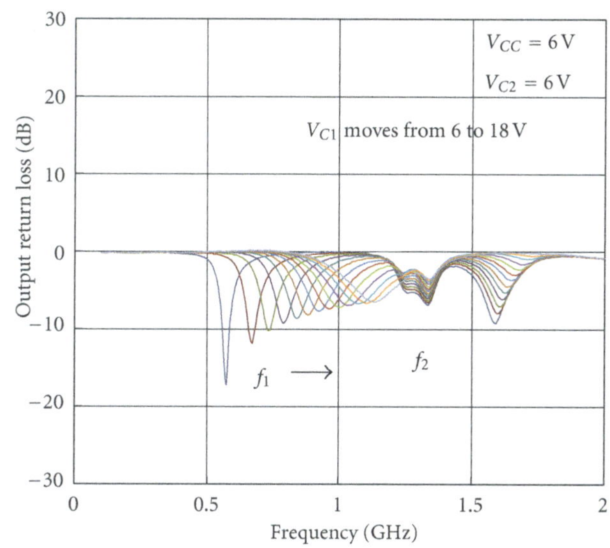

FIGURE 12: Measured output return loss for a fixed $V_{C2}(C_{V2})$ and a variable $V_{C1}(C_{V1})$.

f_2 of 0.97 GHz. The phase variation at 0.63 GHz was 99°. The phase variation at 0.97 GHz, however, was less than 2.1°. V_{CC} and V_{C2} were 6 V and 11 V, respectively.

4.3. Noise Figure Performances. Noise figure performances of the dual-band frequency-tunable and phase-shifting differential amplifier have been measured. The measured noise figures are plotted in Figure 15 for a fixed $V_{C1}(C_{V1})$ and a variable $V_{C2}(C_{V2})$ and in Figure 16 for a variable $V_{C1}(C_{V1})$ and a fixed $V_{C2}(C_{V2})$. V_{CC}, V_{C1}, and V_{C2} were 6 V, 7 V and 9–13 V in Figure 15. Meanwhile, V_{CC}, V_{C1}, and V_{C2} were 6 V, 5–9 V, and 11 V in Figure 16. The measured noise figures were less than 4.1 dB at 0.63 GHz and less than 5.3 dB at 0.97 GHz in Figure 15. Meanwhile, the measured noise figures were better than 4.4 dB at 0.63 GHz and better than 4.3 dB at 0.97 GHz in Figure 16. As compared with Figures 7 and 10, a narrow variable range was chosen for V_{C1} and V_{C2} since it becomes difficult to show all noise figure data at the same time and make clear the difference. Overall the noise figure performances are not seriously affected by the frequency tuning.

4.4. IIP$_3$. IIP$_3$ performances were measured with two tones of a 10 MHz separation for a fixed V_{C1} (6 V) and a variable V_{C2} (6–18 V) as well as a variable V_{C1} (6–18 V) and a fixed V_{C2} (6 V). V_{CC} was 6 V. The amplifier has shown the minimum IIP$_3$ of −8.7 dBm for a fixed V_{C1} (6 V) and a variable V_{C2} (6–18 V) as well as −6.5 dBm for a variable V_{C1}

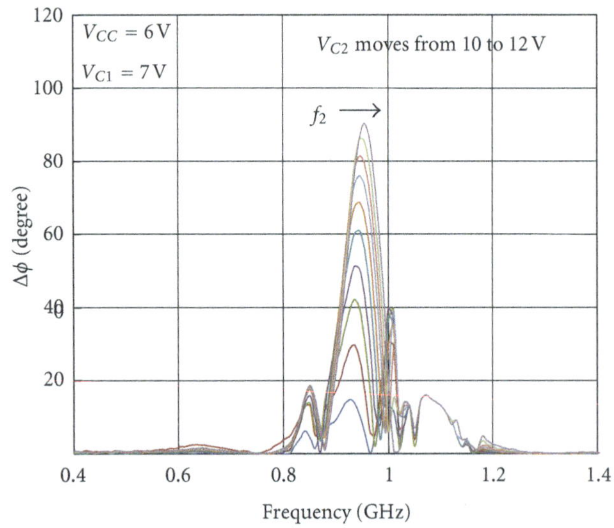

FIGURE 13: Measured phase variations for a fixed $V_{C1}(C_{V1})$ and a variable $V_{C2}(C_{V2})$.

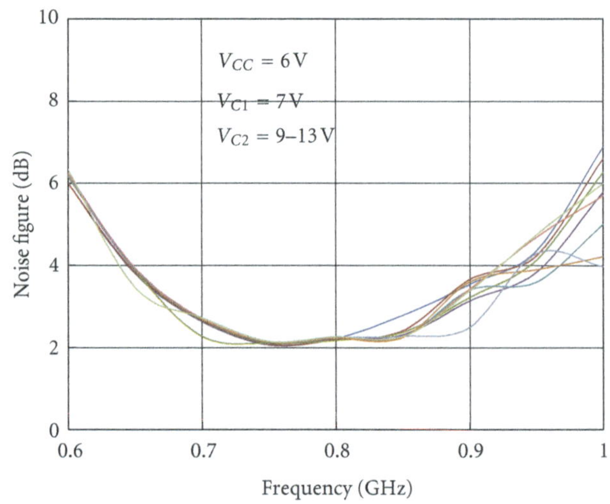

FIGURE 15: Measured noise figure performances for a fixed $V_{C1}(C_{V1})$ and a variable $V_{C2}(C_{V2})$.

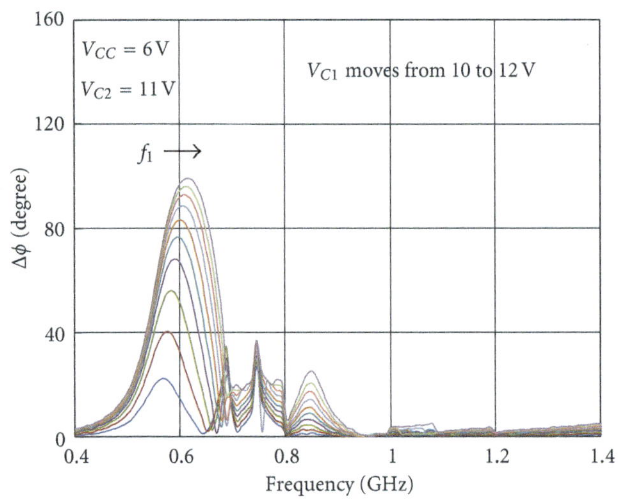

FIGURE 14: Measured phase variations for a fixed $V_{C2}(C_{V2})$ and a variable $V_{C1}(C_{V1})$.

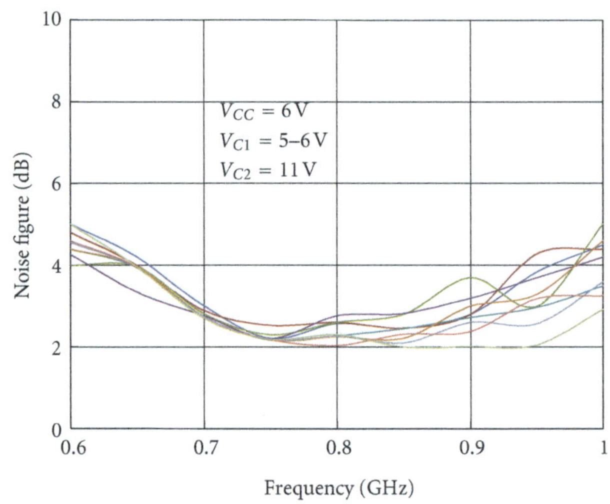

FIGURE 16: Measured noise figure performances for a fixed $V_{C2}(C_{V2})$ and a variable $V_{C1}(C_{V1})$.

(6–18 V) and a fixed V_{C2} (6 V). A total collector current I_C appeared in all measurements and ranged from 9 to 12 mA.

5. Conclusions

The dual-band SiGe HBT frequency-tunable and phase-shifting differential amplifier has been presented for use in the future active phased array antennas with a multiband, multibeam, and multi-target tracking operation. To realize multifunction capabilities including frequency-tunable and phase-shifting capabilities, the novel design approach to incorporate varactor-loaded, stacked LC resonators with different resonant frequencies into the circuit design has been proposed. Then to make sure the usefulness of the design approach, the dual-band frequency-tunable and phase-shifting differential amplifier using SiGe HBT and Si varactor diode have been actually designed, fabricated, and tested. It is confirmed from the measured results that the design approach proposed in this paper would be one of the candidates for the multifunctional amplifier for use in the current and future wireless, aerospace, microwave, and millimeter-wave sensors and radar systems.

References

[1] A. R. Rofougaran, M. Rofougaran, and A. Behzad, "Radios for next-generation wireless networks," *IEEE Microwave Magazine*, vol. 6, no. 1, pp. 38–43, 2005.

[2] R. S. Tahim, "Multi-band antenna technology," in *Proceedings of the IEEE Antennas and Propagation Society Symposium*, vol. 4, pp. 3968–3971, June 2004.

[3] H. Hashemi and A. Hajimiri, "Concurrent multiband low-noise amplifiers-theory, design, and applications," *IEEE Transactions on Microwave Theory and Techniques*, vol. 50, no. 1, pp. 288–301, 2002.

[4] Y. Itoh, "L-Band SiGe HBT differential amplifiers using stacked parallel-resonant circuits," *Contemporary Engineering Sciences*, vol. 1, no. 3, pp. 127–138, 2008.

[5] M. Shirata, T. Shinohara, M. Sato, and Y. Itoh, "An L-band SiGe HBT differential amplifier with frequency and rejection-level tunable, multiple stopband," *International Journal of Microwave and Wireless Technologies*, vol. 1, no. 4, pp. 285–292, 2009.

[6] C. S. Lin, S. F. Chang, C. C. Chang, and Y. H. Shu, "Design of a reflection-type phase shifter with wide relative phase shift and constant insertion loss," *IEEE Transactions on Microwave Theory and Techniques*, vol. 55, no. 9, pp. 1862–1868, 2007.

[7] J. C. Wu, T. Y. Chin, S. F. Chang, and C. C. Chang, "2.45-GHz CMOS reflection-type phase-shifter MMICs with minimal loss variation over quadrants of phase-shift range," *IEEE Transactions on Microwave Theory and Techniques*, vol. 56, no. 10, pp. 2180–2189, 2008.

[8] K. Miyaguchi, M. Hieda, K. Nakahara et al., "An ultra-broad-band reflection-type phase-shifter MMIC with series and parallel LC circuits," *IEEE Transactions on Microwave Theory and Techniques*, vol. 49, no. 12, pp. 2446–2452, 2001.

[9] X. Tang and K. Mouthaan, "Dual-band Class III loaded-line phase shifters," in *Proceedings of the Asia-Pacific Microwave Conference (APMC '10)*, pp. 1731–1734, December 2010.

[10] A. Ocera, E. Sbarra, R. V. Gatti, and R. Sorrentino, "An innovative reconfigurable reflection-type phase shifter for dual band WLAN applications," in *Proceedings of the 36th European Microwave Conference (EuMC '06)*, pp. 64–67, September 2006.

[11] D. R. Banbury, N. Fayyaz, S. Safavi-Naeini, and S. Nikneshan, "A CMOS 5.5/2.4 GHz dual-band smart-antenna transceiver with a novel RF dual-band phase shifter for WLAN 802.11 a/b/g," in *Proceedings of the IEEE Radio Frequency Integrated Circuits Symposium (RFIC '04)*, pp. 157–160, June 2004.

3

Wide Range Temperature Sensors Based on One-Dimensional Photonic Crystal with a Single Defect

Arun Kumar,[1] Vipin Kumar,[2] B. Suthar,[3] A. Bhargava,[4] Kh. S. Singh,[2] and S. P. Ojha[5]

[1] *AITTM, Amity University, NOIDA 201303, India*
[2] *Department of Physics, Digamber Jain (P.G.) College, Baraut 250611, India*
[3] *Department of Physics, Govt. College of Engineering & Technology, Bikaner 334004, India*
[4] *Nanophysics Laboratory, Department of Physics, Govt. Dungar College, Bikaner 334001, India*
[5] *Director General, IIMT Group of Colleges, Noida 201303, India*

Correspondence should be addressed to Vipin Kumar, vrpcommon@gmail.com

Academic Editor: Yeou Song (Brian) Lee

Transmission characteristics of one-dimensional photonic crystal structure with a defect have been studied. Transfer matrix method has been employed to find the transmission spectra of the proposed structure. We consider a Si/air multilayer system and refractive index of Si layer has been taken as temperature dependent. As the refractive index of Si layer is a function of temperature of medium, so the central wavelength of the defect mode is a function of temperature. Variation in temperature causes the shifting of defect modes. It is found that the average change or shift in central wavelength of defect modes is 0.064 nm/K. This property can be exploited in the design of a temperature sensor.

1. Introduction

Since the last two and half decades, investigations on various properties of photonic crystals, particularly photonic bandgap materials, have become an area of interest for many researchers [1–6]. It was observed that periodic modulation of the dielectric functions significantly modifies the spectral properties of the electromagnetic waves. The transmission and reflection spectra of such structures are characterized by the presence of allowed and forbidden photonic bands bands similar to the electronic band structure of periodic potentials. For this reason, such a new class of artificial optical material with periodic dielectric modulation is known as photonic bandgap (PBG) material [3]. Fundamental optical properties like band structure, reflectance, group velocity and the rate of spontaneous emission, and so forth can be controlled effectively by changing the spatial distribution of the dielectric function [4, 5]. This fact has opened up important possibilities for the design of novel optical and optoelectronic devices. Conventional photonic crystals have periodic modulation of homogeneous refractive indices, and they are artificially fabricated with periods that are comparable to the wavelength of the electromagnetic waves. These photonic crystals lead to formation of photonic bandgaps or stop bands, in which propagation of electromagnetic waves of certain wavelengths is prohibited. A one-dimensional photonic crystal (1D PC) structure has many interesting applications such as dielectric reflecting mirrors, optical switches, filters, and optical limiters. It has also been demonstrated theoretically and experimentally that 1D PCs can have absolute omnidirectional PBGs [7–11].

In addition to the existence of wide bandgaps in some properly designed PCs, the feature of a tunable PBG is an interesting property of such PCs. The PBG can be tuned by means of some external agents [12]. For instance, it can be changed by the operating temperature and we call it T-tuning. A superconductor/dielectric PC belongs to this class of PC. This happens because of the temperature-dependent London perturbation length in the superconducting materials [13–15]. Using a liquid crystal as one of the constituents in a PC, the T-tuning optical response is also obtainable [16]. Recently, PCs containing semiconductor as one of the constituents have also been investigated by many researchers. PCs with intrinsic semiconductor belong to T-tuning devices because the dielectric constant of an intrinsic semiconductor is strongly dependent on temperature [17].

Although these applications can be realized using pure PCs, but doped or defective PCs may be more useful, just as semiconductor doped by impurities is more important than the pure ones for various applications. The idea of doped PCs comes from the consideration of the analogy between electromagnetism and solid state physics, which lead to the study of band structures of periodic materials and further to the possibility of the occurrence of localized modes in the bandgap when a defect is introduced in the lattice. These defect-enhanced structures are called doped photonic crystals and present some resonant transmittance peaks in the bandgap corresponding to the occurrence of the localized states [18], due to the change of the interference behavior of the incident waves. Defect(s) can be introduced into perfect PCs by changing the thickness of the layer [19], inserting another dielectric into the structure [20], or removing a layer from it [21, 22].

The introduction of the defect states within PCs has been perceived as a new dimension in the study of photonic crystals, especially in 2D and 3D PCs, due to numerous possible applications that can be achieved by using them. In 2D or 3D PCs, it has been known that a point defect can act as a micro cavity, a line defect like a waveguide, and a planar defect like a perfect mirror [6]. Similar to 2D or 3D PCs, the introduction of the defect layers in 1D PCs can also create localized defect modes within the PBGs. Due to the simplicity in 1D PC fabrications over 2D and 3D PCs, the defect mode can be easily introduced within1D PCs. Such PCs with defect(s) can be exploited for the design of various applications such as in the design of TE/TM filters and splitters [23], in the fabrication of lasers [24], and in light emitting diodes [25]. The existence of defect mode in 1D PCs depends upon a number of parameters such as refractive indices of the layers, filling fraction, the thickness of defect layer, and the angle of incidence. If all other parameters are kept constant and the refractive index of the material is changed, then any change in refractive index of a material results to a change or shift the position of the defect mode in the reflection/transmission spectra. In this present communication, a wide range temperature sensor based on one-dimensional photonic crystal with a single defect has been proposed.

Here, we consider the Si/air multilayer system with a defect in the form of a single Si layer in the middle. Si is taken as one of the constituents of a one-dimensional photonic crystal. Since the refractive index of Si depends on temperature [17], so any change in the temperature of the constituent material will change the refractive index of the material, and hence the wavelength corresponding to the middle of the defect mode will also change or shift. By measuring this change or shift in the position of defect mode, the change in the temperature can be measured, so it can work as a temperature sensor. In this present work, we will restrict our study for normal incidence only.

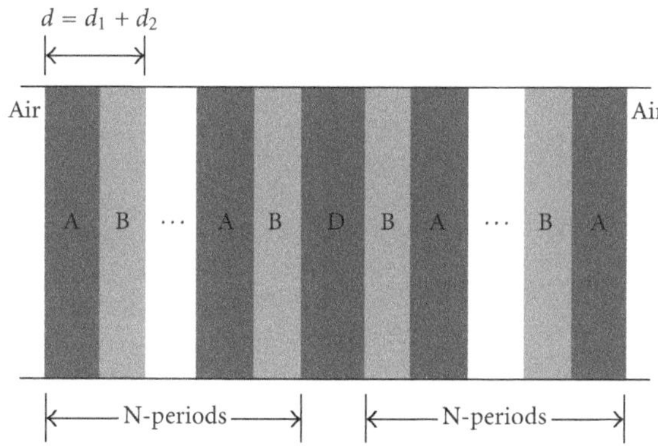

Figure 1: Schematic diagram of 1-D photonic crystal with defect.

2. Theoretical Analysis

The schematic representation of one-dimensional photonic crystal with defect is represented in Figure 1.

We consider air/$(AB)^N D(BA)^N$/air, in which A and B represent the layers of high and low refractive index materials, and D is the defect layer. To compute the defect mode in the transmission spectrum, we employ the transfer matrix method (TMM) [26, 27]. In this method, the transfer matrix for each layer can be written as

$$M_j = Z_j P_j Z_j^{-1}, \tag{1}$$

where j stands for A, B, or D layers and Z_j and P_j are called the dynamical matrix, and the propagation matrix, respectively. The dynamical matrix is given by the following equations:

$$Z_j = \begin{pmatrix} 1 & 1 \\ n_j \cos\theta_j & -n_j \cos\theta_j \end{pmatrix}, \text{ for TE mode of polarization,}$$
$$Z_j = \begin{pmatrix} \cos\theta_j & \cos\theta_j \\ n_j & -n_j \end{pmatrix}, \text{ for TM mode of polarization,} \tag{2}$$

where θ_j is the ray angle in the layers and n_j is the refractive index of the layers. Also, the propagation matrix P_j can be defined as

$$P_j = \begin{pmatrix} e^{i\delta_j} & 0 \\ 0 & e^{-i\delta_j} \end{pmatrix}, \tag{3}$$

where the phase is written as

$$\delta_j = \frac{2\pi d_j}{\lambda} n_j \cos\theta_j. \tag{4}$$

The transfer matrix, for the structure (Figure 1), embedded in air, air/$(AB)^N D(BA)^N$/air can be written as

$$M = \begin{pmatrix} M_{11} & M_{12} \\ M_{21} & M_{22} \end{pmatrix} = Z_0^{-1} (M_A M_B)^N M_D (M_A M_B)^N Z_0, \tag{5}$$

where Z_0 is called the dynamical matrix for air.

The transmittance (T) and reflectance (R) can be calculated by using the matrix elements M_{11} and M_{21} given in (5) as follows:

$$T = \left| \frac{1}{M_{11}} \right|^2, \qquad R = \left| \frac{M_{21}}{M_{11}} \right|^2. \tag{6}$$

In the next section, structural parameters of the device have been defined.

3. Proposed Structure and Structural Parameters

We choose Si for the materials A and D, air for material B, and $N = 5$ in Figure 1. Silicon and air have been taken as the high and the low refractive index materials, respectively, and an additional silicon layer (D) as the defect layer. The geometrical parameters are so chosen that the thicknesses of high and low refractive index materials are $d_1 = 600$ nm, $d_2 = 1200$ nm and the thickness of defect layer is taken 1800 nm. The thermal expansion coefficient for silicon layer is 2.6×10^{-6}/K, which is negligible small quantity. So we neglect the thermal expansion of Si layer in our computation.

The refractive index of air is independent of temperature and wavelength. But the refractive index of silicon layer is taken as a function of both wavelength and temperature. The refractive index of Silicon (Si) in the ranges 1.2 to 14 μm and 20–1600 K is represented as [17]

$$n^2(\lambda, T) = \varepsilon(T) + \frac{e^{-3\Delta L(T)/L_{293}}}{\lambda^2} \times \left(0.8948 + 4.3977 \times 10^{-4} T + 7.3835 \times 10^{-8} T^2\right), \tag{7}$$

where

$$\begin{aligned}
\varepsilon(T) &= 11.4445 + 2.7739 \times 10^{-4} T + 1.7050 \times 10^{-6} T^2 - 8.1347 \times 10^{-10} T^3, \\
\frac{\Delta L(T)}{L_{293}} &= -0.071 + 1.887 \times 10^{-6} T + 1.934 \times 10^{-9} T^2 - 4.554 \times 10^{-13} T^3 \quad \text{for } 293\,\text{K} \le T \le 1600\,\text{K}, \\
\frac{\Delta L(T)}{L_{293}} &= -0.021 - 4.149 \times 10^{-7} T - 4.620 \times 10^{-10} T^2 + 1.482 \times 10^{-11} T^3 \quad \text{for } 20\,\text{K} \le T \le 293\,\text{K}.
\end{aligned} \tag{8}$$

The plot of the refractive index as the function of wavelength and temperature is shown in Figure 2. From Figure 2, it is clear that the refractive index of silicon layer increases with the increase in temperature. Therefore, the refractive index contrast increases with temperature.

In the next section, we will compute the transmission spectra of photonic crystal with defect by using (6) as shown in Figure 1.

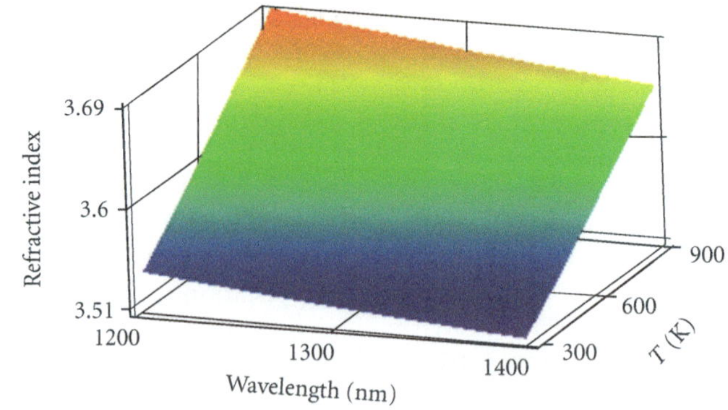

FIGURE 2: Variation of refractive index of Si as a function of wavelength and temperature.

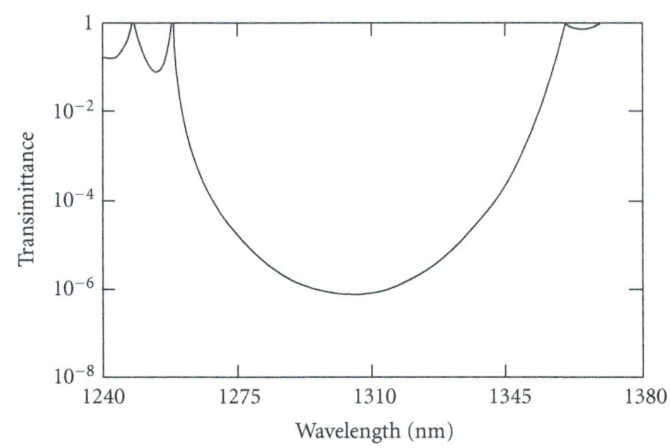

FIGURE 3: Transmittance for an ideal PC of $(\text{Si/air})^{10}$ at 100 K temperature.

4. Result and Discussion

In this section, we present the working principle of the proposed temperature sensor. The transmission spectra of the ideal PC $(\text{Si/air})^{10}$ at 100 K temperature for normal incidence are shown in Figure 3. If a defect which is in the form of silicon layer is introduced in this PC, then a defect mode appears inside the bandgap. The transmission spectra of the PC with a defect $[(\text{Si/air})^5\text{Si}(\text{air/Si})^5]$ at 100 K temperature are shown in Figure 4. As shown in this figure a defect mode has a mid-wavelength centered at 1288.3 nm. Thus, from this figure it is clear that if the incident radiation of wavelengths ranging from 1275 nm to 1350 nm is incident normally on the proposed structure at 100 K, then it will pass only 1288.3 nm wavelength and all other wavelengths will be reflected. More interestingly, this defect mode can be shifted to the longer wavelength region by the variation of temperature. The transmission spectra of the proposed structure at various temperature are shown in Figure 5, and the corresponding data is tabulated in Table 1. From Figure 5 and Table 1, it is clear that the transmission peak centered at 1288.3 nm at 100 K has been shifted to 1300.08 nm, 1312.9 nm, and 1326.8 nm corresponding to temperatures at 300 K, 500 K, and 700 K, respectively. Further, it is also clear that as the temperature of the structure increases,

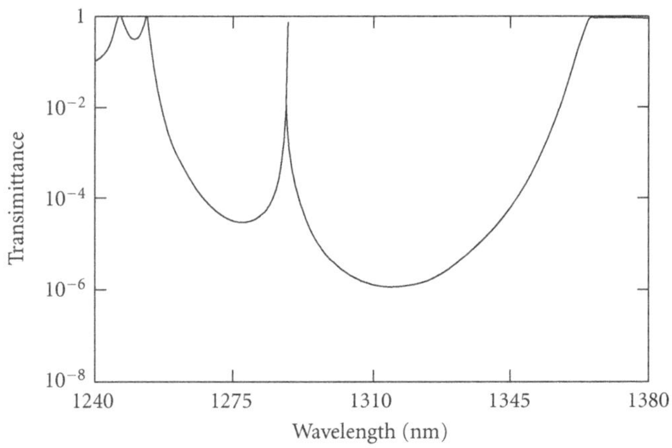

FIGURE 4: The defect mode for a PC with defect of $[(Si/air)^5Si(air/Si)^5]$ at 100 K temperature.

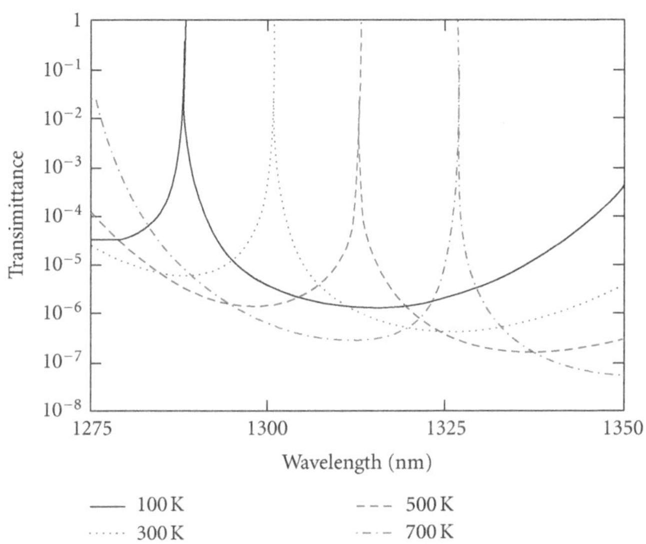

FIGURE 5: Transmission spectra of proposed structure at different temperature.

the defect mode of transmission shifts towards the higher wavelength region. The dependence of defect modes on temperature is due to the dependence of refractive index on temperature as given in (7). The shifting behavior can be explained by using the phase relation given by (4). According to this equation, as refractive index $n(\lambda, T)$ increases with temperature, the wavelength must increase accordingly to keep the phase δ unchanged. The variation of central wavelength of defect modes with temperature is shown in Figure 6. It is found that the central wavelength of defect modes changes approximately linearly with temperature. The average change in central wavelength of defect mode is 0.064 nm/K. This change in central wavelength is almost 9 times more than the previous reported work [28]. So the proposed structure is more sensitive to temperature variation. Earlier studies on defect filters reported response time of the order of ~ms to ~ps range with different techniques [29, 30]. Gan et al. achieved a response rise time

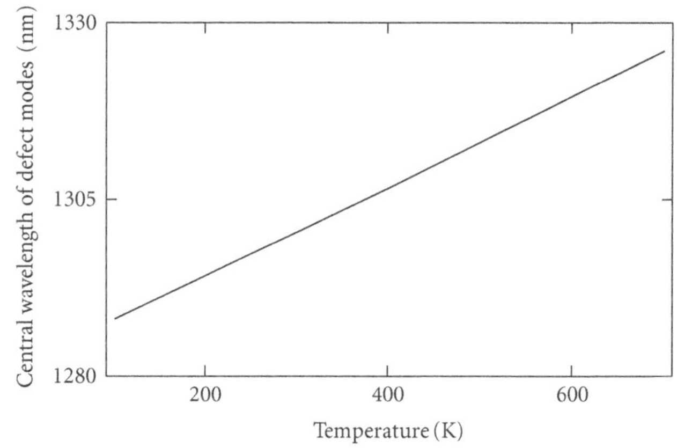

FIGURE 6: The variation of central wavelength of defect modes with temperature.

TABLE 1: Central wavelength of defect modes.

Temperature (K)	Central wavelength of defect mode (nm)
100	1288.3
300	1300.8
500	1312.9
700	1326.8

of 7 μs and a fall time of 14 μs for silicon photonic structures [31]. For the proposed temperature sensor, the response time for silicon will decide slow or fast response. Thus, by measuring the change or shift in the central wavelength of the defect mode, we can measure the temperature in a wide range. Hence, the proposed structure may be used as a temperature sensor in a wide range.

5. Conclusion

In the present communication, a simple design of temperature sensor has been proposed. The proposed structure is based on the one-dimensional photonic structure with defect. It is found that introduction of defect in the conventional one-dimensional photonic crystal creates a defect mode. Silicon has been chosen as one of the constituents of the device, because refractive index of the Si depends on the temperature. It is found that centre of this defect mode depends on the temperature of the device and it change linearly with temperature. An average change or shift in the central wavelength of defect mode is 0.064 nm/K. So, we can use the proposed device as a temperature sensor. The proposed device may also have many applications in different optical systems.

References

[1] S. John, "Strong localization of photons in certain disordered dielectric superlattices," *Physical Review Letters*, vol. 58, no. 23, pp. 2486–2489, 1987.

[2] E. Yablonovitch, "Inhibited spontaneous emission in solid-state physics and electronics," *Physical Review Letters*, vol. 58, no. 20, pp. 2059–2062, 1987.

[3] C. Soukoulis, *Photonic Band Gap Materials*, Kluwer Academic, Dordrecht, The Netherlands, 1996.

[4] J. A. M. Rojas, J. Alpuente, J. Piñeiro, and R. Sánchez, "Rigorous full vectorial analysis of electromagnetic wave propagation in ID inhomogeneous media," *Progress in Electromagnetics Research*, vol. 63, pp. 89–105, 2006.

[5] E. Yablonovitch and T. J. Gmitter, "Photonic band structure: the face-centered-cubic case," *Physical Review Letters*, vol. 63, no. 18, pp. 1950–1953, 1989.

[6] H. Y. Lee and T. Yao, "Design and evaluation of omnidirectional one-dimensional photonic crystals," *Journal of Applied Physics*, vol. 93, no. 2, pp. 819–830, 2003.

[7] J. P. Dowling, "Mirror on the wall: you're omnidirectional after all?" *Science*, vol. 282, no. 5395, pp. 1841–1842, 1998.

[8] E. Yablonovitch, "Engineered omnidirectional external-reflectivity spectra from one-dimensional layered interference filters," *Optics Letters*, vol. 23, no. 21, pp. 1648–1649, 1998.

[9] D. N. Chigrin, A. V. Lavrinenko, D. A. Yarotsky, and S. V. Gaponenko, "Observation of total omnidirectional reflection from a one-dimensional dielectric lattice," *Applied Physics A*, vol. 68, no. 1, pp. 25–28, 1999.

[10] B. Suthar, V. Kumar, Kh. S. Singh, and A. Bhargava, "Tuning of photonic band gaps in one dimensional chalcogenide based photonic crystal," *Optics Communications*, vol. 285, no. 6, pp. 1505–1509, 2012.

[11] V. Kumar, Kh. S. Singh, S. K. Singh, and S. P. Ojha, "Broadening of omnidirectional photonic band gap in Si-based one dimensional photonic crystals," *Progress in Electromagnetics Research M*, vol. 14, pp. 101–111, 2010.

[12] B. Suthar, V. Kumar, A. Kumar, Kh. S. Singh, and A. Bhargava, "Thermal expansion of photonic band gap for one dimensional photonic crystal," *Progress in Electromagnetics Research Letters*, vol. 32, pp. 81–90, 2012.

[13] O. L. Berman, Y. E. Lozovik, S. L. Eiderman, and R. D. Coalson, "Superconducting photonic crystals: numerical calculations of the band structure," *Physical Review B*, vol. 74, no. 9, Article ID 092505, 2006.

[14] H. Takeda and K. Yoshino, "Tunable photonic band schemes in two-dimensional photonic crystals composed of copper oxide high-temperature superconductors," *Physical Review B*, vol. 67, no. 24, Article ID 245109, 6 pages, 2003.

[15] W. H. Lin, C. J. Wu, T. J. Yang, and S. J. Chang, "Terahertz multichanneled filter in a superconducting photonic crystal," *Optics Express*, vol. 18, no. 26, pp. 27155–27166, 2010.

[16] P. Halevi, J. A. Reyes-Avendaño, and J. A. Reyes-Cervantes, "Electrically tuned phase transition and band structure in a liquid-crystal-infilled photonic crystal," *Physical Review E*, vol. 73, no. 4, Article ID 040701(R), 4 pages, 2006.

[17] H. H. Li, "Refractive index of silicon and germanium and its wavelength and temperature derivatives," *The Journal of Physical Chemistry*, vol. 9, no. 3, pp. 561–658, 1980.

[18] G. Guida, A. de Lustrac, and P. Priou, "An introduction to Photonic Band Gap (PBG) materials," *Progress in Electromagnetics Research Letters*, vol. 41, pp. 1–20, 2003.

[19] B. Suthar and A. Bhargava, "Temperature-dependent tunable photonic channel filter," *IEEE Photonics Technology Letters*, vol. 24, no. 5, Article ID 6096365, pp. 338–340, 2012.

[20] C. M. Soukoulis, "Photonic band gaps and localization," in *Proceedings of the NATO Advanced Research Workshop*, Plenum Press, London, UK, May 1993.

[21] D. R. Smith, R. Dalichaouch, N. Kroll, S. Schultz, S. L. McCall, and P. M. Platzman, "Photonic band structure and defects in one and two dimensions," *Journal of the Optical Society of America B*, vol. 10, no. 2, pp. 314–321, 1993.

[22] V. Kumar, Kh. S. Singh, and S. P. Ojha, "Abnormal behaviour of one-dimensional photonic crystal with defect," *Optik*, vol. 122, no. 13, pp. 1183–1187, 2011.

[23] Z. M. Jiang, B. Shi, D. T. Zhao, J. Liu, and X. Wang, "Silicon-based photonic crystal heterostructure," *Applied Physics Letters*, vol. 79, no. 21, pp. 3395–3397, 2001.

[24] W. D. Zhou, J. Sabarinathan, P. Bhattarcharya et al., "Characteristics of a photonic bandgap single defect microcavity electroluminescent device," *IEEE Journal of Quantum Electronics*, vol. 37, no. 9, pp. 1153–1160, 2001.

[25] M. W. Feise, I. V. Shadrivov, and Y. S. Kivshar, "Bistable diode action in left-handed periodic structures," *Physical Review E*, vol. 71, no. 3, Article ID 037602, 4 pages, 2005.

[26] P. Yeh, *Optical Waves in Layered Media*, John Wiley and Sons, New York, NY, USA, 1988.

[27] M. Born and E. Wolf, *Principle of Optics*, Pergamon, Oxford, UK, 4th edition, 1970.

[28] W. C. L. Hopman, P. Pottier, D. Yudistira et al., "Quasi-one-dimensional photonic crystal as a compact building-block for refractometric optical sensors," *IEEE Journal on Selected Topics in Quantum Electronics*, vol. 11, no. 1, pp. 11–16, 2005.

[29] S. Toyoda, N. Ooba, A. Kaneko, M. Hikita, T. Kurihara, and T. Maruno, "Wideband polymer thermo-optic wavelength tunable filter with fast response for WDM systems," *Electronics Letters*, vol. 36, no. 7, pp. 658–660, 2000.

[30] X. Hu, Q. Gong, Y. Liu, B. Cheng, and D. Zhang, "Ultrafast tunable filter in two-dimensional organic photonic crystal," *Optics Letters*, vol. 31, no. 3, pp. 371–373, 2006.

[31] F. Gan, T. Barwicz, M. A. Popović et al., "Maximizing the thermo-optic tuning range of silicon photonic structures," in *Proceeding of the IEEE International Conference on Photonics in Switching*, August 2007.

Analysis and Design of Ultra-Wideband 3-Way Bagley Power Divider Using Tapered Lines Transformers

Khair Al Shamaileh,[1] Abdullah Qaroot,[2] Nihad Dib,[3] Abdelfattah Sheta,[4] and Majeed A. Alkanhal[4]

[1] *Waseela for Integrated Telecommunication Solutions, P.O. Box 962487, Amman 11196, Jordan*
[2] *Electrical Engineering Department, King Faisal University, P.O. Box 400, Al-Ahsa 31982, Saudi Arabia*
[3] *Electrical Engineering Department, Jordan University of Science and Technology, P.O. Box 3030, Irbid 22110, Jordan*
[4] *Electrical Engineering Department, King Saud University, P.O. Box 800, Riyadh 11421, Saudi Arabia*

Correspondence should be addressed to Khair Al Shamaileh, khair.shamaileh@waseela-net.com

Academic Editor: Walter De Raedt

An ultra-wideband (UWB) modified 3-way Bagley polygon power divider (BPD) that operates over a frequency range of 2–16 GHz is presented. To achieve the UWB operation, the conventional quarter-wave transformers in the BPD are substituted by two tapered line transformers. For verification purposes, the proposed divider is simulated, fabricated, and measured. The agreement between the full-wave simulation results and the measurement ones validates the design procedure.

1. Introduction

Recently, the need for microwave components that have the capability of operating over a wide range of frequencies has motivated many researchers. Thus, many papers that investigate the ultra-wide operation for different microwave devices such as the Wilkinson power dividers (WPDs), branch line couplers (BLCs), and antennas were proposed. In [1], an UWB directional coupler that operates over a frequency range of 3.1–10.6 GHz was presented. To realize such an UWB coupler, two elliptically shaped microstrip lines, which are broadside coupled through an elliptically shaped slot, were used. In [2], a similar approach was used to design a slot-coupled multisection quadrature hybrid coupler for UWB applications. A novel approach for the design of UWB 3 dB couplers, out-of-phase equal-split power dividers, omnidirectional monopole antennas, and directional tapered slot antennas was proposed in [3]. In [4], a novel UWB WPD with modifications on the traditional divider by adding an extra open stub on each branch was proposed. In [5], an UWB WPD that consists of two branches of impedance transformers, each one consisting of two sections of transmission lines with different characteristic impedances and different lengths, was proposed. A modified UWB WPD formed by implementing one delta stub on each branch was proposed in [6]. In [7], and based on the theory presented in [8], a WPD that operates over a frequency range of 2–10.2 GHz was designed by substituting its conventional quarter-wave arms by tapered lines. Three resistors were added along the tapered lines to achieve an acceptable isolation between the output ports.

One of the power dividers, which has been a new area of research, is the Bagley polygon power divider (BPD) [9–17]. Compared to other power dividers, such as the Wilkinson power divider, Bagley polygon power divider does not use lumped elements, such as resistors, and can be easily extended to any number of output ports. However, the output ports for such dividers are not matched, and the isolation between them is not as good as that of the Wilkinson power divider. In [9], reduced size 3-way and 5-way Bagley power dividers (BPDs), using open stubs, were presented. In [10], an optimum design of a modified 3-way Bagley rectangular power divider was investigated. In [11, 12], a general design of compact multiway dividers based

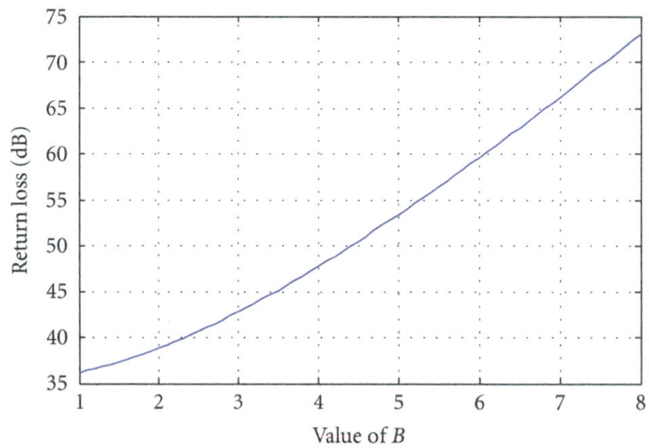

FIGURE 1: The obtained input return loss (in dB) versus B for $Z_s = 100\,\Omega$ and $Z_l = 33.333\,\Omega$.

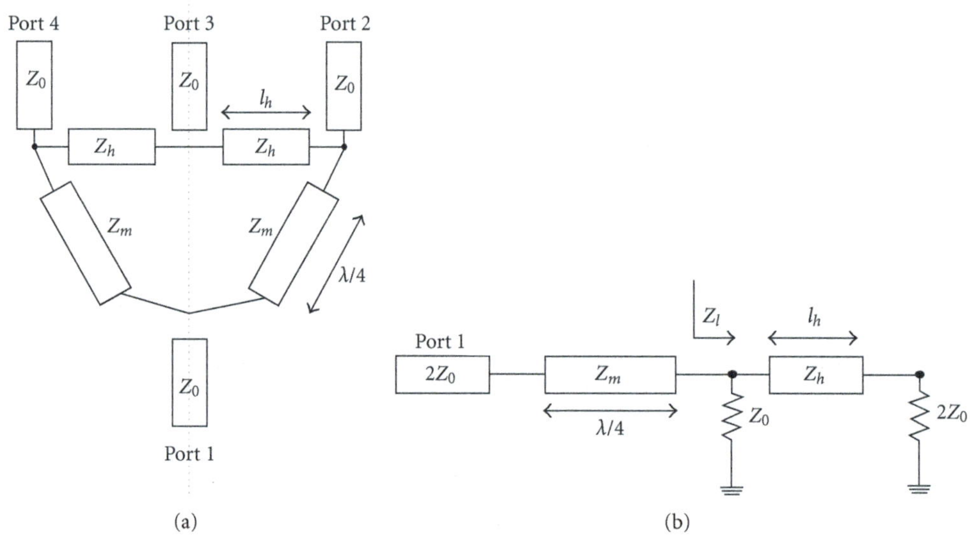

FIGURE 2: Schematic diagram of the 3-way BPD and its equivalent circuit.

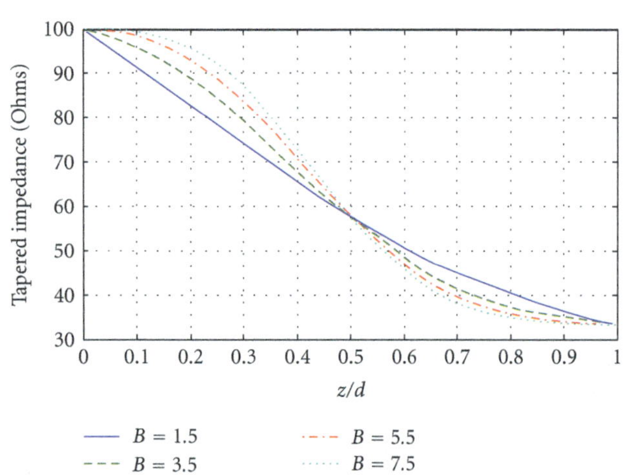

FIGURE 3: The tapered impedance variation for different values of B.

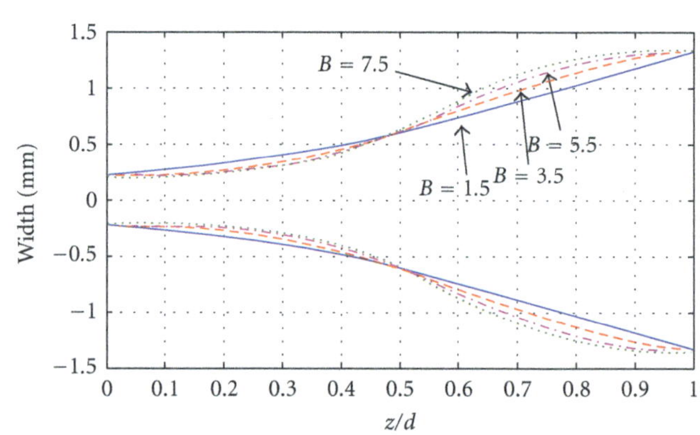

FIGURE 4: The tapered line width variation for different values of B.

on BPDs was introduced. In [13], a compact dual-frequency 3-way BPD using composite right-/left-handed (CRLH)

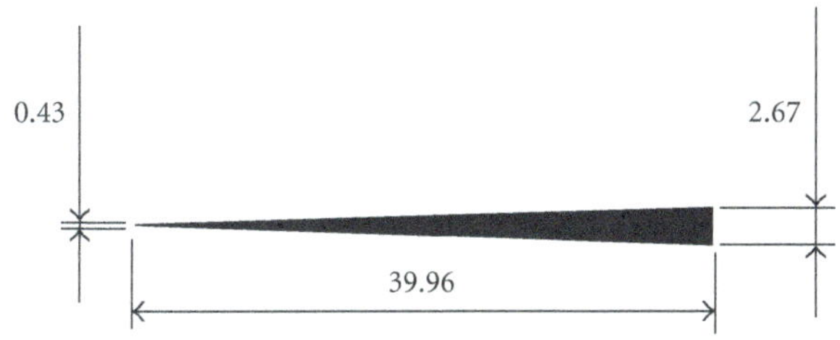

(a)

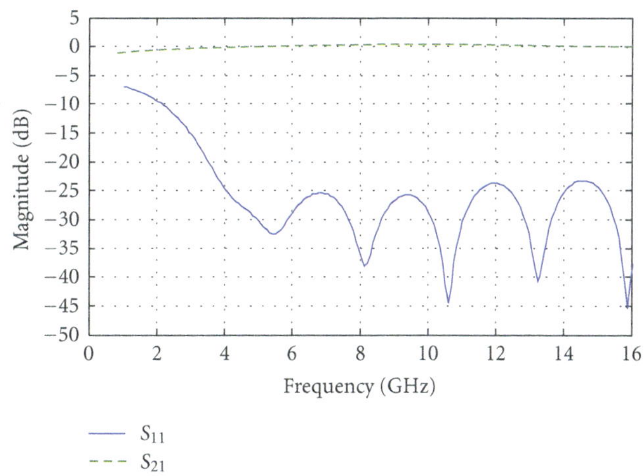

(b)

FIGURE 5: (a) The designed UWB tapered line transformer (dimensions are in mm). (b) The input port matching parameter S_{11} and the transmission loss parameter S_{21}.

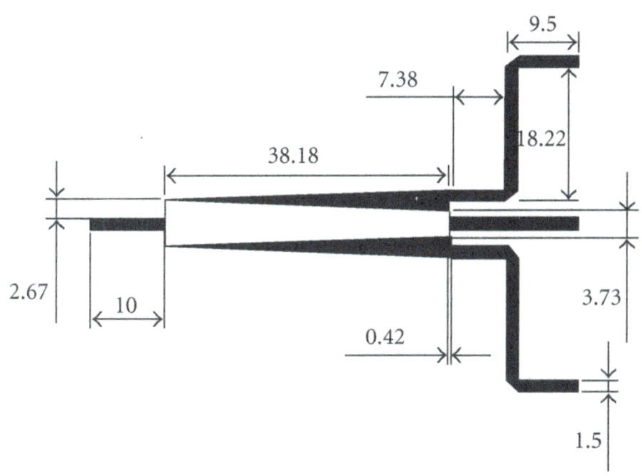

FIGURE 6: The layout of the proposed UWB 3-way BPD (dimensions are in mm).

transmission lines was implemented. Recently, and based on the generalized 3-way Bagley polygon power divider, dual-passband filter section was presented in [14]. Moreover, compact 5-way BPD for dual-band (or wide-band) operation was proposed in [15]. Dual-band modified 3-way BPDs based on substituting the quarter-wave sections of the conventional design by their equivalent dual-band matching networks were presented in [16]. Very recently, multiband miniaturized 3-way and 5-way BPDs were proposed in [17]. It should be mentioned here that all of the BPDs investigated in [9–17] have an odd number of output ports. In [18], a novel approach for the design of modified BPDs with even number of output ports was proposed.

In this paper, an UWB modified 3-way BPD that operates over the frequency range of 2–16 GHz is presented. To have such a divider, the quarter-wave sections are substituted by their equivalent UWB tapered lines. The designed UWB divider is simulated using two full-wave EM simulators. Moreover, the divider is fabricated and measured, and the simulation and measurement results are in a good agreement.

2. Tapered Line Design

According to [7, 8], the maximum input return loss (in dB) for a given tapered line that is used in order to match a source impedance Z_s to a load impedance Z_l is given by the following equation:

$$\left|\mathrm{RL_{input}}\right|_{\max} = -20\log\left[\tanh\left(\frac{B}{\sinh B}(0.21723)\ln\left(\sqrt{\frac{Z_l}{Z_s}}\right)\right)\right], \quad (1)$$

where B is a predefined design parameter used to determine the tapered line curve. Figure 1 shows the effect of increasing B on the obtained input return loss.

As seen from Figure 1, larger values of B result in lower reflection at the input port. However, increasing B will demand wider tapered line width and longer length.

After choosing the value of B in order to achieve a desired input return loss, the exponential tapered line characteristic impedance is calculated using the following equations [7, 8]:

$$\ln\left(\frac{Z(z)}{Z_s}\right) = 0.5\ln\left(\frac{Z_l}{Z_s}\right)\left[1 + G\left(B, 2\left(\frac{z}{d} - 0.5\right)\right)\right], \quad (2a)$$

where

$$G(B,\xi) = \frac{B}{\sinh B}\int_0^{\xi} I_0\left(B\sqrt{1-\xi'^2}\right)d\xi'. \quad (2b)$$

It should be mentioned here that $Z(z)$ in (2a) represents the characteristic impedance of the tapered line at point z, and $I_0(x)$ represents the modified zero-order Bessel function. The tapered line length d is a predefined variable chosen appropriately to achieve the desired maximum return loss.

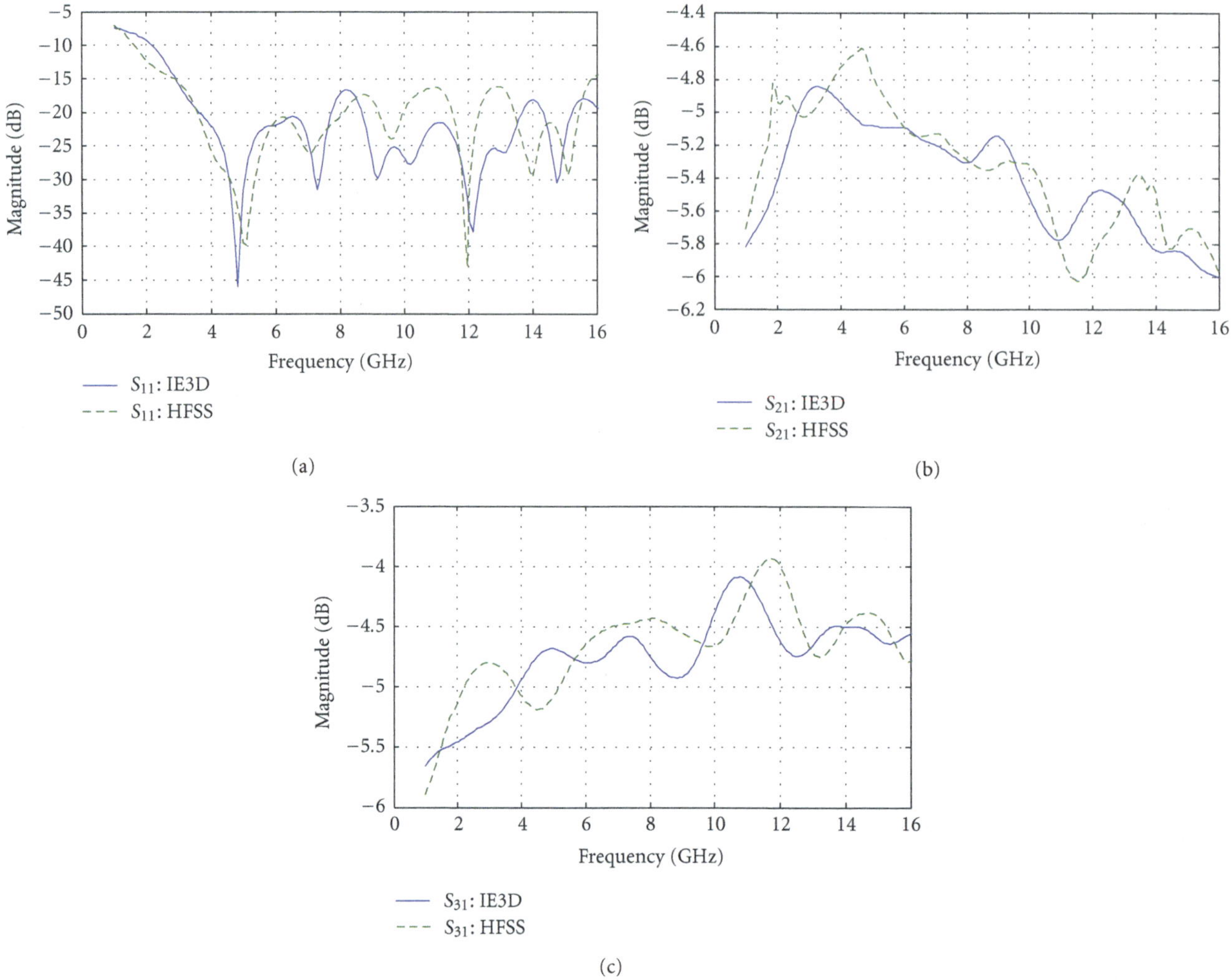

FIGURE 7: Simulated scattering parameters for the designed UWB BPD.

3. UWB 3-Way BPD Design

In this section, the design of a modified UWB 3-way BPD is presented. Figure 2(a) shows the schematic diagram of the 3-way modified BPD [11]. Noting that this divider is symmetric around its center line, an equivalent circuit (looking from port 1 to the right or left side) can be drawn as shown in Figure 2(b).

Referring to the equivalent circuit, it can be easily realized that choosing $Z_h = 2Z_0$ makes the design of this BPD independent of the length l_h. In this case, the characteristic impedance (Z_m) of the quarter wave section is $Z_m = \sqrt{(2Z_0)Z_l}$, where $Z_l = 2Z_0/3$. This gives

$$Z_m = \frac{(2Z_0)}{\sqrt{3}}. \qquad (2)$$

Thus, each quarter-wave section matches a load impedance of $Z_l = 2Z_0/3$ to a source impedance of $Z_s = 2Z_0$, resulting in a perfect match at port 1 (the input port) and equal split power division to the three output ports. As noted in the Introduction, the BPD does not contain any lumped elements, and it can be easily extended to any number of output ports.

Now, considering a characteristic impedance of 50 Ω, the values of Z_s and Z_l are 100 Ω and 33.333 Ω, respectively. These values will be incorporated in the tapered line design equations given in (1) and (2a). Then, the resulting tapered line will replace each conventional quarter-wave transmission line transformers in the BPD presented in Figure 2 in order to obtain an UWB operation. Figure 3 shows the variation of the tapered impedance for different values of B, which can be translated into microstrip width variation as shown in Figure 4. It is worth mentioning here that the substrate used in order to obtain the tapered line width for all cases is Duroid RT5870 with a relative permittivity $\varepsilon_r = 2.33$, a thickness of 0.508 mm, and a loss tangent of 0.0012.

It can be seen from Figures 3 and 4 that larger values of B result in a wider microstrip line width. In our design, B is chosen to be 5.5, which corresponds to a maximum input return loss of 56.44 dB. The length of the designed tapered line is set to 40 mm, which is about 1.48 times the length of the conventional transmission line transformer at the lower

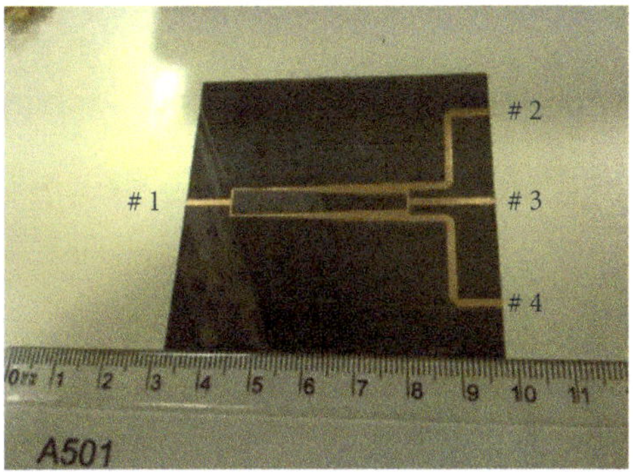

FIGURE 8: The photograph of the fabricated divider.

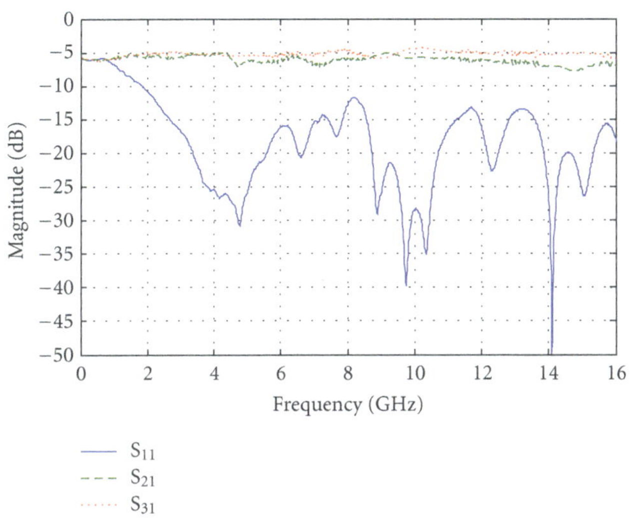

FIGURE 9: Measured scattering parameters of the divider shown in Figure 8.

frequency (2 GHz). However, such slight increase in the circuitry size leads to obtaining the desired electrical performance, especially the input port matching and transmission parameters performances, not only at a single frequency, but also over a considerable wide range of 2–16 GHz. Figure 5(a) shows the designed tapered transformer that matches a source impedance of 100 Ω to a load impedance of 33.333 Ω along with its obtained input port matching parameter (S_{11}) and transmission loss parameter (S_{21}) shown in Figure 5(b). An input return loss better than 10 dB is obtained over a frequency range of 2–16 GHz for the designed transformer. Moreover, the transmission coefficient S_{21} equals to −0.2 dB over the entire frequency range. It is worth to point out here that these results were obtained using the full-wave simulator IE3D [19].

4. Simulation Results

Figure 6 shows the layout of the designed UWB modified 3-way BPD. This proposed divider is simulated using two different full-wave electromagnetics simulators: IE3D [19]; which solves Maxwell's equations using the method of moments (MoM), and HFSS [20]; which solves the same equations using the finite element method. Figure 7 shows the obtained scattering parameters.

Figure 7(a) shows that an input return loss better than 10 dB is achieved over the frequency range of 2–16 GHz. Moreover, the resulting transmission parameter S_{21} (which is equals to S_{41} because of the symmetry of the structure) is close to its theoretical value of −4.7 dB ± 1 dB over the same frequency range except for the increase in the losses at higher frequencies. Such losses can be decreased through the use of low-loss tangent substrates. The transmission parameter S_{31} is also close to its theoretical value (−4.7 dB ± 0.8 dB) over the frequency band 2–16 GHz. The discrepancies between the results of the two simulators are thought to be due the different technique each simulator follows to solve Maxwell's equations, and the way the structure was divided in the meshing process during the simulations.

5. Measurement Results

The circuit layout shown in Figure 6 is implemented on the same substrate mentioned in Section 3 (Duroid RT5870 with a relative permittivity $\varepsilon_r = 2.33$ and a thickness of 0.508 mm). The extended ports in the circuit layout have been chosen to allow accurate S-parameter measurements using universal test fixture (GigaLane) without soldering. A photograph of the fabricated circuit is shown in Figure 8. The measurements have been performed using Anritsu 37369C network analyzer. The measured results are shown in Figure 9. The measured return loss is better than 10 dB from 2 to 16 GHz. The measured S_{31} is almost flat, around −5 dB, in the entire band. It changes at a few bands to −6 dB and some others to −4.5 dB. On the other hand, the measured S_{21} is approximately −5.6 ± 0.7 dB from 2 GHz to 12 GHz except at 4.8 GHz and 7 GHz, where it reaches −7.2 dB. From 12.5 GHz to 16 GHz, S_{21} changes from −6 dB to −7 dB; except for a small notch at about 15 GHz at which S_{21} is about −7.7 dB.

6. Conclusions

In this paper, an UWB 3-way BPD using tapered line transformers was designed, simulated, fabricated, and measured. Simulation results show a very good performance of the designed divider over a frequency range of 2–16 GHz. Measurement results show an acceptable performance with little discrepancies from the simulation ones. These differences could be mainly due to the fabrication process, as well as, measurement errors.

References

[1] A. M. Abbosh and M. E. Bialkowski, "Design of compact directional couplers for UWB applications," *IEEE Transactions on Microwave Theory and Techniques*, vol. 55, no. 2, pp. 189–194, 2007.

[2] A. Moscoso-Martir, J. G. Wanguemert-Perez, I. Molina-Fernandez, and E. Marquez-Segura, "Slot-coupled multisection quadrature hybrid for UWB applications," *IEEE Microwave*

and Wireless Components Letters, vol. 19, no. 3, pp. 143–145, 2009.

[3] M. Bialkowski, A. Abbosh, and H. Kan, "Design of compact components for ultra wideband commiunication front ends," in *Proceedings of the NEWCOM-ACoRN Joint Workshop*, Vienna, Austria, 2006.

[4] X. P. Ou and Q. X. Chu, "A modified two-section UWB Wilkinson power divider," in *Proceedings of the International Conference on Microwave and Millimeter Wave Technology (ICMMT '08)*, pp. 1258–1260, April 2008.

[5] L. Yang and Q. X. Chu, "Design of a compact UWB Wilkinson power divider," in *Proceedings of the International Conference on Microwave and Millimeter Wave Technology (ICMMT '08)*, pp. 360–362, April 2008.

[6] B. Zhou, H. Wang, and W.-X. Sheng, "A modified UWB Wilkinson power divider using delta stub," *Progress In Electromagnetics Research Letters*, vol. 19, pp. 49–55, 2010.

[7] C.-T. Chiang and B.-K. Chung, "Ultra wideband power divider using tapered line," *Progress in Electromagnetics Research*, vol. 106, pp. 61–73, 2010.

[8] R. P. Hecken, "A near optimum matching section without discontinuities," *IEEE Transactions on Microwave Theory and Techniques*, vol. 20, no. 11, pp. 734–739, 1972.

[9] T. Wuren, K. Taniya, I. Sakagami, and M. Tahara, "Miniaturization of 3- and 5- way Bagley Polygon power dividers," in *Proceedings of the Asia-Pacific Microwave Conference (APMC '05)*, vol. 5, December 2005.

[10] H. Oraizi and S. A. Ayati, "Optimum design of a modified 3-way bagley rectangular power divider," in *Proceedings of the 10th Mediterranean Microwave Symposium (MMS '10)*, pp. 25–28, August 2010.

[11] I. Sakagami, T. Wuren, M. Fujii, and M. Tahara, "Compact multi-way power dividers similar to the Bagley Polygon," in *Proceedings of the IEEE MTT-S International Microwave Symposium (IMS '07)*, pp. 419–422, June 2007.

[12] I. Sakagami, T. Wuren, M. Fujii, and Y. Tomoda, "A new type of multi-way microwave power divider based on Bagley Polygon power divider," in *Proceedings of the Asia-Pacific Microwave Conference (APMC '06)*, pp. 1353–1356, December 2006.

[13] D. S. Elles and Y. K. Yoon, "Compact dual band three way bagley polygon power divider using Composite Right/Left Handed (CRLH) transmission lines," in *Proceedings of the IEEE MTT-S International Microwave Symposium (IMS '09)*, pp. 485–488, June 2009.

[14] R. Gómez-García and M. Sánchez-Renedo, "Application of generalized Bagley-polygon four-port power dividers to designing microwave dual-band bandpass planar filters," in *Proceedings of the IEEE MTT-S International Microwave Symposium (MTT '10)*, pp. 580–583, May 2010.

[15] I. Sakagami and T. Wuren, "Compact multi-way power dividers for dual-band, wide-band and easy fabrication," in *Proceedings of the IEEE MTT-S International Microwave Symposium (IMS '09)*, pp. 489–492, June 2009.

[16] A. M. Qaroot, K. Shamaileh, and N. Dib, "Design and analysis of dual-frequency modified 3-way Bagley power dividers," *Progress In Electromagnetics Research C*, vol. 20, pp. 67–81, 2011.

[17] K. A. Al Shamaileh, A. Qaroot, and N. Dib, "Non-uniform transmission line transform-ers and their application in the design of compact multi-band bagley power dividers with harmonics suppression," *Progress in Electromagnetics Research*, vol. 113, pp. 269–284, 2011.

[18] K. A. Al Shamaileh, A. Qaroot, and N. Dib, "Design of N-way power divider similar to the Bagley polygon divider with an even number of output ports," *Progress In Electromagnetics Research C*, vol. 20, pp. 83–93, 2011.

[19] 2006, http://www.mentor.com/electromagnetic-simulation/.

[20] HFSS: High Frequency Structure Simulation based on Finite Element Method, V. 10, Ansoft Corporation, http://www.ansys.com/.

5

Microwave-Osmotic Dehydration of Cranberries under Continuous Flow Medium Spray Conditions

Derek Wray and Hosahalli S. Ramaswamy

Department of Food Science and Agricultural Chemistry, McGill University, Macdonald Campus, 21,111 Lakeshore Road, Ste. Anne de Bellevue, Montreal, QC, Canada H9X 3V9

Correspondence should be addressed to Hosahalli S. Ramaswamy; hosahalli.ramaswamy@mcgill.ca

Academic Editor: Tanmay Basak

Microwave-osmotic dehydration of cranberries was evaluated under continuous flow medium spray (MWODS) conditions after some pretreatments. A central composite rotatable design was used with three input variables at five levels (temperature, 33°C–67°C; sucrose concentration, 33°B–67°B; and contact time, 5–55 min). Responses were moisture loss (ML), solids gain (SG), and weight reduction (WR) as well as color and texture parameters. The responses were related to process variables using response surface methodology and statistical analysis: each model was tested for lack of fit to assure nonsignificance ($P > 0.05$) and each process variable was tested for significance ($P < 0.05$ or $P < 0.0001$). Temperature was found to have the most prominent effect as it was significant with all drying (ML, SG, and WR) and quality (hardness and chewiness) parameters, while contact time was found to be significant with ML and WR. Concentration wasn't found to be significant for any response. Increasing skin pretreatment severity generally promoted SG but had little effect on ML. The exception was chemical peeling, which favored ML but had no effect on SG. Overall, MWODS enables food dehydration in a much faster period of time than conventional osmotic dehydration (COD), while specifically promoting moisture loss over solids gain.

1. Introduction

Dehydration is one of the oldest methods of food preservation and is still prevalent in the food industry. The main objective of the dehydration process is to remove moisture to a point where the product is microbiologically and enzymatically stable and limit product deterioration during storage [1]. There is also an interest in producing dried ingredients which can then be incorporated into other products like breakfast cereals or baked goods. Drying often results in degradation of flavor volatiles, deterioration of color and texture, and an overall decrease in nutritional value, largely due to exposure to high temperatures for long periods of time in the presence of air [2]. Therefore, the key to improving the quality of dried products is to limit changes to the aforementioned quality characteristics during processing.

Osmotic dehydration (OD) represents a mild processing step in which texture is only moderately affected, nutritional value is well maintained, and the product quality, especially the color, can often be enhanced throughout the process [3]. In osmotic dehydration, the driving force is the osmotic pressure across a semipermeable membrane, which in food products is the cell wall. In addition to water exiting the plant tissue there is also a counter-current flow of solutes from the osmotic solution into the product. Furthermore, since the plant cell wall does not represent a perfect membrane, there is usually also a small leakage of low molecular weight substances such as vitamins, minerals, or organic acids which diffuse out of the cell along with the water [4]. This last diffusion tends to be insignificant in terms of mass balance but can be important in terms of quality aspects; it has been noted that anthocyanin content in cranberries after OD was 10% less than in fresh berries [5].

OD can easily be adapted to industrial applications and represents an energy efficient means to obtain an intermediate moisture food product. The energy efficiency is attributed to both the use of moderate temperatures and to the fact that there is no need to provide the latent heat of vaporization because water is removed by physical diffusion instead of as a vapor [6]. One drawback of conventional osmotic dehydration is the relatively slow process and therefore there has been a push to develop techniques that can be applied in conjunction with osmotic dehydration in order to speed up the process without negatively affecting the product quality [7]. The methods are applied either before or during osmotic dehydration and examples include application of vacuum [8, 9], ultrasound [8] as well as pulsed electric field and high hydrostatic pressure [10].

In general, microwaves have been more commonly applied in conjunction with air drying in various other studies [11]. Microwave energy has proven to be effective in drying for several reasons. First, microwaves generate heat by exciting dipolar molecules (mainly water) and polarizing ionic salts; both of which try to orient themselves to the microwave field, causing rapid heating from within the sample itself, an effect known as volumetric heating [11]. During microwave-assisted dehydration there is a rapid and differential heat generated within the food product as a result of the microwave absorption, largely because the main component in the food product is water. This results in a pressure buildup within the food which in turn promotes moisture loss by forcing moisture out of the product [12]. In the case of osmotic dehydration, the increased rate of moisture loss under microwave conditions tends to reduce the inward flow of solids gain. This effect was demonstrated by Azarpazhooh and Ramaswamy, [6, 13, 14] and Li and Ramaswamy [15–17]. In these studies, apple cylinders were treated by combined osmotic and microwave drying with a continuous flow of the osmotic medium either in an immersion (MWODI) or spray mode (MWODS). As compared to conventional osmotic dehydration, these techniques were demonstrated to provide high moisture loss, weight reduction, and ML/SG ratio along with the low solids gain and short dehydration times.

Cranberries (*Vaccinium macrocarpon*) represent an important cash crop in North America. The United States and Canada are the first and the second in world production and together represent about 98% of worldwide yield [5]. The vast majority of that crop is juiced, frozen, or dried prior to use, most often with sugar being incorporated because naturally cranberries are generally considered too sour and tart for direct consumption. As such, cranberries represent ideal candidates for osmotic drying because the incorporation of sucrose during the process sweetens the product as it is being dehydrated. Cranberries are also garnering interest for their health benefits which could include decreased risk of coronary heart disease [18], stroke [19], and lung cancer [20]. Moderate drying techniques are ideal in this case as high temperatures could destroy the heat labile flavonoid compounds responsible for these activities including various anthocyanins and catechins [21].

The objective of this study was to evaluate the potential for the MWODS process to be applied to a new product as each of the prior projects had been completed using apple cylinders. Further, as an extension to previous studies the focus was to quantify the effects of skin pretreatment and more specifically, MWODS process variables (temperature, contact time, sucrose concentration) on moisture loss, solids gain, and weight reduction, as well as texture and color changes in cranberry samples using a central composite rotatable design (CCRD) for the experiments and response surface methodology (RSM) for analysis of results.

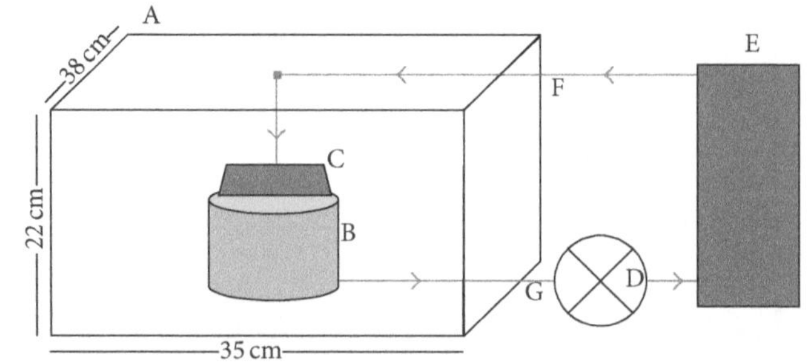

Figure 1: Schematic of MWODS setup: A: microwave oven cavity, B: microwave transparent sample chamber, C: spray head, D: peristaltic pump, E: water bath (containing heat exchanging coils, not pictured), and F and G are thermocouple measuring points immediately before and after the solution enters and leaves the microwave cavity, respectively.

2. Materials and Methods

2.1. Raw Materials. Frozen whole cranberries (Ocean Spray Canada) were obtained from a local food service supply company and kept frozen (−21 to −27°C) until use. Prior to use the cranberries were thawed for one hour at room temperature water (approximately 20°C), following similar parameters to those already described for cranberries destined for OD [22]. Commercial grade sucrose (Lantic Sugar Ltd., Montreal, Qc, Canada) was used in conjunction with tap water for the osmotic solution, and the concentration was determined using a handheld refractometer (Model N2-E, ATAGO Company, Tokyo, Japan).

2.2. MWODS Setup. The experimental setup used is illustrated in Figure 1 and was the same as previously described by Azarpazhooh and Ramaswamy [6, 13].

3. Experimental Procedure

3.1. Sample Determination. In this study some initial tests were performed in order to select a sample type because previous authors [23] made reference towards difficulty in working with cranberries for an osmotic dehydration process because of the resistance against water and solids transfer provided by the waxy skin that encases the fruit. For this series of tests frozen berries were thawed in room temperature water as previously described. Pretreatments were applied when the berries were completely thawed and immediately before the samples were placed in the MWODS setup. Each of these tests was done at the same conditions (50°C, 50°B, and 30 minutes), to demonstrate the effects of the skin pretreatments only. The treatments were performed as outlined here. The 7-hole treatment consisted of poking through the skin 7 times with a fine tipped needle. There were

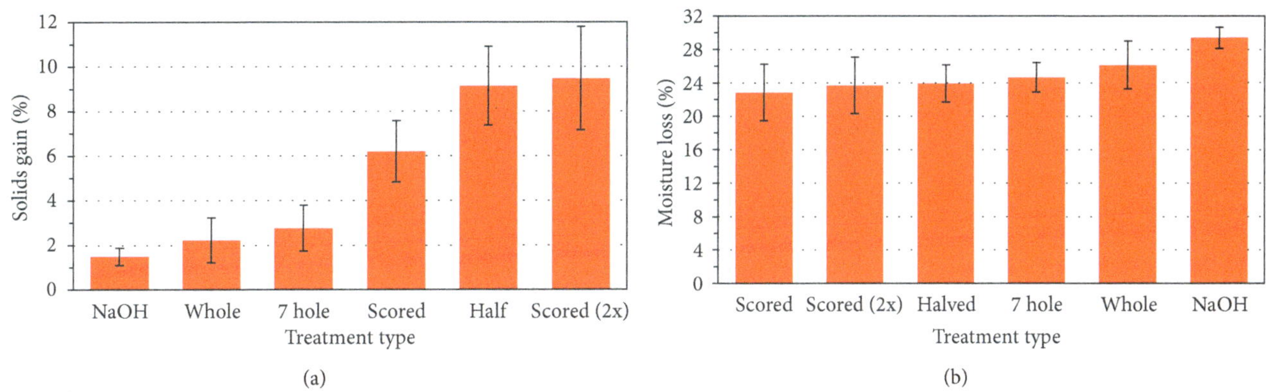

FIGURE 2: Moisture loss and solids gain for various skin pretreatments of cranberries. Average values with standard deviation shown.

5 holes places around the equator of the berry and one at each end. For scored berries the tip of a knife was moved around the long end of the berry (starting and ending at the stem) to create either a single U-shaped cut around the fruit for the "scored" type or two intersecting U-shaped cuts for the double-scored-type pretreatment. Halved berries were cut in half along the equator by hand just prior to treatment. Finally, for chemical peeling the berries were placed in 0.5% NaOH for 3 min at 25°C, quickly rinsed in tap water, blotted dry, and then placed in the MWODS setup as outlined by Grabowski et al. [23]. Each sample was prepared in triplicate, and the results averaged for Figure 2.

3.2. MWODS Experiments. Individual samples of approximately 20 g (17-18 berries) were weighed and placed in a Nylon mesh bag to contain the sample. The system was setup and solution was preheated according to the prescribed temperature of the run type. The sample was then placed on the acrylic stage in the sample chamber in a single layer, the pump was turned on and the solution allowed to flow, and then the microwave was turned on. After the allotted time the pump was stopped and the sample was removed. The berries were then rinsed twice in a container of room temperature tap water to remove excess sugar solution from the surface of the product. The cranberries were blotted with a moist paper towel to remove surface moisture, weighed again, and were then either examined for quality parameters or dried to constant weight in an oven set at 105°C for approximately 24 hours [24]. The moisture content of the frozen-thawed (untreated) berries was determined to be 89.08% (wb) on average.

3.3. Experimental Design. Design Expert v6.1 (Stat-Ease Inc, Minneapolis, Minnesota, USA) was used to create a central composite rotatable design (CCRD) with three factors (temperature, sucrose concentration, and contact time) at five coded levels each (−1.68, −1, 0, 1, and 1.68). The complete experimental design with both real and coded values can be seen in Table 1.

TABLE 1: CCRD experimental design for MWODS in coded (in brackets) and real values.

Run #	Temperature (°C)	Concentration (°B)	Contact time (min)
1	50 (0)	50 (0)	30 (0)
2	50 (0)	67 (+1.68)	30 (0)
3	50 (0)	50 (0)	30 (0)
4	40 (−1)	60 (+1)	15 (−1)
5	50 (0)	50 (0)	30 (0)
6	40 (−1)	60 (+1)	45 (+1)
7	60 (+1)	60 (+1)	15 (−1)
8	40 (−1)	40 (−1)	15 (−1)
9	40 (−1)	40 (−1)	45 (+1)
10	50 (0)	33 (−1.68)	30 (0)
11	60 (+1)	60 (+1)	45 (+1)
12	67 (+1.68)	50 (0)	30 (0)
13	60 (+1)	40 (−1)	15 (−1)
14	33(−1.68)	50 (0)	30 (0)
15	50 (0)	50 (0)	30 (0)
16	50 (0)	50 (0)	5 (−1.68)
17	50 (0)	50 (0)	30 (0)
18	60 (+1)	40 (−1)	45 (+1)
19	50 (0)	50 (0)	30 (0)
20	50 (0)	50 (0)	55 (+1.68)

4. Quality Analysis

For each run type, partial samples were taken for quality analysis, while the rest of the samples were dried until completion for the determination of solids gain and moisture loss.

4.1. Color Measurement. Color values of the MWODS-treated samples were determined in the L^*, a^*, b^* system using a tristimulus Minolta Chroma Meter (Minolta Corp., Ramsey, NJ, USA). The Chroma Meter was warmed up 20

TABLE 2: CCRD run numbers with results for post-MWODS drying and quality parameters (average values with standard deviation shown).

Run #	ML (%)	SG (%)	WR (%)	Hardness (N)	Chewiness (mJ)
1	22.2 (3.10)	2.73 (0.53)	22.4 (2.68)	162 (41.4)	89.9 (41.3)
2	21.7 (1.16)	2.74 (0.16)	21.9 (1.10)	92.2 (25.8)	24.4 (14.5)
3	23.8 (0.60)	2.75 (0.51)	24.0 (0.31)	154 (14.8)	64.8 (18.4)
4	14.3 (1.29)	1.92 (0.22)	15.3 (1.30)	206 (61.5)	79.4 (28.4)
5	22.3 (2.35)	2.28 (0.27)	22.9 (2.32)	157 (22.2)	62.6 (33.1)
6	17.6 (1.77)	2.31 (0.21)	18.2 (1.94)	205 (27.9)	94.2 (22.5)
7	22.5 (2.20)	1.99 (0.66)	23.5 (1.43)	162 (24.9)	66.6 (14.0)
8	19.8 (1.26)	2.65 (0.86)	14.3 (2.75)	155 (24.5)	67.4 (13.9)
9	22.0 (3.55)	1.79 (0.32)	23.2 (3.71)	257 (33.6)	184 (31.0)
10	19.8 (1.26)	2.75 (0.31)	20.0 (1.56)	178 (38.0)	73.0 (30.0)
11	29.7 (1.89)	3.27 (1.03)	29.4 (2.83)	125 (14.1)	36.0 (6.26)
12	21.3 (3.09)	5.91 (1.30)	18.4 (2.86)	156 (66.8)	66.5 (48.0)
13	23.7 (1.18)	2.20 (0.13)	24.5 (1.05)	180 (39.4)	66.3 (32.1)
14	17.0 (3.65)	2.36 (2.34)	17.6 (3.62)	239 (36.4)	143 (23.6)
15	22.1 (2.37)	2.34 (0.98)	22.6 (3.32)	156 (27.1)	49.9 (19.3)
16	17.3 (2.11)	2.10 (0.58)	18.3 (2.50)	208 (20.4)	76.5 (12.5)
17	21.7 (1.08)	2.49 (0.40)	22.1 (0.72)	183 (24.8)	105 (28.0)
18	30.8 (2.87)	5.41 (2.62)	28.3 (1.49)	108 (17.6)	34.8 (10.6)
19	25.1 (2.07)	1.93 (0.42)	26.1 (1.68)	194 (33.3)	77.0 (32.0)
20	24.0 (1.14)	2.34 (0.52)	24.6 (0.78)	154 (17.2)	78.9 (35.2)
Fresh				293 (71.9)	121 (69.7)

minutes prior to use and calibrated against a white standard. Eight measurements were taken with each sample, and the values were averaged in order to obtain the L^* (lightness), a^* (green (−) to red (+)), and b^* (blue (−) to yellow (+)) values of the individual trials. The ΔE (total color change), C^* (chroma), and H° (hue angle) were also determined according to the following equations [25]:

$$\Delta E = \sqrt{(L_0 - L)^2 + (a_0 - a)^2 + (b_0 - b)^2},$$
$$\text{Chroma } (C^*) = \sqrt{a^2 + b^2}, \quad (1)$$
$$\text{Hue Angle } (H^\circ) = \tan^{-1}\frac{b}{a},$$

where L_0, a_0, and b_0 and L, a, and b represent the values of the fresh and processed samples, respectively.

4.2. Texture Analysis. Texture profile analysis (TPA) of both the MWODS-treated samples and fresh berries was performed using a TA.XT*Plus* Texture Analyser (Stable Microsystems, Surrey, UK). TPA allows for the determination of a wide range of responses such as adhesiveness, chewiness, cohesiveness, fracturability, gumminess, hardness, and springiness [26]. In this study hardness and chewiness were selected as parameters of interest as they represent quality indicators one would experience during mastication of a raisin-type semidried product. Hardness was taken as the peak of the first curve (N), while chewiness (mJ) is calculated as the gumminess multiplied by springiness, as defined by Bourne [26]. The analysis was performed with eight replicates, and the average values (with standard deviation) are presented in Table 2. Analysis was performed by means of TPA using a flat bottom probe (25 mm diameter) with pretest speed of 1 mm/sec, test speed of 5 mm/sec and posttest speed of 5 mm/sec. The target was a distance of 10 mm, which represents the 50% of the average diameter of the cranberries tested. These settings were used with guidance from Rahman and Al-Farsi [27], who used TPA on a similar dried product (dates).

5. Osmotic Dehydration Responses and Data Analysis

To analyze the mass exchange that took place during osmotic dehydration, the parameters of moisture loss (ML), solids gain (SG), and weight reduction (WR) were calculated according to the following equations:

$$\%\text{ML} = 100\frac{M_0 x_0 - M_t x_t}{M_0},$$
$$\%\text{SG} = 100\frac{M_0 s_0 - M_t s_t}{M_0}, \quad (2)$$
$$\%\text{WR} = 100\frac{M_0 - M_t}{M_0},$$

where M_0 and M_t are the total sample mass as time 0 and time t, respectively; x_0 and x_t are the moisture fractions (kg/kg wet basis) at time 0 and time t, respectively; s_0 and s_t are the solid fractions (kg/kg wet basis) at time 0 and time t, respectively. These equations assume a one-way mass transfer of sucrose into the product (i.e., there is no significant leeching of solids from the sample into the solution). These parameters were collected by weighing the samples before MWODS treatment (M_0), after treatment (M_t). The initial solids fraction (s_0) was calculated by difference from the moisture content of fresh cranberries, and the posttreatment solids content (s_t) was defined as the mass gained after drying to constant weight. Each run was completed in triplicate with average values used for data analysis.

6. Results and Discussion

6.1. Pretreatment. For most of the treatment types the moisture loss for the various skin treatments remained near constant and certainly within the error of the samples (Figure 2). The only exception is the chemically (NaOH) peeled sample, which did have a slightly higher moisture loss when compared to the rest. Interestingly, the halved berries provided nearly the same amount of moisture loss when compared to the whole berry. This is in contrast to Grabowski et al. [23], who found that cutting the berries in half increased the moisture diffusion by about 100-fold when compared to the mass transfer through the cranberry skin and overall by about a 4.5% increase in moisture loss over the length of the process. There are a few possible explanations for this change, including a smaller solution-sample contact area in a spray setup when compared to a traditional immersion setup. This is particularly important when you consider the positioning of the cranberry halves under the spray head. While each sample was placed facing "up" to sit like a bowl on the sample stage, if a stream of spray did not exactly enter and fill that piece, it would not have nearly the same, effect as if the sample was completely submerged in the solution.

The solids gain for the most part increased with the level of destruction to the skin. The lone exception is that halved berries accounted for less solids gain, on average, than the double scored samples, although the variation is within the error. Also of interest is the fact that the chemically peeled samples showed no increased absorption of sucrose when compared to untreated samples. This suggests that while removing the waxy layer allows water to more easily traverse the skin, the skin tissue that remains may still be a physical barrier that prevents solids uptake.

Since any of the previous pretreatments were considered to be fully satisfactory, it was decided to use the untreated whole berries for two key reasons. First, the unaltered skin on the whole sample provided an interesting property in that it would limit the uptake of solute. This is of particular interest in osmotic dehydration where it is optimal to provide a high ML/SG ratio, that is, providing the highest possible moisture loss while limiting solids gain [6]. Secondly, while the NaOH treatment increased the moisture loss and potential ML/SG over the untreated berries, it was concluded that the effect was not enough to justify the use of synthetic chemicals in the process. These results agree with another conclusion from a similar study where it was determined that chemical pretreatment provided no significant influence on water transfer during osmotic dehydration of cranberries [28]. However, throughout the study it was observed that there were random breaks in the skin likely due to the freeze-thaw process. This possibly contributed to the variation of the results and would have therefore reduced the accuracy of the predicting model. In future studies it may be advantageous to use a skin pretreatment in order to minimize these effects.

7. Response Surface Methodology (MWODS)

The experimental design for this set of experiments differed from traditional studies of osmotic dehydration. Because conventional osmotic dehydration is a relatively slow process, studies tend to fix variables such as temperature, concentration, and flow rate and measure samples at a given interval until a targeted equilibrium is reached [6]. Because of the requirement to test the statistical significance of each of the process variables (temperature, concentration and contact time) in a manageable number of experiments, a central composite rotatable design (CCRD) was employed and provided a set of 20 individual run types to perform that independently tests the effect of that parameter on the response (Table 1). By design, the program does not require duplicates of each individual run, as there are six repeated run types (center points) which determine the variability of the process and the lack of fit of the model. However, due to the intrinsic variability of microwaves in drying, each run type was performed in triplicate and the mean value entered into Design Expert for analysis (Table 2).

In working with a CCRD design one of the most important steps is to determine the statistical significance of the model and each of the parameters uses analysis of variance (ANOVA). The models selected for each of the responses in this study were selected according to their statistical significance and lack of fit values (Table 3). For moisture loss and weight reduction, this represented the linear model (both $P < 0.0001$), while solids gain and hardness used a quadratic model (both $P < 0.05$), and finally chewiness was best represented by an interaction (2FI) model ($P < 0.05$). For each of the models it is critical to ensure that the lack of fit is not significant ($P > 0.05$); this value for each model is shown along with correlation coefficients in Table 3. Lack of fit is determined by testing the repeated center points, and an insignificant lack of fit tells us that the model is able to adequately predict the response variables. Using the predicting equations determined by Design Expert (Table 3), we can predict the outcome for responses such as moisture loss, solids gain, weight reduction, hardness, or chewiness depending on a function of input process variables.

8. Effects of Individual Process Variables

8.1. Moisture Loss. Beginning with moisture loss, it was found that only the linear terms of temperature ($P < 0.0001$) and contact time ($P < 0.05$) had a significant effect on ML. No interactions or quadratic terms were found to have any significant effect. Based on the sum of squares produced by ANOVA (not shown), it can be stated that temperature had a larger effect than contact time in terms of determining

TABLE 3: Selected model, predicting equation, and model evaluation for each response.

Response	Model	Prob > F	Equation in terms of actual factors	Lack of fit	R^2
ML	Linear	<0.0001	ML (%) = $0.649 + 0.336T + 0.180t - 0.0232C$	0.0527 (NS)	0.7304
SG	Quadratic	0.0018	SG (%) = $9.550-0.367T-0.128t+0.0936C+5.456\times10^{-3}T^2-5.531\times10^{-4}t^2+6.838\times10^{-4}C^2+3.300\times10^{-03}Tt-3.918\times10^{-03}TC+2.617\times10^{-4}tC$	0.0556 (NS)	0.8749
WR	Linear	<0.0001	WR (%) = $0.0272 + 0.355T + 0.158t - 4.56 \times 10^{-3}C$	0.2856 (NS)	0.8230
Hardness	Quadratic	0.0233	Hardness (N) = $56.705 - 0.602T + 9.662t + 11.433C + 0.109T^2 + 0.024t^2 - 0.108C^2 - 0.174Tt + 3.71 \times 10^{-04}TC - 0.056tC$	0.0839 (NS)	0.7765
Chewiness	Interaction (2FI)	0.0006	Chewiness (mJ) = $155.644 - 3.194T + 12.625t - 3.598C - 0.161Tt + 0.099TC - 0.084tC$	0.5006 (NS)	0.8003

Where T is Temperature (°C), t is contact time (minutes), and C is concentration (°B); $\alpha = 0.05$.

moisture loss during MWODS. Since the coefficients for both of these terms were positive, it can be further stated that an increase in either of the variables would result in higher moisture loss. This makes sense in the context of previous studies as osmotic dehydration has been previously described as a temperature-dependant process [15]. Response surface plots showing the effect on moisture loss of increasing temperature (Figure 3(a)) and concentration (Figure 3(d)) as a function of a time show that while increasing temperature has a large effect on the ML whole cranberries over the length of the process, concentration has little to no effect. In this case, moisture loss remains relatively stable as concentration increases and changes mainly with longer contact times. Many studies have formed the general conclusion that increasing temperature or contact time increased moisture loss whether it is in an immersion or spray setup and with or without microwaves being applied [1, 6, 13, 15–17]. The effects of temperature on moisture loss can be largely attributed to three key points. First, higher temperatures reduce the solution viscosity, allowing better contact between the sample and the osmotic solution, higher mobility of water molecules, and overall higher extraction ability of the solution. This effect is particularly important at high sucrose concentrations where viscosity can cause issues in a spray-based system [6]. Second, higher temperatures cause the individual cells in the sample to swell, thereby increasing the moisture permeability across the cell wall, which normally moderates moisture movement. Finally, thermal energy promotes moisture diffusion within the sample itself, which means that the water molecules can more easily reach the sample surface and be removed into the osmotic solution. These three causes were determined by several studies [6, 15]. While contact time was found to provide a significant effect on the moisture loss of the sample, it was lower than that of temperature. This is in contrast to some previous studies on both conventional [1, 15] and microwave-assisted [6] osmotic dehydration; all of which concluded that contact time had the largest influence on moisture loss. Similarly, past studies concluded that concentration does provide a significant impact on moisture loss, just not as large as that of temperature [29]. These incongruities with past results were perhaps caused by the effect of the cranberry skin, which is naturally intended to keep water inside the fruit as well as foreign substances out. If on the skin, which is at the junction of the fruit flesh and the osmotic solution, there was a layer of sucrose that formed near the beginning of the run; it is possible that this would provide enough resistance to moisture removal that longer run times or increased concentration had no appreciable effect on the overall moisture loss.

8.2. Solids Gain. In terms of solids (sucrose) gain by the cranberry samples only the temperature had a significant effect, where it was found to be significant ($P < 0.05$) at both the linear and the quadratic levels. The interaction effect between temperature and contact time was also found to be significant ($P < 0.05$). Like moisture loss, all of the significant terms for solids gain had positive coefficients, signifying that increasing temperature or contact time would results in higher solids gain. This effect can be visualized in Figure 3(b), where the interaction effect between contact time and temperature shows an increase in either of these variables resulting in increasing solids gain. These results agree with the general conclusions of several studies where increased temperature also increases the solids gain of the product [1, 6, 15, 30]. The concentration of sucrose in the osmotic solution was found to be insignificant and actually slightly decreased the solids gain. This trend can be seen in Figure 3(e), where higher concentrations actually result in lower solids gain throughout the range, an effect that partially disagrees with the majority of published reports. It has been seen in past studies that solids gain increases with solute concentration until extremely high concentrations at which point solids gain is often lower than at more moderate concentrations. This effect is likely caused by the formation of a dense superficial layer of solute at the surface of the sample, which then blocks further uptake of the solute [15]. While normally only evident at higher concentrations in the case of whole cranberries it is likely that the skin provides an additional barrier which blocks solids uptake in the fruit and facilitates the formation of the dense surface layer of solute, thereby slowing the uptake of solids over a broader range of concentrations instead of only at high concentrations.

8.3. Weight Reduction. Since weight reduction is determined by the moisture loss mediated by the solids gain and the moisture loss is typically larger than the solids gain by an order of magnitude, it is expected that the responses of weight reduction and moisture loss will follow a similar trend [6]. Therefore, it is not surprising to see that for weight reduction, like moisture loss, only the linear terms of temperature ($P < 0.0001$) and contact time ($P < 0.05$) were significant, while all of the quadratic and interaction effects proved to be insignificant. For the linear terms, the positive coefficients confirm that increasing either the temperature or contact time would increase the level of weight reduction, much like moisture loss. In general, increasing temperature will increase the weight reduction over the length of the MWODS process (Figure 3(c)), which shows that higher temperature tends to favour moisture loss over solids gain, thereby promoting moisture loss. It can also be concluded that concentration had little to no effect on weight reduction (Figure 3(f)), where the increase was due to increased contact time alone. This is nearly identical to response of the same interaction for moisture loss (Figure 3(d)).

8.4. Color. In this study, none of the parameters (or models) had any significant effect on the change in color measurements of the MWODS samples (in terms of $L^*a^*b^*$, Chroma (C^*), Hue Angle (H°), or total color change (ΔE) values). As such, none of these values are presented here. For these responses, statistical analysis dictated that the mean change in the value better represented the predicted outcome than any linear (or higher) model. There are two possible conclusions from this outcome; first there is the possibility that there was no significant change in the color throughout the MWODS

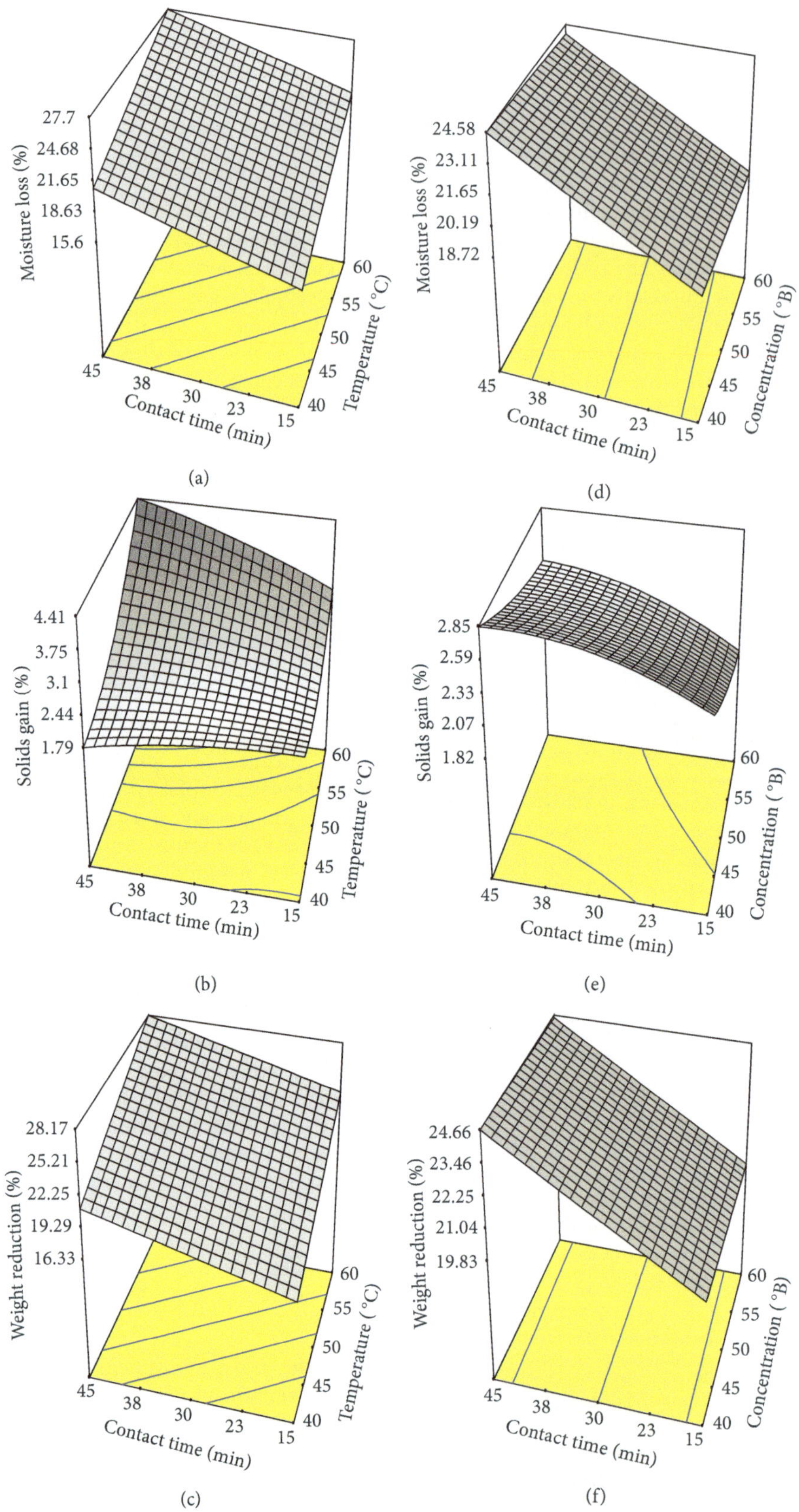

FIGURE 3: Response surface plots for dehydration parameters (ML, SG, and WR). For variable temperature plots (a–c), concentration was kept constant at its center point (50°B), while for variable concentration plots (d–f), temperature was kept at its center point (50°C).

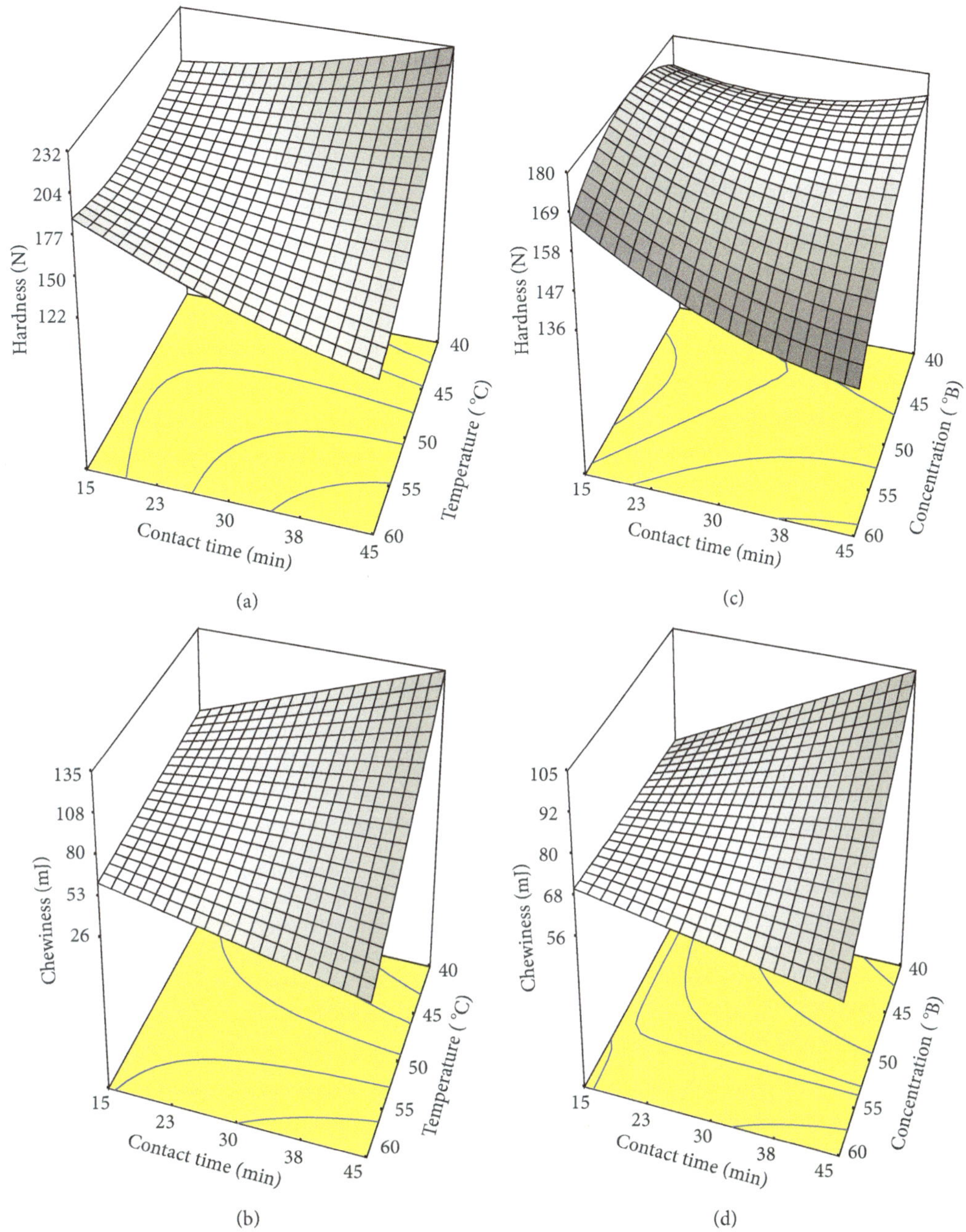

FIGURE 4: Response surface plots for quality parameters (Hardness and Chewiness). For variable temperature plots (a, b), concentration was kept constant at its center point (50°B), while for variable concentration plots (c, d), temperature was kept at its center point (50°C).

process. This is quite possible as typically these quality tests have only been performed on finished products (after secondary drying). The second possible conclusion is that even the least severe MWODS treatment made some change to the color properties, and this change did not increase with increased treatment severity.

8.5. Texture. Like color measurements, mechanical properties are typically only tested after the finished drying step, but the interest here was to see if the MWODS process caused any significant change in the textural attributes as a standalone process. For both chewiness and hardness the linear effect of temperature provided the only significant process variable effect, where $P < 0.0001$ and $P < 0.05$ for hardness and chewiness, respectively. For both of these responses the interaction effect between temperature and contact time proved to be significant on a $P < 0.05$ level; these interactions can be seen in Figures 4(a) and 4(b). The variable coefficients for these effects are negative, indicating that increasing either temperature or contact time actually decreases the hardness and chewiness when compared to the fresh product. This agrees with previous studies that have concluded that osmotically dried samples become softer and more plastic as samples losing these characteristics will become softer and less elastic, respectively [31]. It has been described that the mechanical properties of dried products are closely related to the physical state and structure developed as a result of typical deformations such as cell shrinking or swelling, changes in intercellular spaces (volume), and rupturing of cell bonds [32]. These changes influence mechanical behaviour in several fashions; first, air and liquid volumes within the sample change as solute is taken into the food, and the size of the product can be altered as well [9]. Secondly, detachment of the middle lamella of plant cells results in the loss of cell turgor which in turn affects the cell wall and puncture strength [33]. In examining the relationship between cell structure and texture, it would be a logical conclusion that higher temperatures and concentrations would exacerbate cell damage and therefore further affect the mechanical properties, which concur with the trends seen here. The effects of concentration of textural responses were more mixed.

9. Conclusions

Overall the level of destruction of cranberry skin increased moisture and solids diffusion, however, not to a significant extent. Therefore, the decision to leave the cranberries as a whole with no skin pretreatment was done in part to omit chemical peeling techniques and ease of processing but also to observe the ability of the MWODS system to overcome an the effect of a natural moisture barrier that is prevalent in small berries such as cranberries. A CCRD model combined with RSM analysis was used to determine the effect of individual process parameters (temperature, sucrose concentration, and contact time) on a series of responses (ML, SG, WR, hardness, and chewiness). The CCRD model was used to reduce the number of experiments required while still testing each variable independently. Overall, it was found that temperature was a significant factor for every response, where higher temperatures produced higher moisture loss, solids gain, and weight reduction along with reduced hardness and chewiness. Additionally, contact time was found to be a significant positive independent variable for ML, SG, and WR; implying increasing process time would in turn enhance these responses, while longer process times tended to decrease textural characteristics. Concentration effects tended to be minimal for each of the responses, which can be largely attributed to the unique properties of the cranberry skin. No model was found to be significant for any color change response, and therefore the color destruction during MWODS was insignificant. Overall it was found that the MWODS process is suitable for products like small berries; however, removal of the skin would make it easier to further model the process.

References

[1] N. H. Van Nieuwenhuijzen, M. R. Zareifard, and H. S. Ramaswamy, "Osmotic drying kinetics of cylindrical apple slices of different sizes," *Drying Technology*, vol. 19, no. 3-4, pp. 525–545, 2001.

[2] I. Alibas, "Microwave, air and combined microwave-air-drying parameters of pumpkin slices," *LWT—Food Science and Technology*, vol. 40, no. 8, pp. 1445–1451, 2007.

[3] C. Lerici, G. Pinnavaia, M. Rosa, and L. Bortolucci, "Osmotic dehydration of fruit influence of osmotic agents on drying behaviour and product quality," *Journal of Food Science*, vol. 50, no. 5, pp. 1217–1219, 1985.

[4] D. Torreggiani, "Osmotic dehydration in fruit and vegetable processing," *Food Research International*, vol. 26, no. 1, pp. 59–68, 1993.

[5] S. Grabowski, M. Marcotte, M. Poirier, and T. Kudra, "Drying characteristics of osmotically pretreated cranberries—energy and quality aspects," *Drying Technology*, vol. 20, no. 10, pp. 1989–2004, 2002.

[6] E. Azarpazhooh and H. S. Ramaswamy, "Evaluation of factors influencing microwave osmotic dehydration of apples under continuous flow medium spray (MWODS) conditions," *Food and Bioprocess Technology*, vol. 5, no. 4, pp. 1265–1277, 2010.

[7] N. K. Rastogi, K. S. M. S. Raghavarao, K. Niranjan, and D. Knorr, "Recent developments in osmotic dehydration: methods to enhance mass transfer," *Trends in Food Science and Technology*, vol. 13, no. 2, pp. 48–59, 2002.

[8] Y. Deng and Y. Zhao, "Effects of pulsed-vacuum and ultrasound on the osmodehydration kinetics and microstructure of apples (Fuji)," *Journal of Food Engineering*, vol. 85, no. 1, pp. 84–93, 2008.

[9] P. Fito, "Modelling of vacuum osmotic dehydration of food," *Journal of Food Engineering*, vol. 22, no. 1–4, pp. 313–328, 1994.

[10] K. A. Taiwo, A. Angersbach, B. I. O. Ade-Omowaye, and D. Knorr, "Effects of pretreatments on the diffusion kinetics and some quality parameters of osmotically dehydrated apple slices," *Journal of Agricultural and Food Chemistry*, vol. 49, no. 6, pp. 2804–2811, 2001.

[11] V. Orsat, W. Yang, V. Changrue, and G. S. V. Raghavan, "Microwave-assisted drying of biomaterials," *Food and Bioproducts Processing*, vol. 85, no. 3, pp. 255–263, 2007.

[12] M. E. Sosa-Morales, L. Valerio-Junco, A. López-Malo, and H. S. García, "Dielectric properties of foods: reported data in the 21st century and their potential applications," *LWT—Food Science and Technology*, vol. 43, no. 8, pp. 1169–1179, 2010.

[13] E. Azarpazhooh and H. S. Ramaswamy, "Microwave-osmotic dehydration of apples under continuous flow medium spray conditions: comparison with other methods," *Drying Technology*, vol. 28, no. 1, pp. 49–56, 2010.

[14] E. Azarpazhooh and H. S. Ramaswamy, "Evaluation of diffusion and Azuara models for mass transfer kinetics during microwave-osmotic dehydration of apples under continuous flow medium-spray conditions," *Drying Technology*, vol. 28, no. 1, pp. 57–67, 2010.

[15] H. Li and H. Ramaswamy, "Osmotic dehydration of apple cylinders: I. Conventional batch processing conditions," *Drying Technology*, vol. 24, no. 5, pp. 619–630, 2006.

[16] H. Li and H. Ramaswamy, "Osmotic dehydration of apple cylinders: II. Continuous medium flow heating conditions," *Drying Technology*, vol. 24, no. 5, pp. 631–642, 2006.

[17] H. Li and H. S. Ramaswamy, "Osmotic dehydration of apple cylinders: III. Continuous medium flow microwave heating conditions," *Drying Technology*, vol. 24, no. 5, pp. 643–651, 2006.

[18] P. Knekt, R. Järvinen, A. Reunanen, and J. Maatela, "Flavonoid intake and coronary mortality in Finland: a cohort study," *British Medical Journal*, vol. 312, no. 7029, pp. 478–481, 1996.

[19] S. O. Keli, M. G. L. Hertog, E. J. M. Feskens, and D. Kromhout, "Dietary flavonoids, antioxidant vitamins, and incidence of stroke: the Zutphen study," *Archives of Internal Medicine*, vol. 156, no. 6, pp. 637–642, 1996.

[20] P. Knekt, R. Järvinen, R. Seppänen et al., "Dietary flavonoids and the risk of lung cancer and other malignant neoplasms," *American Journal of Epidemiology*, vol. 146, no. 3, pp. 223–230, 1997.

[21] W. Zheng and S. Y. Wang, "Oxygen radical absorbing capacity of phenolics in blueberries, cranberries, chokeberries, and lingonberries," *Journal of Agricultural and Food Chemistry*, vol. 51, no. 2, pp. 502–509, 2003.

[22] J. Yongsawatdigul and S. Gunasekaran, "Microwave-vacuum drying of cranberries: part I. Energy use and efficiency," *Journal of Food Processing and Preservation*, vol. 20, no. 2, pp. 121–143, 1996.

[23] S. Grabowski, M. Marcotte, D. Quan et al., "Kinetics and quality aspects of Canadian blueberries and cranberries dried by osmo-connective method," *Drying Technology*, vol. 25, no. 2, pp. 367–374, 2007.

[24] AOAC, "Moisture in dried fruits," in *Official Methods of the Association of Official Analytical Chemists*, AOAC International, Maryland, Md, USA, 17th edition.

[25] N. Maftoonazad and H. S. Ramaswamy, "Effect of pectin-based coating on the kinetics of quality change associated with stored avocados," *Journal of Food Processing and Preservation*, vol. 32, no. 4, pp. 621–643, 2008.

[26] M. Bourne, "Principles of objective texture measurement," in *Concept and Measurement*, pp. 114–117, Academic Press, San Diego, Calif, USA, 1982.

[27] M. S. Rahman and S. A. Al-Farsi, "Instrumental texture profile analysis (TPA) of date flesh as a function of moisture content," *Journal of Food Engineering*, vol. 66, no. 4, pp. 505–511, 2005.

[28] P. S. Sunjka, T. J. Rennie, C. Beaudry, and G. S. V. Raghavan, "Microwave-convective and microwave-vacuum drying of cranberries: a comparative study," *Drying Technology*, vol. 22, no. 5, pp. 1217–1231, 2004.

[29] F. Nsonzi and H. S. Ramaswamy, "Osmotic dehydration kinetics of blueberries," *Drying Technology*, vol. 16, no. 3–5, pp. 725–741, 1998.

[30] H. N. Lazarides, E. Katsanidis, and A. Nickolaidis, "Mass transfer kinetics during osmotic preconcentration aiming at minimal solid uptake," *Journal of Food Engineering*, vol. 25, no. 2, pp. 151–166, 1995.

[31] A. Monsalve-Gonzalez, V. Gustavo, and P. Ralph, "Mass transfer and textural changes during processing of apples by combined methods," *Journal of Food Science*, vol. 58, no. 5, pp. 1118–1124, 1993.

[32] C. Contreras, M. E. Martín-Esparza, N. Martínez-Navarrete, and A. Chiralt, "Influence of osmotic pre-treatment and microwave application on properties of air dried strawberry related to structural changes," *European Food Research and Technology*, vol. 224, no. 4, pp. 499–504, 2007.

[33] A. Chiralt and P. Talens, "Physical and chemical changes induced by osmotic dehydration in plant tissues," *Journal of Food Engineering*, vol. 67, no. 1-2, pp. 167–177, 2005.

6

Compact Multiband Printed IFA on Electromagnetic Band-Gap Structures Ground Plane for Wireless Applications

Dalia Elsheakh and Esmat Abdallah

Microstrip Department, Electronics Research Institute, ELtahrir Street, Dokki, Giza 12622, Egypt

Correspondence should be addressed to Dalia Elsheakh; daliaelsheakh@gmail.com

Academic Editor: Ramon Gonzalo

The fourth mobile generation requires of multistandard operating handsets of small physical size as well as has an increasing demand for higher data rates. Compact multiband printed inverted-F antennas (IFAs) for available wireless communications are proposed in this paper. A new design of a printed IFA based on a uniplanar compact EBG concept is proposed. An L-loaded printed IFA shaped over an artificial ground plane is designed as the main antenna to cover the GSM, LTE, UMTS, bluetooth, and WLAN. The multi-band is created by means of an electromagnetic band-gap (EBG) structure that is used as a ground plane. Different shapes of uniplanar EBG as ring, split ring resonator, and a spiral rather than mushroom-like structure are investigated. The proposed antenna is built on the uniplanar EBG ground plane with a size of $35 \times 45\ \text{mm}^2$, which is suitable for most of the mobile devices.

1. Introduction

The current upsurge in wireless communication systems has forced antenna engineering to face new challenges, which include the need for wide bandwidth, small-size, high-performance, robustness, ease of mounting on a host surface and low-cost antennas. To have compatible uses for a wide range of applications by fulfilling mobility requirement and holding up the performance as well as the capability of obtaining dual and triple frequency operations are challenges [1].

Printed IFAs offer an attractive solution to a compact and ease-low-cost design of modern wireless communication systems due to having the above advantages. Printed inverted-F antenna is a simple and compact radiator; however, it cannot radiate efficiently near a perfect electric conductor (PEC) ground plane due to the reverse image currents. Also, printed IFAs suffer from a number of disadvantages as compared to conventional nonprinted antennas. Some of their major drawbacks are the narrow bandwidth, $\lambda/4$ length, low gain, and surface wave excitation that reduce radiation efficiency.

To solve the problem of their narrow bandwidth, several techniques can be used [2–4]. A thicker substrate with a low dielectric constant or a ferrite composition provides a wider bandwidth but this approach leads to no low-profile designs and an increase in size. Noncontacting feeding methods such as proximity/aperture coupled can be used to improve the impedance bandwidth, but this is difficult to fabricate. Another possibility is multiresonator stack configuration with the inconvenience of resulting a large thickness prototype [5, 6].

To overcome the problem of large size, the basic antenna miniaturization techniques can be classified into lumped-element loading, material loading, the use of ground planes, short circuits, and the antenna environment, as applied antenna on electromagnetic band-gap structures. Based on this, many attempts to design compact antennas have been studied in the past [7]. Among various EBG structures, uniplanar EBG designs such as the mushroom-like structure [8] received a great potential for wireless communication systems because of their low-profile configurations. One important application of EBG structures is to serve as the

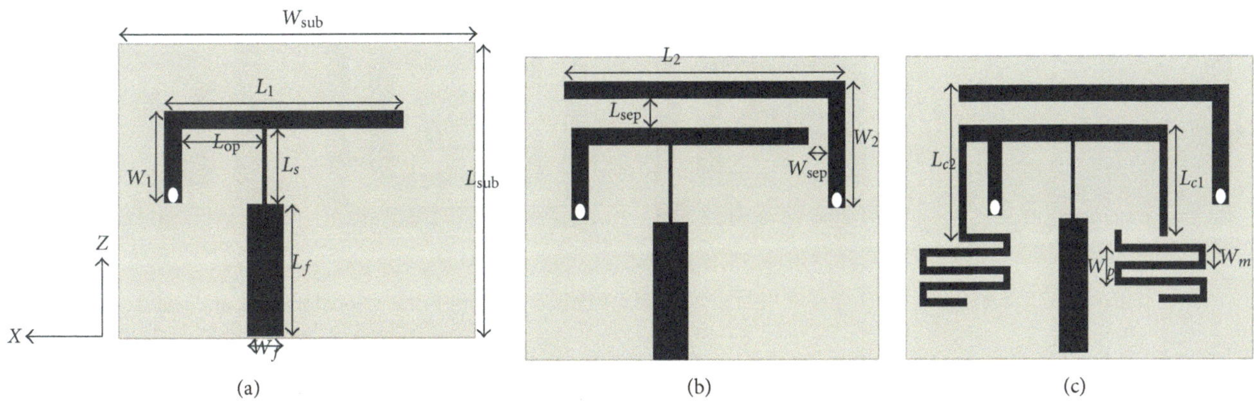

FIGURE 1: The configuration of (a) conventional IFA, (b) L-loaded IFA and (c) meandered L-loaded IFA.

ground plane for low-profile antennas [9–12]. The EBG structures are usually periodic arrangements of metallic or dielectric elements that exhibit band-gap characteristics which are sensitive to the permittivity, thickness of the substrate, and most importantly to the EBG geometry. The important properties of EBG are the ability to guide and control the propagation of electromagnetic waves, the suppression of surface waves within the band-gap range, and hence the improvement of the antenna performance by reducing cross-polarization, back radiation, and mutual coupling [13, 14]. Most of researches were focused on the three-dimensional EBG structures which are complicated, heavy, and difficult to integrate with microwave circuits. To avoid complexity, planar 2D EBG-like mushroom shape, split-ring-resonator-like shape, spiral shape, and so forth were suggested [15–17].

In this paper, the performance of L-loaded printed IFAs with meandering shape at the open end near the edge of a PEC ground plane is first investigated. Then different shapes of mushroom-like high impedance surface (HIS) EBG structures are used to improve the antenna parameters such as broaden bandwidth and efficiency and reduce the electrical size. Satisfactory results are obtained. All simulations are carried out using the EM commercial simulator, HFSS version 13.0.

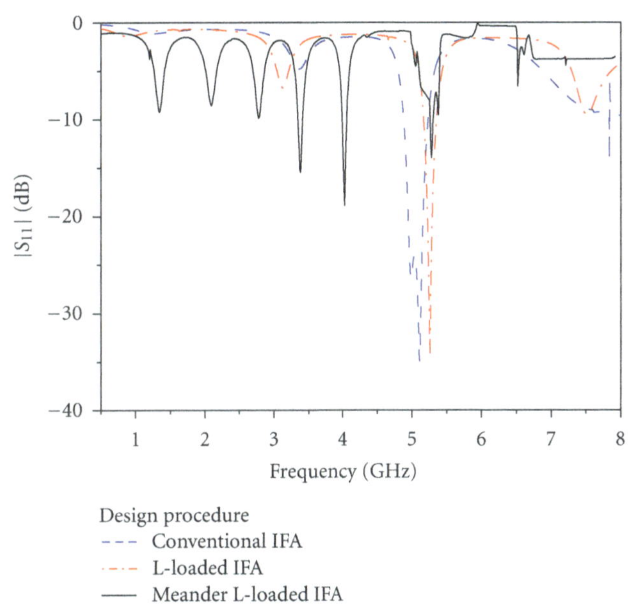

FIGURE 2: Simulated $|S_{11}|$ of the design antenna procedures with conventional ground.

2. Design of Printed IFAs

This section gives the details of the design procedure of the proposed multiband printed IFA antenna. It starts by the design of a conventional printed IFA that operates at 5.2 GHz, over a conventional ground plane as shown in Figure 1(a). The printed IFA dimensions are length $L_1 = 21.75$ mm with shorting wall size $W_1 = 7.4$ mm and ground plane size $L_{sub} \times W_{sub} = 35 \times 45\ \text{mm}^2$. The distance between the feed and the short end $L_{op} = 16.5$ mm with L_{sep} and W_{sep} are $6.5{\times}4.5\ \text{mm}^2$ and all transmission line widths are 2 mm. Then, two different approaches were investigated to introduce more bands to the printed IFA. The first approach is using L-loaded printed IFA as shown in Figure 1(b) with length and width $L_2 \times W_2 = 30 \times 12\ \text{mm}^2$. The antenna resonates at dual resonant frequencies 3.1 GHz and 5.2 GHz. It may be noted that the dual resonance frequency independent. Finally, to reduce the electrical size of L-loaded IFA, additional inductors are added. Two meandering shape with two unit cells at both open end of the L-loaded IFA as shown in Figure 1(c) are added. Multiband with the reduction in electrical antenna size is created with approximately same antenna parameters. Three-antenna design steps are shown in Figure 2. Figure 2 shows that the conventional IFA resonates at 5.2 GHz while an additional resonant frequency added by loading L shaped at 3 GHz and kept the first resonant slightly unchanged. The antenna is resonant at multiband by adding a meander line at the open ends of L-loaded IFA at 1.4 GHz, 2 GHz, 2.8 GHz, 3.2 GHz, 4 GHz, and 5.2 GHz with poor antenna matching.

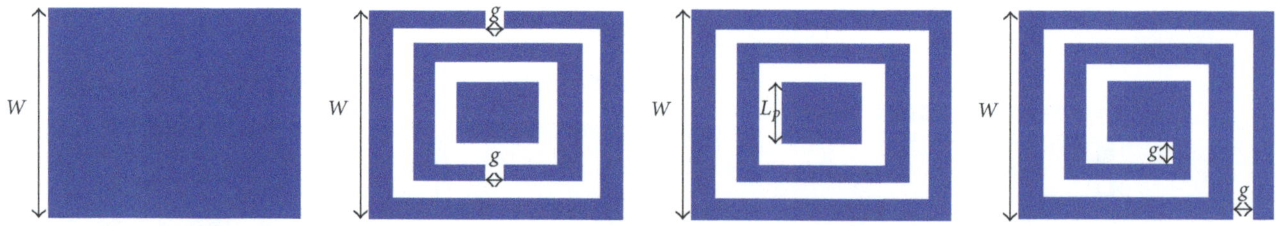

FIGURE 3: Different unit cell shapes of EBG ground plane mushroom, split ring resonator, ring, and spiral.

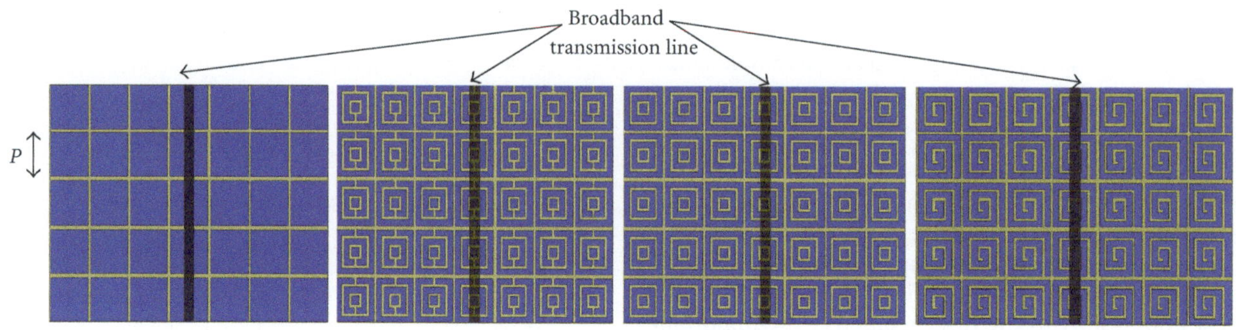

FIGURE 4: Different shapes of EBG structures (a) mushroom, (b) split ring resonator (SRR), (c) ring, and (d) spiral.

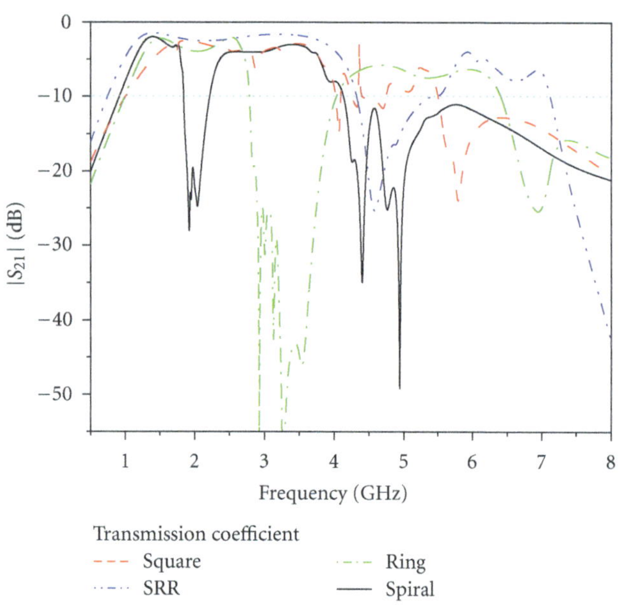

FIGURE 5: The transmission response for EBG ground, mushroom, split ring resonator, ring, and spiral.

3. Design of Different Shapes of EBG

The design of a uniplanar structure of a high impedance surface with compactness is a challenge nowadays. The absence of ground connecting makes the challenge significant, but not impenetrable. Because of the frequency sensitiveness of the surface impedance of the periodic structures like EBG, the configuration must be designed carefully.

Miniaturization and good performance are difficult to achieve simultaneously, particularly at lower frequencies. Some studies have been reported, which improve the performance of patch antennas using periodic structures [8]. In fact, it is well known that a patch antenna on a high dielectric constant substrate is a highly inefficient radiator due to surface wave losses and has a very narrow bandwidth. However, a significant effort has been recently made to realize high-performance antennas employing electromagnetic band-gap (EBG) structures printed on high permittivity substrates.

High-impedance electromagnetic surfaces have been studied by Sievenpiper [7]. In his approach, high-impedance surfaces (in general) consist of a lattice of metal plates, connected to a solid metal sheet

$$2W = \lambda_{\text{guid}}$$
$$\frac{W}{P} = 0.8 \sim 0.9, \quad (1)$$

where W, P, and λ_{guid} are side length, periodicity, and guided wavelength, respectively [7–12].

In this paper, different shapes of high-impedance surface structures were introduced as mushroom, ring, split ring resonator, and spiral shape cells, respectively, as shown in Figure 3 to produce wide or dual band gaps and as a means to suppress surface waves effectively [18–20]. The configurations of transmission line over the 7 × 5 cell arrangement used to test the proposed EBG structures are shown in Figure 4. The design consists of a one-layer substrate; printed on FR4

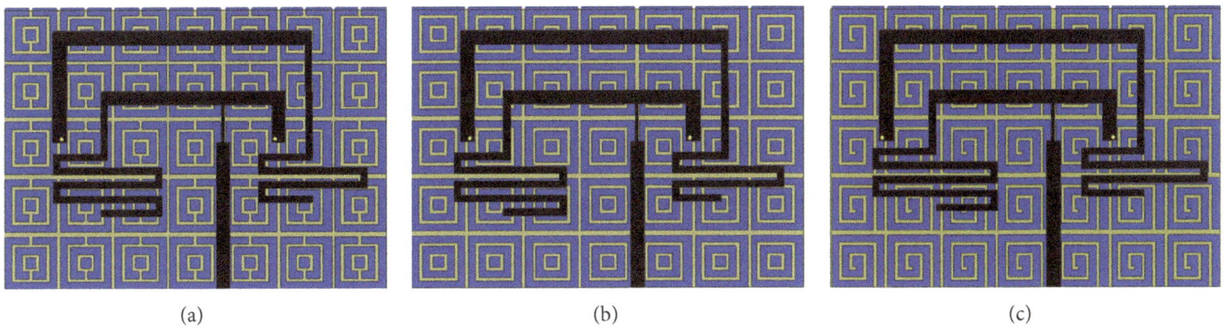

FIGURE 6: Printed meander L-loaded IFA with different EBG configurations (a) split ring resonator, (b) ring, and (c) spiral.

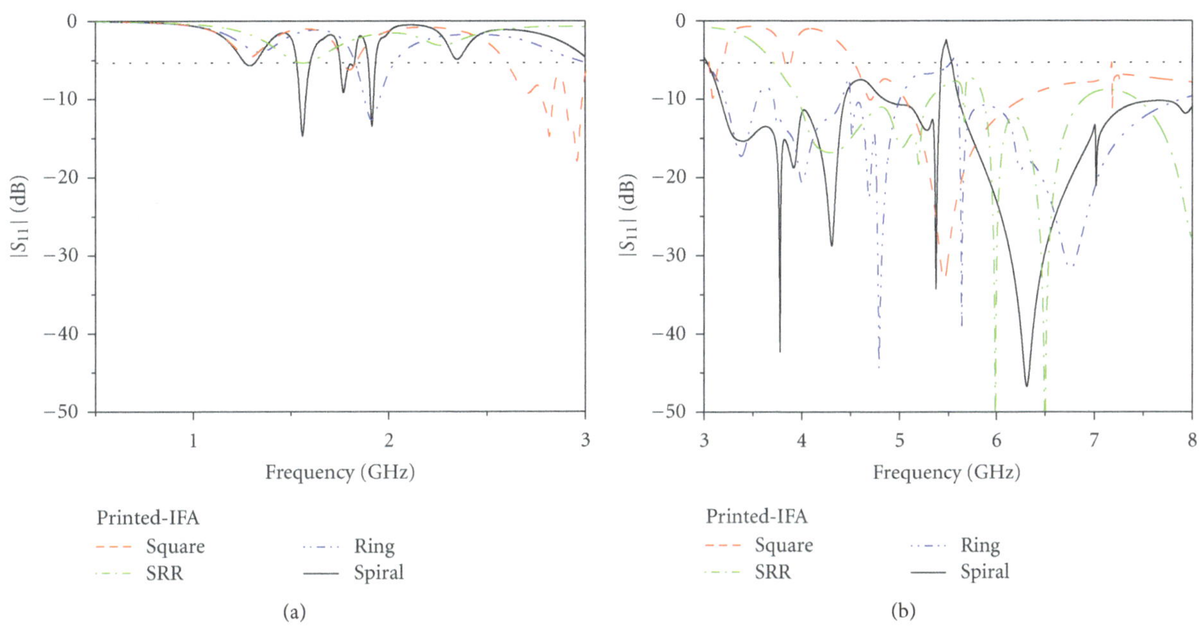

FIGURE 7: $|S_{11}|$ of the printed IFA antenna on different uniplanar EBG ground plane shapes.

board of thickness 1.6 mm. This figure represents the metallic periodic structure, which is etched on the dielectric substrate. The main element of this EBG square lattice has side length $W = 7.5$ mm and periodicity $P = 8$ mm. The split ring resonator is etched with gap width and length $g = 1$ mm. Then ring shape is etched with strip width equal to separation air gap $g = 1$ mm. Finally, one arm spiral is created to introduce the second band gap. The bandwidth of band gap depends on width of the arm, dielectric constant, and height of the substrate when using 50 Ω broadband transmission lines. The transmission responses of these configurations are shown in Figure 5. This figure shows that the square shape has one band gap started from 4.5 up to 8 GHz, while split ring resonator creates two band gaps at 4.5 to 5.5 GHz and from 7.5 to 8 GHz, the two-ring EBG shape has a dual band gap started from 2.5 to 4 GHz and from 6.5 to 8 GHz. Finally spiral EBG shapes create multiband gap, one at lower frequency from 1.7 to 2.3 GHz and the other started from 4 GHz to 8 GHz.

4. Printed IFA with EBG Ground Plane

The last step in the proposed antenna design is replacing the conventional antenna ground by 5×7 patche arrays of high-impedance surface EBG structure to suppress the surface waves and to increase the number of resonant frequencies by means of coupled resonators effect. The EBG lattice is arranged under the radiator. Inductive and capacitive values are added to reduce the antenna size and create multiband. Different shapes of uniplanar EBG ground plane, split ring resonator shape, ring shape, spiral shape, and mushroom shape are applied as a ground plane; all cells have the same dimensions and periodicity. The design started by applying these shapes of EBG ground plane on single printed IFA shown in Figure 1(a). The simulated reflection coefficient is shown in Figure 7. Multibands are generated at a low frequency especially less than 3 GHz by using spiral EBG, while at high frequency the spiral and ring shape give a good performance. Then L-loaded IFA is applied on different

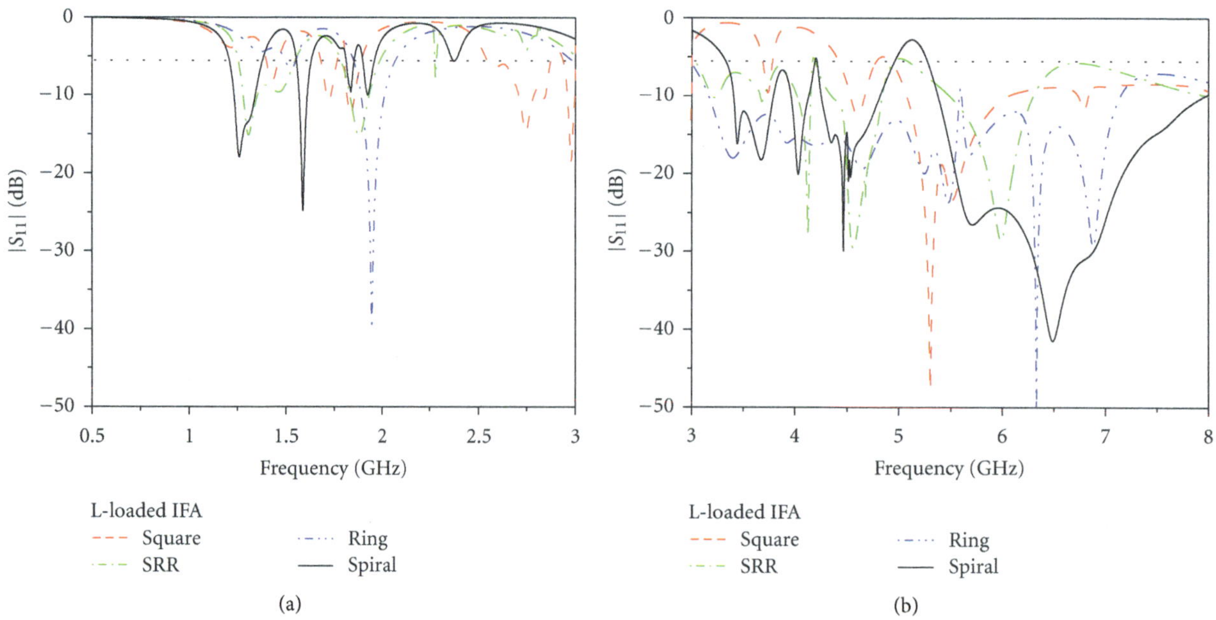

FIGURE 8: $|S_{11}|$ of the printed L-loaded IFA antenna on different uniplanar EBG ground plane shapes.

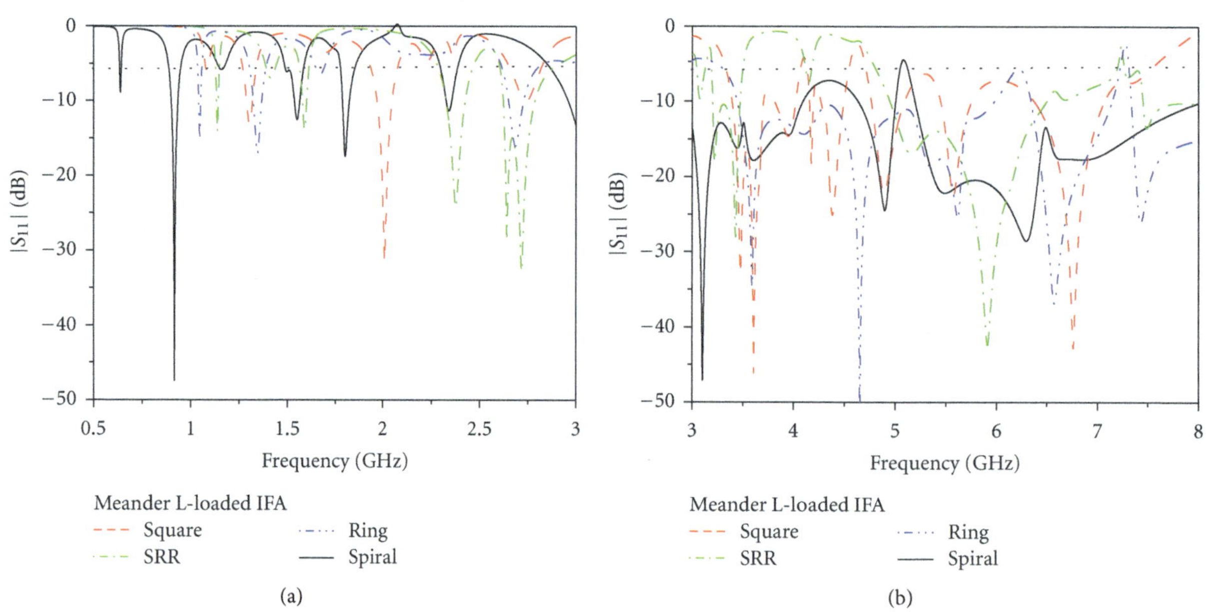

FIGURE 9: Comparison of $|S_{11}|$ of the printed meandered L-loaded IFA antenna on different uniplanar EBG ground plane shapes.

TABLE 1: Simulated antenna parameters.

Antenna Ch/cs.	Meandered L-loaded IFA Conventional ground	Final proposed antenna with ring EBG ground plane	Final proposed antenna with spiral EBG ground plane
Resonante Frequency (GHz)	2.65, 3, 3.6, 5.2	0.7, 1, 1.5, 2.5, 3, 3.6, 4.1, 5 and 6	0.55, 0.9, 1.5, 1.7, 2, 2.5, 3, 3.6, 4.1, 5 and 6
Average −6 dB BW	2%	5%, and extended from 3 to 8 GHz	Extended from 0.5 up to 8 GHz
Average Radiation Efficiency	0.8	0.83	0.85
Average gain (dBi)	3.5	4.2	4.8

TABLE 2: Simulated radiation patterns, antenna with conventional ground —, antenna with ring EBG ground - - and spiral EBG ground - ·· -. Axes are shown in Figure 1.

Frequancy	Φ = 0° (Z⊠)	Φ = 90° (ZY)	Θ = 90° (⊠Y)
1.8 GHz	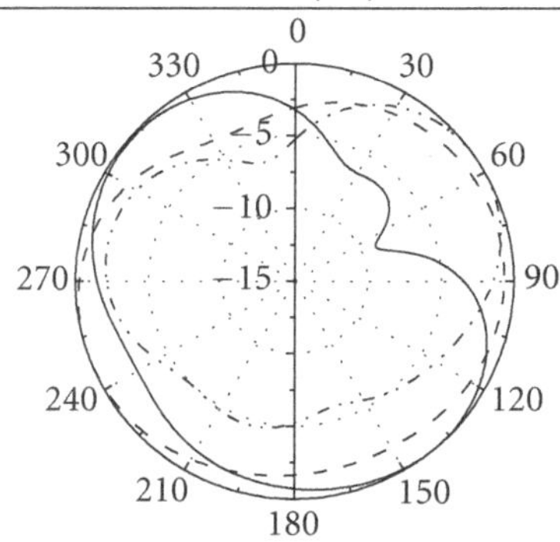	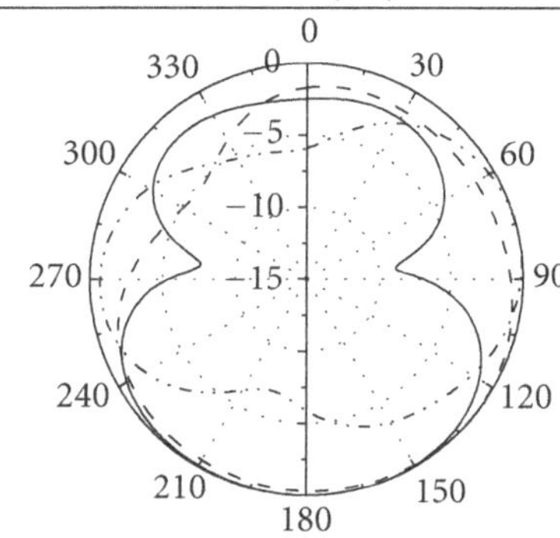	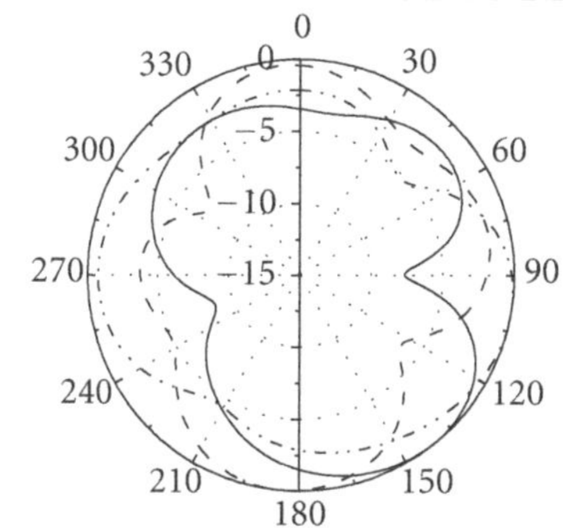
2.4 GHz	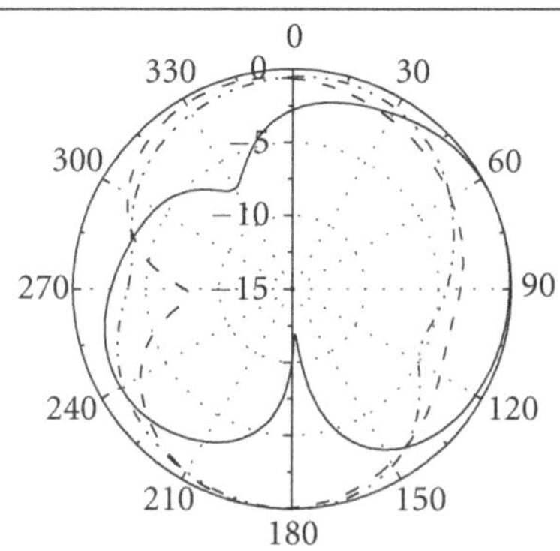	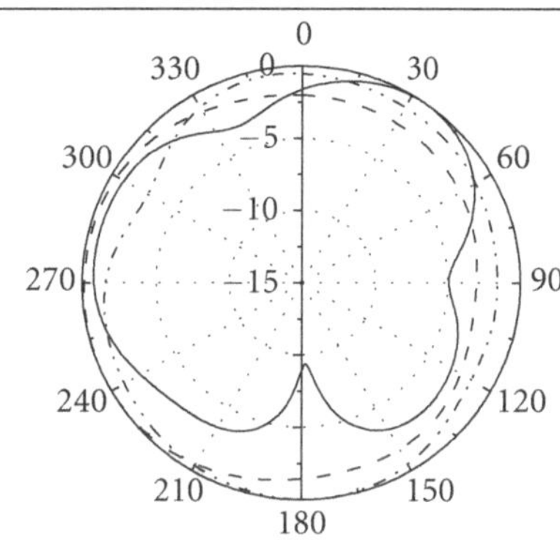	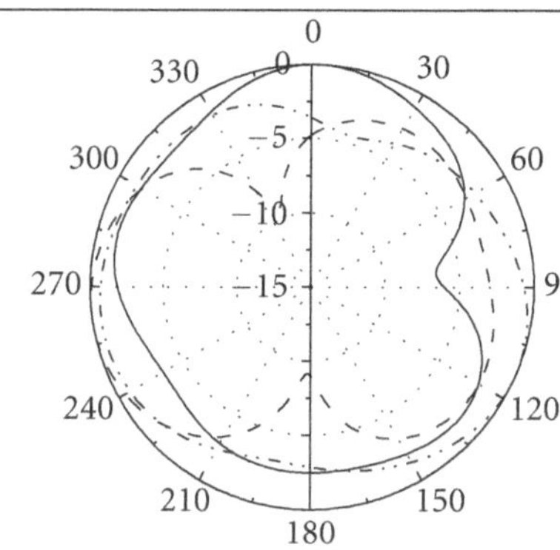
3.5 GHz	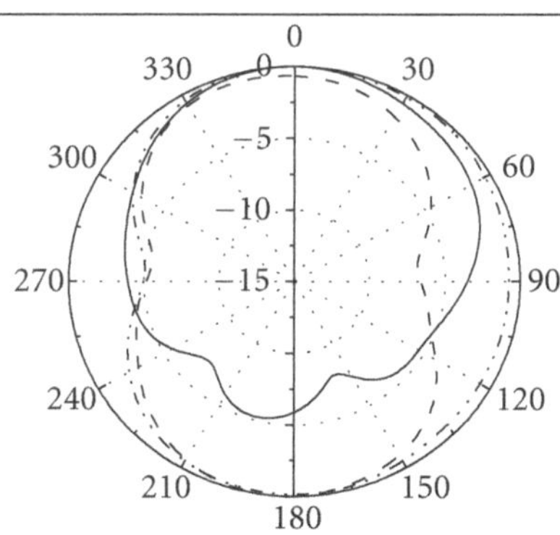	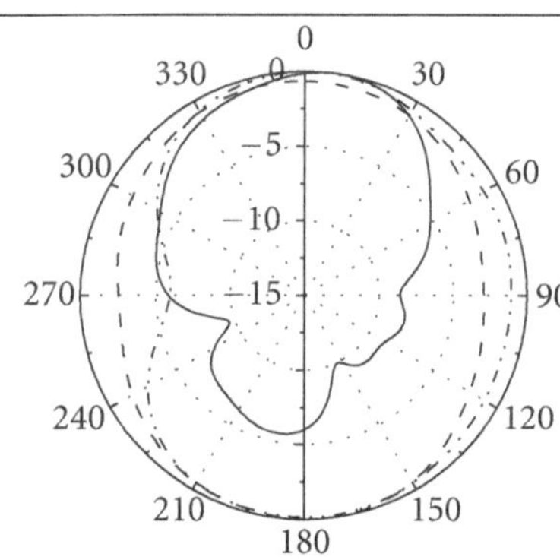	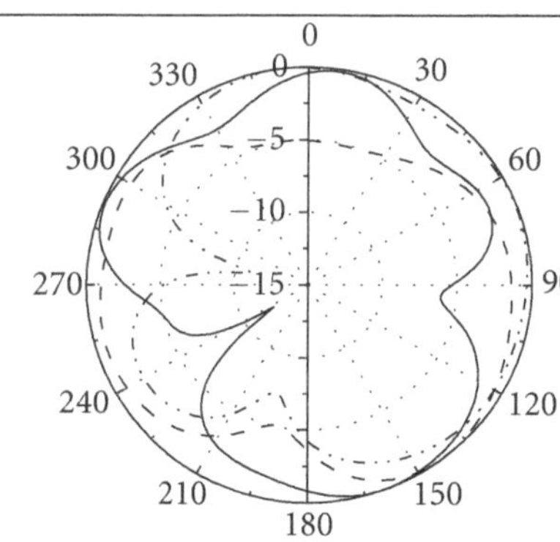
5.2 GHz	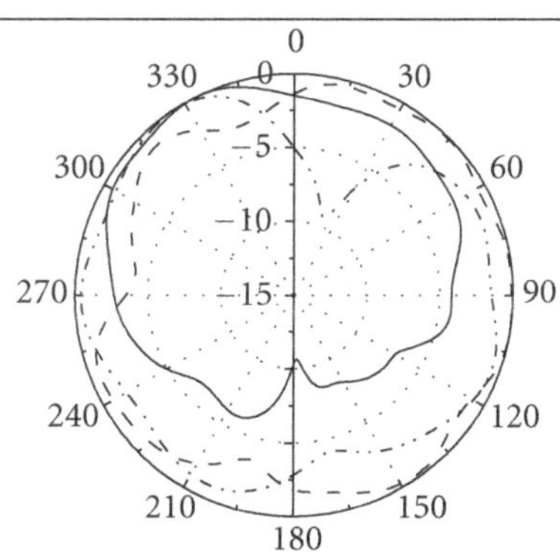	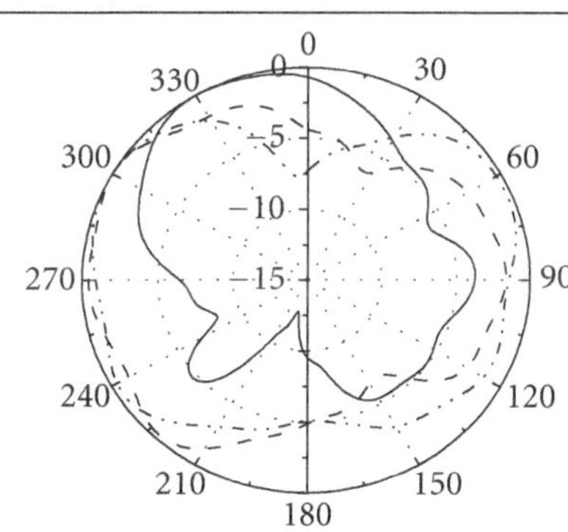	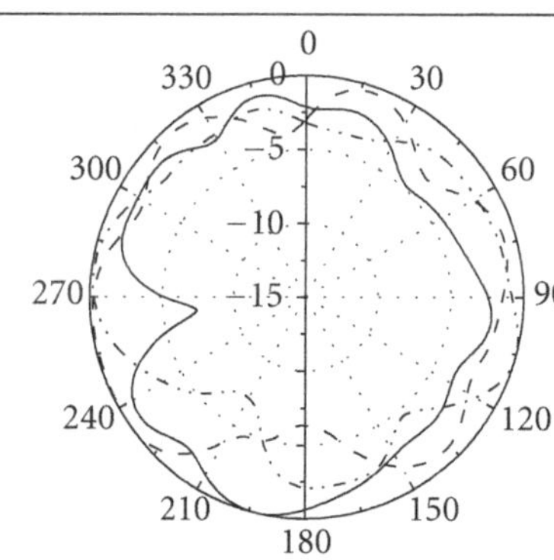

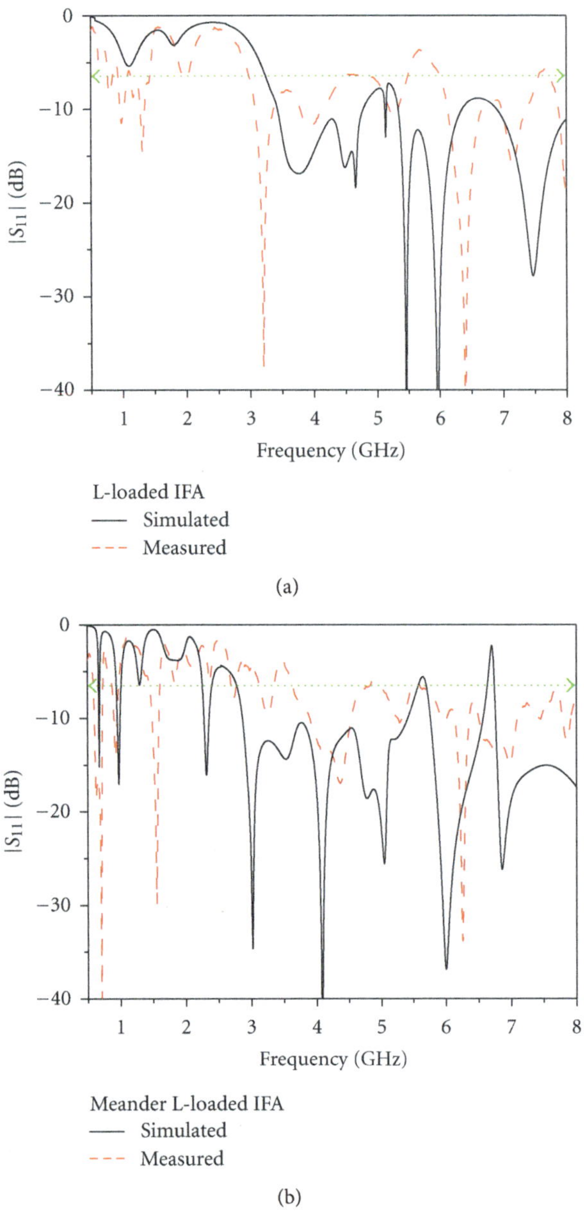

Figure 10: Comparison between measured and simulated $|S_{11}|$ of L-loaded IFA antenna with and without meandered shape on ring EBG ground plane.

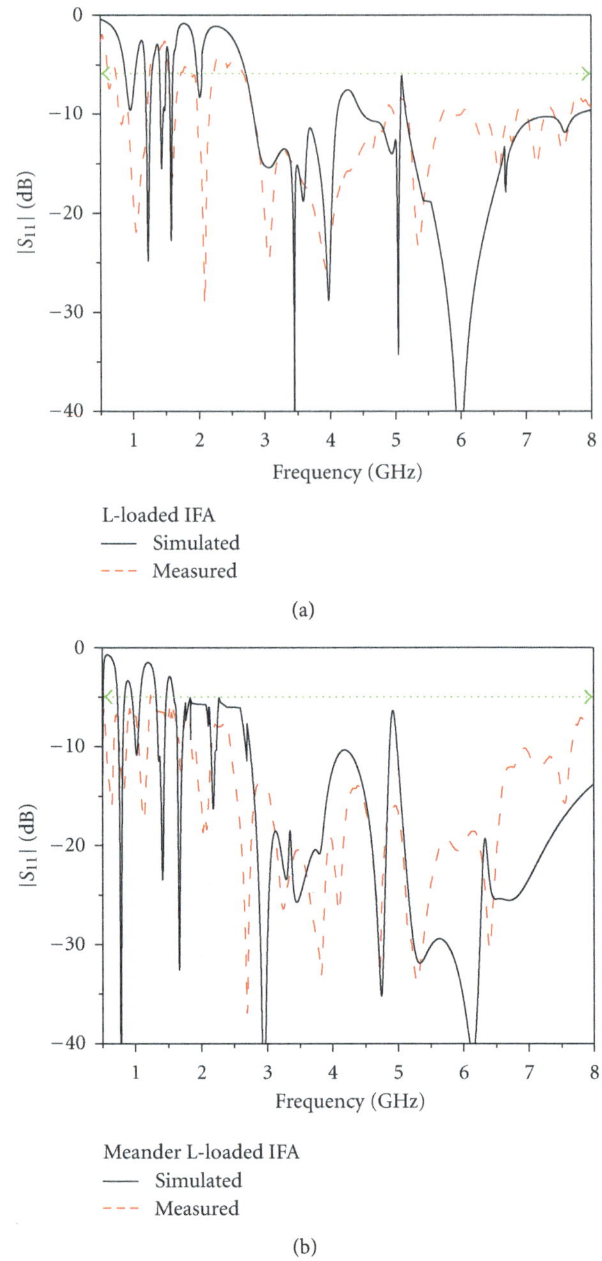

Figure 11: Comparison between measured and simulated $|S_{11}|$ of L-loaded IFA antenna with and without meandered shape on a spiral EBG ground plane.

EBG shapes ground plane shown in Figure 1(b). Simulated S_{11} is shown in Figure 8, the number of antenna resonances is increased especially by using spiral and ring shapes which give a good performance.

By loading meander shapes at the open end of L-loaded IFA antenna, the number of resonances is increased and the antenna matching is improved as shown in Figure 9. Spiral, ring, and split ring resonator give a good performance, respectively, with a good antenna matching. These results could be explained as a spiral shape creates band gap at a lower frequency from 1.5 to 2.5 GHz and band gap opened from 4 GHz to 8 GHz, while ring EBG shape ground plane creates a broad band gap from 2.5 to 4 GHz and another band gap opened from 6 GHz. So, the number of resonant frequencies is increased in this region and by adding inductance and capacitance under the radiator, the antenna size is reduced and the antenna matching is improved.

5. Simulation and Measurement Results

The optimized multiband proposed antennas, presented in a previous section, were fabricated on FR4 substrate ($\varepsilon_r = 4.4$, $h = 1.6$ mm, and $\tan\delta = 0.02$). From the above results, ring and spiral shapes ground plane were selected due to their

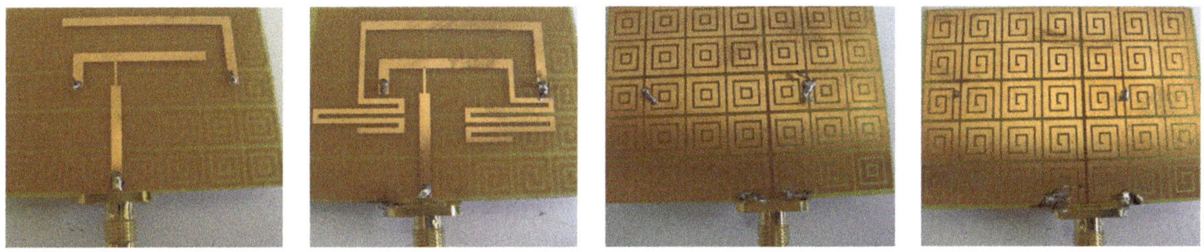

(a) (b) (c) (d)

FIGURE 12: Fabricated antennas, top layer: (a) L-printed-IFA, (b) meander loaded printed IFA, bottom layers, (c) ring-EBG ground plane, and (d) spiral-EBG ground plane.

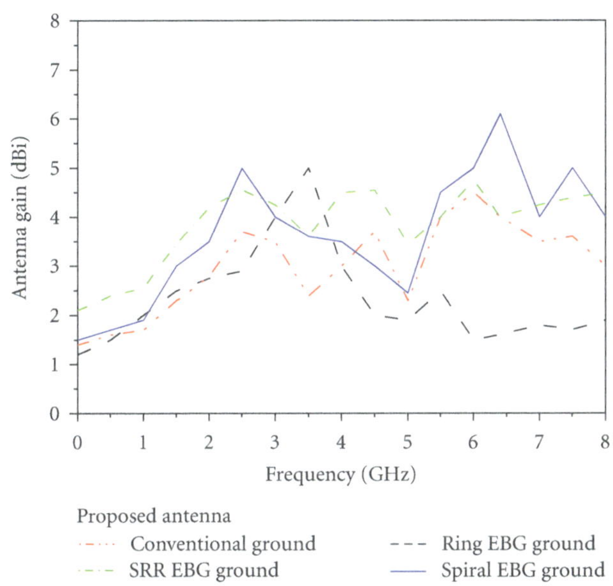

FIGURE 13: Simulated antenna gain for the meandered L-loaded printed IFA with and without a EBG ground plane.

good performance to be fabricated for both configurations, L-loaded IFA with and without meander shape. The fabricated antenna structures for both upper and lower layers are shown in Figure 6. The comparison between simulated and measured reflection coefficient is shown in Figure 10.

The measurements agree well with EM simulations and show that the operating bands have a return loss better than −6 dB. When these EBG structures were applied as printed IFA ground plane, the antenna resonant frequencies were reduced due to the capacitive effects that were generated between the two combined structures. Multi-resonant frequencies are achieved at 1.1 GHz, 1.58 GHz, and 2.49 GHz and the bandwidth extended from 3.5 GHz to 5.5 GHz due to number of band-gap stops as shown in Figure 11 for both L-loaded PIFA and meandered L-loaded IFA, respectively. The fabricated antenna is shown in Figure 12. The simulated antenna gain with and without EBG ground plane is shown in Figure 13 and Table 1. Figure 13 shows that the antenna with spiral EBG ground plane has an average gain greater than the other two shapes. However, the antenna with ring EBG ground plane has gain higher at frequency band from 3 to 4 GHz.

Finally, simulated radiation patterns for three antenna planes (Z⊠, ZY, and ⊠Y) are also studied. Comparison of radiation pattern between proposed antenna with and without a EBG ground plane at four different resonant frequencies 1.8 GHz, 2.4 GHz, 3.5 GHz, and 5.2 GHz is presented in Table 2.

6. Conclusion

Multiband compact printed IFAs for the internal mobile phone antenna and wireless applications have been proposed and studied. Three different shapes of EBG rather than mushroom were studied as a printed antenna ground plane. It can be concluded that spiral and ring EBG structures prototype improved the antenna matching, reduced the antenna size, and generated an extra number of resonant frequencies. The operating frequencies are 0.9 GHz, 1.57 GHz, and 2.48 GHz, resonant frequencies extended from 3.5 GHz up to 8 GHz for L-loaded printed-IFA spiral EBG ground plane while the lower frequency improved and the higher frequency was slightly unchanged by using meandered L-loaded IFA. EBG-printed IFA provides around an 80% size reduction, while the antenna characteristics remain almost unchanged. The proposed technique in this paper can be extended to create more operating bands.

Acknowledgment

This paper is funded by the National Telecommunication Regularity Authority (NTRA), the Ministry of Communication and Information Technology, Egypt.

References

[1] J. L. Volakis, C. Chih, and K. Fujimoto, *Small Antennas: Miniaturization Techniques and Applications*, Mc-Graw Hill, New York, NY, USA, 2010.

[2] M. N. M. Kehn, O. Quevedo-Teruel, and E. Rajo-Iglesias, "Reconfigurable loaded planar inverted-F antenna using varactor diodes," *IEEE Antennas and Wireless Propagation Letters*, vol. 10, pp. 466–469, 2011.

[3] R. A. Bhatti, Y. T. Im, and S. O. Park, "Compact PIFA for mobile terminals supporting multiple cellular and non-cellular standards," *IEEE Transactions on Antennas and Propagation*, vol. 57, no. 9, pp. 2534–2540, 2009.

[4] Q. Quevedo-Teruel, E. Pucci, and E. Rajo-Iglesias, "Compact loaded PIFA for multifrequency applications," *IEEE Transactions on Antennas and Propagation*, vol. 58, no. 3, pp. 656–664, 2010.

[5] K. L. Wong, *Compact and Broadband Microstrip Antennas*, Wiley-Inter Science, NewYork, NY, USA, 2002.

[6] S. Chaimool, K. L. Chung, and P. Akkaraekthalin, "Bandwidth and gain enhancement of microstrip patch antennas using reflective metasurface," *IEICE Transactions on Communications B*, vol. E93, no. 10, pp. 2496–2503, 2010.

[7] D. Sievenpiper, L. Zhang, R. F. Broas, N. G. Alexöpolous, and E. Yablonovitch, "High-impedance electromagnetic surfaces with a forbidden frequency band," *IEEE Transactions on Microwave Theory and Techniques*, vol. 47, no. 11, pp. 2059–2074, 1999.

[8] F. Yang and Y. Rahmat-Samii, *Electromagnetic Band-Gap Structures in Antenna Engineering: The Cambridge RF and Microwave Engineering Series*, Cambridge University Press, Cambridge, Mass, USA, 2008.

[9] F. Yang and Y. Rahmat-Samii, "Curl antennas over electromagnetic band-gap surface: a low profiled design for CP applications," in *Proceedings of the IEEE Antennas and Propagation. Society International Symposium*, vol. 3, pp. 372–375, July 2001.

[10] J. Y. Park, C. C. Chang, Y. Qian, and T. Itoh, "An improved low-profile cavity-backed slot antenna loaded with 2D UC-PBG reflector," in *Proceedings of the IEEE Antennas and Propagation. Society International Symposium*, pp. 194–197, July 2001.

[11] F. Yang, V. Demir, D. A. Elsherbeni, A. Z. Elsherbeni, and A. A. Eldek, "Planar dipole antennas near the edge of an EBG ground plane for WLAN applications," in *Proceedings of the IEEE Antennas and Propagation Society International Symposium*, vol. 1A, pp. 750–753, July 2005.

[12] D. Nashaat, H. A. Elsadek, E. A. Abdallah, M. F. Iskander, and H. M. Elhenawy, "Ultrawide bandwidth 2 × 2 microstrip patch array antenna using electromagnetic band-gap structure (EBG)," *IEEE Transactions on Antennas and Propagation*, vol. 59, no. 5, pp. 1528–1534, 2011.

[13] F. Yang and Y. Rahmat-Samii, "Microstrip antennas integrated with electromagnetic band-gap (EBG) structures: a low mutual coupling design for array applications," *IEEE Transactions on Antennas and Propagation*, vol. 51, no. 10, pp. 2936–2946, 2003.

[14] S. D. Assimonis, V. Yioultsis, and C. S. Antonopoulos, "Computational investigation and design of planar EBG structures for coupling reduction in antenna applications," *IEEE Transactions on Magnetics*, vol. 48, pp. 771–774, 2012.

[15] M. Y. Fan, R. Feng, Z. H. Hu, ⊠. ⊠. Zhang, and Q. Hao, "Advance in 2D-EBG research," *Journal of Infrared Millimeter Waves*, vol. 22, pp. 883–887, 2003.

[16] M. F. Abedin, M. Z. Azad, and M. Ali, "Wideband smaller unit-cell planar EBG structures and their application," *IEEE Transactions on Antennas and Propagation*, vol. 56, no. 3, pp. 903–908, 2008.

[17] M. Yamamoto and K. T. Nojima, "Leaf-shaped bowtie antenna backed by a periodic patch loaded grounded slab," in *Proceedings of the IEEE International Symposium on Antennas and Propagation*, pp. 622–625, 2011.

[18] R. C. Hadarig, M. E. de Cos, and F. Las-Heras, "Microstrip patch antenna bandwidth enhancement using AMC/EBG structures," *International Journal of Antennas and Propagation*, vol. 2012, Article ID 843754, pp. 1–6, 2012.

[19] A. Pirhadi, F. Keshmiri, M. Hakkak, and M. Tayarani, "Analysis and design of dual band high directivity EBG resonator antenna using square loop FSS AS superstrate layer," *Progress in Electromagnetics Research*, vol. 70, pp. 1–20, 2007.

[20] Z. Duan, S. Qu, and Y. Hou, "Electrically small antenna inspired by spired split ring resonator," *Progress In Electromagnetics Research Letters*, vol. 7, pp. 47–57, 2009.

7

Nonuniformly Spaced Linear Antenna Array Design Using Firefly Algorithm

Mohammad Asif Zaman and Md. Abdul Matin

Department of Electrical and Electronic Engineering, Bangladesh University of Engineering and Technology, Dhaka 1000, Bangladesh

Correspondence should be addressed to Mohammad Asif Zaman, asifzaman13@gmail.com

Academic Editor: Ramon Gonzalo

A nonuniformly spaced linear antenna array with broadside radiation characteristics is synthesized using firefly algorithm and particle swarm optimization. The objective of the work is to find the optimum spacing between the radiating antenna elements which will create a predefined arbitrary radiation pattern. The excitation amplitudes of all the antenna elements are assumed to be constant. The optimum spacing between the array elements are obtained using firefly algorithm. The minimum allowed distance between the antenna elements is defined in such a way that mutual coupling between the elements can be ignored. Numerical analysis is performed to calculate the far-field radiation characteristics of the array. Two numerical examples are shown to form two different desired predefined radiation patterns. The performance of the firefly algorithm and particle swarm optimization is compared in terms of convergence rate and global best solution achieved. The performances of the optimized nonuniformly spaced arrays are analyzed. Finally, contour plots of the radiated power from the optimized array in the horizontal plane and vertical plane in the far-field region are provided.

1. Introduction

Multiple antennas can be arranged in space in various geometrical configurations to form an antenna array with highly directive radiation pattern [1, 2]. The radiation characteristics of the antenna array depend on some input parameters. These parameters are the relative magnitude and phase of the excitation current of each radiating element, radiation characteristics of each radiating element, the geometrical configuration of the array, and the separation distance between the array elements [2]. An antenna array can be designed to produce almost any arbitrary prescribed pattern by controlling these parameters. For this reason, antenna arrays find application in RADAR and wireless communication systems [3, 4].

Most antenna arrays are designed to produce a directive beam at a particular direction and while keeping the sidelobe level (SLL) small to avoid interference with other radiating sources. In most cases, this is achieved by controlling the magnitude and phase of the excitation amplitudes [5, 6]. In most cases, a relatively simple geometry is considered where the distance between two consecutive radiating elements is constant. However, exact control of phase and magnitude of excitation current of array elements requires complex and expensive electronic circuitry [4]. In case of phased arrays, where the direction of the main beam needs to be controlled electronically in real time [7], uses of such electronic circuits are unavoidable. However, in many applications, a dynamic control of array radiation pattern is not required. It is highly desirable in such cases to design an antenna array that does not require complex circuitry to control the phase and magnitude of excitation currents. Desired radiation characteristics can be achieved by proper placement of each individual radiating element in space while keeping the excitation current constant for all elements. Such arrays are known as nonuniformly spaced antenna arrays.

In linear antenna arrays, the antenna elements are placed along a straight line. Linear array of identical radiating elements are one of the common type of antenna arrays [4, 6]. Design method of linear uniformly spaced antenna arrays are widely covered in literature [5–8]. Most of these methods employ a heuristic optimization algorithm to find the phase and/or magnitude of the excitation currents of the array elements while keeping the separation between the array elements uniform. Particle Swarm Optimization (PSO) [5, 7], Genetic Algorithm (GA) [6] and Artificial Bee Colony (ABC) algorithm [8] have been successfully used to design such arrays. Although, design methods of Nonuniformly Spaced Linear Antenna arrays (NUSLA) exist, they have not received equal attention in the literature.

One of the first major articles describing nonuniformly spaced antenna arrays was published in 1961 [9]. The design method employed perturbation methods and concentrated on sidelobe reduction. An iterative method for sidelobe reduction was developed by Hodjat and Hovanessian [10]. Recently, Fourier transform and window techniques have been applied for designing NUSLA [11]. These methods are suitable for synthesizing radiation pattern with low SLL, but lacks flexibility to synthesize arbitrary radiation pattern.

Optimization-algorithm-based design methods of NUSLA allowed engineers to easily synthesize arbitrary radiation patters. The applications of DE and GA to synthesize NUSLA for low SLL have been reported in the literature [12, 13]. PSO has been successfully used to design NUSLA with low SLL and nulls in arbitrary positions [14]. Ant Colony Optimization (ACO) algorithm has also been used to synthesize NUSLA with arbitrary radiation pattern [15].

Oraizi and Fallahpour have presented an impressive study of NUSLA in their paper [16]. They used modified GA-based analysis to optimize the position of the array elements. However, the work does not mention the number of iterations required by GA to converge or any other indication convergence time. GA usually is not fast converging for antenna array problems and it is often outperformed by PSO and ABC [7, 8]. In this paper, we used a newly developed optimization algorithm called Firefly Algorithm (FA) to synthesize NUSLA and compare the performance of the algorithm with existing methods.

FA is a heuristic numeric optimization algorithm inspired by the behavior of fireflies [17, 18]. FA has been successfully used in many applications. Recently, FA has been used by Basu and Mahanti for designing linear antenna arrays [19]. The work concentrated on designing equally spaced linear array with variable excitation current. No work has yet been published on the application of FA for NUSLA design.

In this paper, FA is used to find the optimum spacing between the array elements to produce a desired broadside radiation pattern. The design examples with numerical results are provided. In the first example, a NUSLA is synthesized with minimum possible SLL. In the second example, a NUSLA is synthesized with low SLL and nulls in arbitrary direction. The paper is organized as follows: array geometry and mathematical formulation of the radiated electric field radiated by the array is presented in Section 2. A brief description of FA is given in Section 3. Numerical results are provided in Section 4 and concluding remarks are given in Section 5.

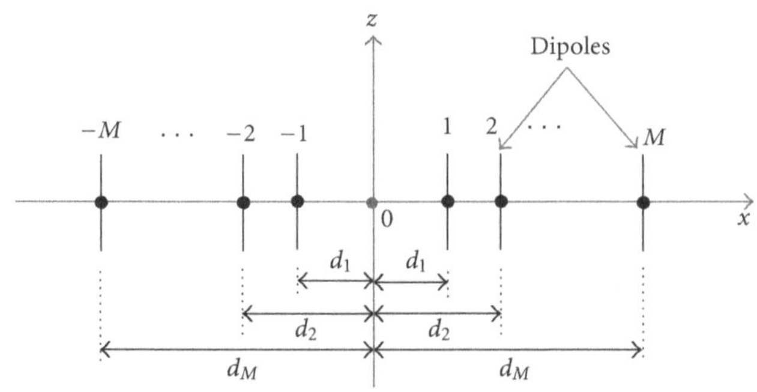

FIGURE 1: Two-dimensional schematic diagram of the array geometry.

2. Array Geometry and Mathematical Formulation

The linear antenna array is assumed to be composed of N identical radiating elements. The radiating elements are taken to be dipoles antenna. The dipoles are positioned symmetrically around the origin on the x-axis. The arms of the dipoles are parallel to the z-axis. A two-dimensional schematic diagram of the array geometry is shown in Figure 1.

The total number of elements, N is assumed to be even. The elements are divided into two groups of M elements where $N = 2M$. The elements are numbered as $-M, -M + 1 \cdots - 1, 1, 2, \ldots M - 1, M$. The distances of elements $1, 2, \ldots M$ from origin is denoted by $d_1, d_2, \ldots d_M$. Due to symmetry, $-1, -2, \ldots -M$ elements have the same distance values. The symmetry around the origin creates symmetrical radiation pattern, which is often desirable. It is possible to create symmetrical radiation pattern with unsymmetrical positioning of the array elements. However, the symmetry condition reduces computational complexities, as now only position of M elements must be optimized instead of position of N elements. To avoid complexities, symmetrical distribution of the elements around the origin is assumed in this paper.

The radiated field of the array depends on the radiation pattern of each array element and relative spacing of the array elements. The far-field radiation pattern of the array is given by [1, 3]

$$\mathrm{FF}(\theta, \phi) = \mathrm{EP}(\theta, \phi) \times \mathrm{AF}(\theta, \phi). \tag{1}$$

Here, $\mathrm{EP}(\cdot)$ is the radiation pattern of individual array elements, $\mathrm{AF}(\cdot)$ is the array factor, and (θ, ϕ) are the zenith and azimuth angle of the spherical coordinate system. The coordinate system and the three dimensional geometry are shown in Figure 2.

The radiation pattern of each array element is assumed to be

$$\mathrm{EP}(\theta, \phi) = \sin \theta. \tag{2}$$

The array factor is given by [16]

$$\mathrm{AF}(\theta, \phi) = \frac{1}{N} \sum_{n=1}^{M} \cos(k d_n \cos \phi). \tag{3}$$

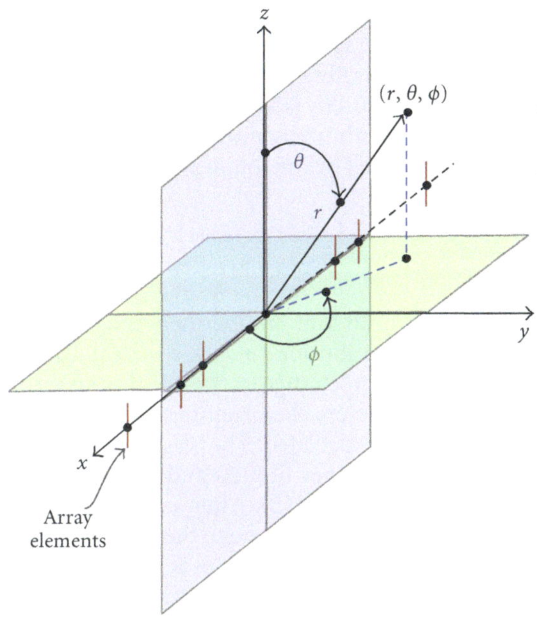

FIGURE 2: Coordinate system and three dimensional geometry of the array.

Here, k is wave number $= 2\pi/\lambda$ and λ is wavelength of radiation.

Using (1), (2), and (3), the overall radiation pattern is given by

$$\mathrm{FF}(\theta, \phi) = \sin\theta \left[\frac{1}{N} \sum_{n=1}^{M} \cos\left(k d_n \cos\phi\right) \right]. \quad (4)$$

In this paper, the objective is to find optimum values of d_n to produce a desired far-field radiation pattern. It is noted that for the given geometry and type of radiating elements chosen, d_n does not affect the field variation with θ. So, for optimization process is carried out at $\theta = 90°$ plane only.

Desired radiation pattern must be defined to measure the performance of the array. Since two different designs are to be implemented using the proposed method, two different desired patterns are defined as follows:

$$\mathrm{FF}_{\mathrm{des},1}(90°, \phi) = \begin{cases} 0\ \mathrm{dB}, & -\dfrac{\mathrm{BW}}{2} < \phi < \dfrac{\mathrm{BW}}{2}, \\ \mathrm{SLL}_{\mathrm{des}}, & \text{otherwise}. \end{cases} \quad (5)$$

$$\mathrm{FF}_{\mathrm{des},2}(90°, \phi) = \begin{cases} 0\ \mathrm{dB}, & -\dfrac{\mathrm{BW}}{2} < \phi < \dfrac{\mathrm{BW}}{2}, \\ \mathrm{NLL}_{\mathrm{des}}, & \dfrac{\phi_{\mathrm{NL}}}{2} < \phi < \dfrac{\phi_{\mathrm{NH}}}{2}, \\ \mathrm{SLL}_{\mathrm{des}}, & \text{otherwise}. \end{cases} \quad (6)$$

Here, BW is desired main beamwidth, $\mathrm{SLL}_{\mathrm{des}}$ is desired sidelobe level, $\mathrm{NLL}_{\mathrm{des}}$ is desired null level, ϕ_{NS} is desired starting angular position of the null, and ϕ_{NE} is desired ending angular position of the null.

Equation (5) represents a desired radiation pattern with a prescribed SLL and beamwidth. Equation (6) represents a desired radiation pattern with a specific SLL and beamwidth along with a predefined null level at an arbitrary direction. The angular extent of the null region is defined by the angles ϕ_{NS} and ϕ_{NE}.

The deviation of the obtained far-field pattern from the desired pattern is evaluated using the cost function. The cost function is defined as

$$f_{\cos t} = \begin{cases} \displaystyle\sum_{\phi=0°}^{180°} \left\{\mathrm{FF}(90°, \phi) - \mathrm{FF}_{\mathrm{des}}(90°, \phi)\right\}^2, \\ \mathrm{FF}(90°, \phi) > \mathrm{FF}_{\mathrm{des}}(90°, \phi), \\ 0, \quad \text{otherwise}. \end{cases} \quad (7)$$

Here, the summation is performed over the discrete field points that are numerically calculated. For the two different desired patterns defined in (5) and (6), two different cost functions can be defined. For each case, the fitness function is defined as

$$f_{\mathrm{fitness}} = -f_{\cos t}. \quad (8)$$

It can be noted that, when the obtained radiation pattern matches with the desired pattern, the cost function and the fitness function have zero values. When the patterns do not match, the cost function has positive values and the fitness function has negative values. Large deviations result is high positive cost values and high negative fitness values. So, the goal of the optimization process is to minimize the cost function or maximize the fitness function by finding appropriate values of d_n.

3. Overview of Firefly Algorithm

FA is an optimization algorithm inspired by behavior and motion of fireflies. It is a population-based optimization algorithm which uses swarm intelligence to converge [17, 18]. It is similar to other optimization algorithms employing swarm intelligence such as PSO and ABC. But FA is found to have superior performance in many cases [19].

FA initially produces a swarm of fireflies located randomly in the search space. The initial distribution is usually produced from a uniform random distribution. The position of each firefly in the search space represents a potential solution of the optimization problem. The dimension of the search space is equal to the number of optimizing parameters in the given problem. The fitness function takes the position of a firefly as input and produces a single numerical output value denoting how good the potential solution is. A fitness value is assigned to each firefly. The FA uses a phenomenon known is *bioluminescent communication* to model the movement of the fireflies through the search space. The brightness of each firefly depends on the fitness value of that firefly. Each firefly is attracted by the brightness of other fireflies and tries to move towards them. The velocity or the pull a firefly towards another firefly depends on the attractiveness. The attractiveness depends on the relative distance between the fireflies. It can be a function of the brightness of the fireflies as well. A brighter firefly far away may not be as attractive as a less bright firefly that is closer. In each iterative step, FA computes the brightness and the relative attractiveness of each firefly. Depending on these values, the positions of the fireflies are updated. After a sufficient amount of iterations, all fireflies converge to the best possible position on the search space.

The number of fireflies in the swarm is known as the population size, P. The selection of population size depends on the specific optimization problem. However, typically a population size of 20 to 40 is used for PSO and FA for most applications [14, 17]. For the current problem, the solution space is M dimensional, where each dimension represents the position of an array element. The position of the nth firefly is denoted by a vector $\mathbf{x_n}$ where,

$$\mathbf{x_n} = \left(x_n^1, x_n^2, x_n^3, \ldots, x_n^m, \ldots, x_n^M\right). \quad (9)$$

Here, $n = 1, 2, 3 \cdots P$ and $m = 1, 2, 3, \ldots, M$.

The search space is limited in mth dimension by the following inequality:

$$x_{\text{Low}}^m < x_n^m < x_{\text{High}}^m. \quad (10)$$

The value of the variables x_{Low}^m and x_{High}^m depend on the optimization problem. For the current NUSLA synthesis problem, these variables represent minimum allowed and maximum allowed separation distance of consecutive array elements. Initially, the positions of the fireflies are generated from a uniform distribution using the following equation:

$$x_n^m = x_{\text{Low}}^m + \left(x_{\text{High}}^m - x_{\text{Low}}^m\right) \times \text{rand}. \quad (11)$$

Here, rand is a uniform random variable with values ranging from 0 to 1. The value of *rand* is different for each value of m and n. Equation (11) generates random values from a uniform distribution within the prescribed range defined by (10). The initial distribution does not significantly affect the performance of the algorithm. Each time the algorithm is executed, the optimization process starts with a different set of initial points. However, in each case, the algorithm finds the optimum solution. In case of multiple possible sets of solutions, the algorithm may converge on different solutions each time. But each of those solutions will be valid as they all will satisfy the design requirements.

The brightness of the nth firefly, I_n is given by

$$I_n = f_{\text{fitness}}(\mathbf{x_n}). \quad (12)$$

The attractiveness between the nth and pth firefly, β_{np} given by [17]

$$\beta_{np} = \beta_o \exp\left(-\gamma r_{np}^2\right). \quad (13)$$

Here, r_{np} is Cartesian distance between $\mathbf{x_n}$ and $\mathbf{x_p}$ given by

$$r_{np} = \left\|\mathbf{x_n} - \mathbf{x_p}\right\| = \sqrt{\sum_{m=1}^{M}\left(x_n^m - x_p^m\right)^2}. \quad (14)$$

β_o is a constant taken to be 1. γ is another constant whose value is related to the dynamic range of the search space.

The position of firefly is updated in each iterative step. If the brightness of pth firefly is larger than the brightness of the nth firefly, then the nth firefly moves towards the pth firefly. The motion is denoted by the following equation:

$$\mathbf{x_{n,new}} = \mathbf{x_{n,old}} + \beta_{mn}\left(\mathbf{x_{p,old}} - \mathbf{x_{n,old}}\right) + \alpha(\text{rand} - 0.5). \quad (15)$$

Here, rand is a random number between 0 and 1, taken from a uniform distribution. α is a constant whose value depends on the dynamic range of the solution space.

At each iterative step, the brightness and the attractiveness of each firefly is calculated. The brightness of each firefly is compared with all other fireflies and the positions of the fireflies are updated using (15). After a sufficient number of iterations, all the fireflies converge to the same position in the search space and the global optimum is achieved.

4. Numerical Simulation and Results

For numerical simulations, a $N = 20$ element NUSLA is considered. As symmetry about the origin is assumed, only the position of $M = 10$ elements located on the positive x-axis needs to be optimized. The population size is taken to be 40 and maximum number of iterations is limited to 65. The search space is bounded by defining the minimum and maximum separation between two consecutive elements to be 0.35λ and 0.9λ, respectively, implying the solution space limiting variables defined in (10) have the following values:

$$\begin{aligned} x_{\text{Low}}^m &= 0.35\lambda, \\ x_{\text{High}}^m &= 0.9\lambda. \end{aligned} \quad (16)$$

For $m = 1, 2, \ldots, M$. The limiting values in all dimensions of the solution space are assumed to be the same. These values are selected so that mutual coupling between the elements remains negligible [16]. The dynamic range of the search space is $0.9\lambda - 0.35\lambda = 0.55\lambda$. α is taken to be 80% of the dynamic range initially and the value is linearly decreased to zero at maximum iteration number. The parameter γ is taken to be equal to the dynamic range. So,

$$\begin{aligned} \alpha &= 0.8 \times \left(x_{\text{High}}^m - x_{\text{Low}}^m\right), \\ \gamma &= x_{\text{High}}^m - x_{\text{Low}}^m. \end{aligned} \quad (17)$$

The parameters of the first desired radiation pattern defined by (5) are selected as

$$\begin{aligned} \text{BW} &< 13.4^\circ, \\ \text{SLL}_{\text{des}} &= -23.5\,\text{dB}. \end{aligned} \quad (18)$$

These values are selected based on typical beamwidth and SLL values of 20 element linear antenna array [16]. From multiple trial runs of the optimization algorithm, it was verified that these were the lowest possible values of beamwidth and SLL.

Table 1: Optimized position of the array elements for the first design example.

Element number, n		1	2	3	4	5	6	7	8	9	10
Position, d_n/λ	FA	0.189	0.540	0.934	1.292	1.726	2.141	2.639	3.173	3.899	4.634
	PSO	0.239	0.857	1.362	1.929	2.576	3.193	3.881	4.762	5.662	6.397

The parameters of the second desired radiation pattern defined by (6) are selected as

$$\begin{aligned} \text{BW} &< 12^\circ \\ \text{SLL}_{\text{des}} &= -20\,\text{dB} \\ \text{NLL}_{\text{des}} &= -40\,\text{dB} \\ \phi_{\text{NS}} &= 46^\circ, 126^\circ \\ \phi_{\text{NE}} &= 54^\circ, 134^\circ. \end{aligned} \tag{19}$$

Two nulls symmetrical spaced around the main beam (located at 90°) extending from 46° to 54° and 126° to 134° with null level of –40 dB are desired. To achieve this, the requirements on desired SLL must be relaxed compared to the previous case. For this reason, desired SLL in this case is set to –20 dB compared to previous value of –23 dB. This relaxation in SLL constraint allows the possibility of reduction of main beamwidth. For this reason, the maximum main beamwidth in this case is selected as 12° compared to 13.4° of the previous case.

The NUSLA for the two desired radiation patterns is designed using FA algorithm. For comparison, the same designs are performed using PSO algorithm also. The value of the parameters of the PSO algorithm is selected the same as the ones used in [14]. However, the population size of 20 was used in [14]. For proper comparison, the population size for the PSO algorithm is taken to be 40, which is identical to the population size used in the FA algorithm.

FORTRAN computer coding (with G95 compiler) is used to implement the FA and PSO algorithm and formulating the far-field pattern. The optimum position of the array elements obtained from FA and PSO for the first desired radiation pattern is shown in Table 1.

It can be observed from Table 1 that the two algorithms produce different sets of solution values. Using these values, the far-field radiation pattern at $\theta = 90^\circ$ plane (xy plane) is calculated. The radiation pattern obtained from the FA is shown in Figure 3 and the radiation pattern obtained from the PSO algorithm is shown in Figure 4. The desired pattern is also highlighted in red in the same figures.

It can be seen that both radiation patterns satisfy most of the design criterion. In [16], the maximum SLL of –22.6 dB was achieved with a 20 element NUSLA of similar geometry. In this paper, using FA algorithm, the SLL value is limited to –23.5 dB using the same number of array elements. So the proposed FA-based outperforms the GA-based method described in [16]. For the PSO algorithm, SLL constraint has not been met by the design obtained from PSO in only a small angular region. So, FA outperforms PSO in terms of output as well. However, the performance of the algorithms cannot be compared form the final output only. The convergence speed of the algorithms can show the difference in performance: the convergence rate observed from the fitness versus iteration plot for the algorithms. This is shown in Figure 5.

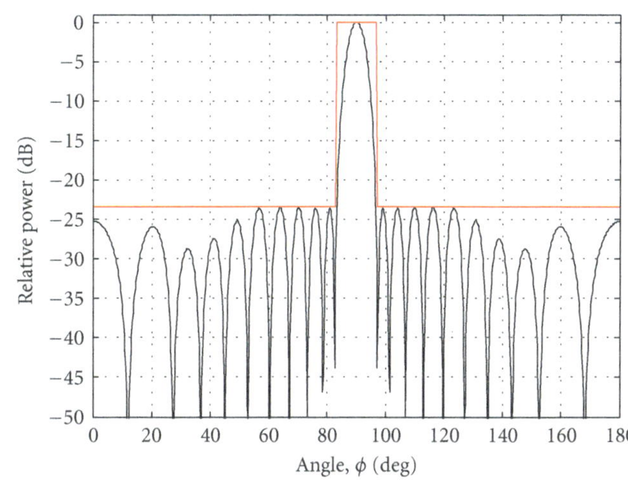

Figure 3: Optimized radiation pattern of the NUSLA at $\theta = 90^\circ$ plane obtained from FA algorithm for the first design example.

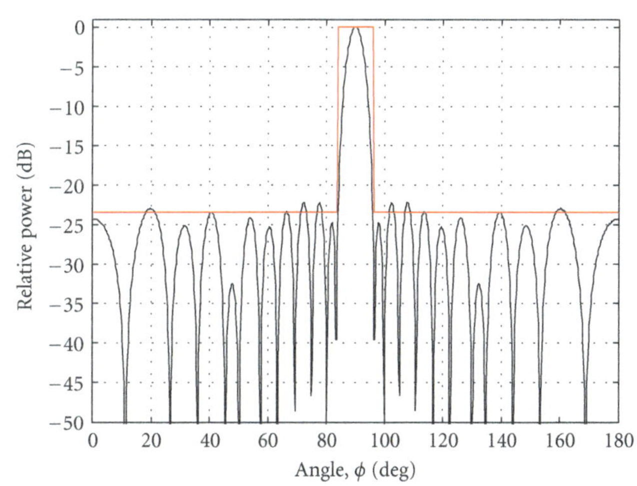

Figure 4: Optimized radiation pattern of the NUSLA at $\theta = 90^\circ$ plane obtained from PSO algorithm for the first design example.

It can be seen that for FA, the fitness function value reaches its maximum possible value zero within 20 iterations. The saturation of fitness value implies convergence. However, for PSO, maximum value of –14.317 is achieved. It took PSO around 50 iterations to reach this value. The fact that zero fitness value was not achieved implies that the obtained radiation pattern does not perfectly match with the desired radiation pattern, which is apparent from Figure 4. The most noticeable fact observed from Figure 5 is the fast converging characteristics of FA compared to PSO indicated

TABLE 2: Optimized position of the array elements for the second design example.

Element number, n		1	2	3	4	5	6	7	8	9	10
Position, d_n/λ	FA	0.214	0.689	1.087	1.504	2.018	2.578	3.045	3.805	4.449	5.106
	PSO	0.219	0.654	1.125	1.539	2.176	2.666	3.236	3.950	4.478	5.083

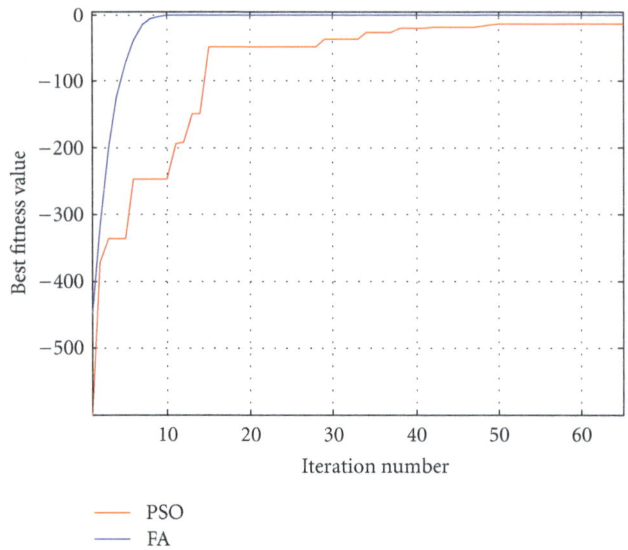

FIGURE 5: Fitness value versus iteration number for FA and PSO algorithm for the first design problem.

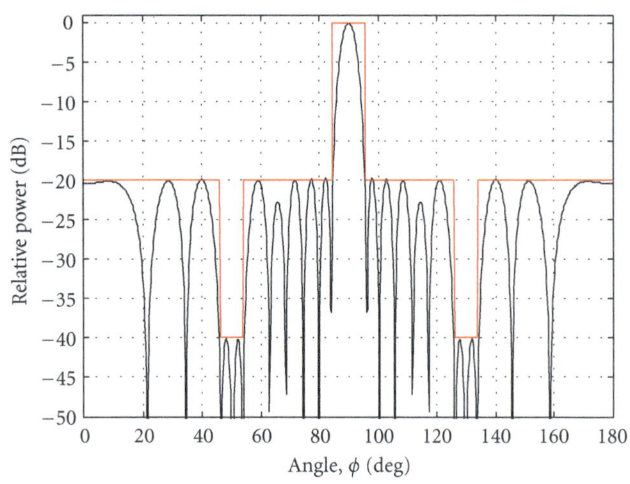

FIGURE 6: Optimized radiation pattern of the NUSLA at $\theta = 90°$ plane obtained from FA algorithm for the second design example.

by the sharper slope of the blue curve compared to the red one.

Similar analysis was performed to synthesize NUSLA for the second desired radiation pattern. The second desired pattern is characterized by two nulls located symmetrically around the main-lobe region. The optimum position of the array elements obtained from FA and PSO for the first desired radiation pattern is shown in Table 2.

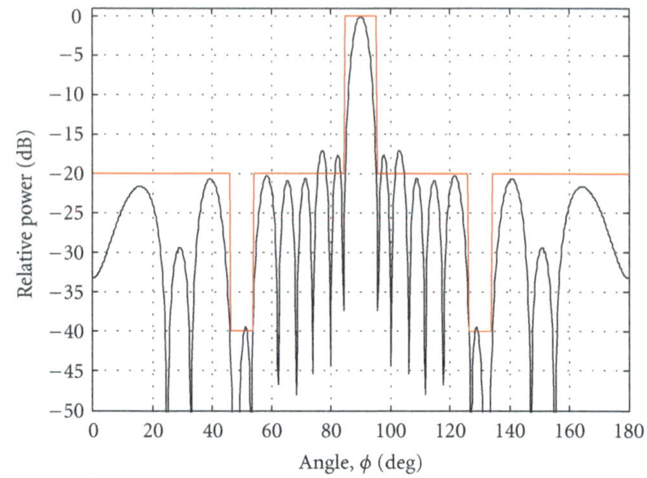

FIGURE 7: Optimized radiation pattern of the NUSLA at $\theta = 90°$ plane obtained from PSO algorithm for the second design example.

It can be observed from Table 2 that the two algorithms produce different sets of solution values. The radiation pattern obtained from the FA and PSO algorithm are shown in Figures 6 and 7, respectively. The desired pattern is also highlighted in red in the same figures.

It can be observed that the results obtained from FA satisfy the design criterion very well. The results from PSO maintain the desired pattern for the most part, with only a few sidelobes exceeding the limit. This implies that PSO has not converged perfectly. This fact can be further illustrated by observing the fitness function values. This is shown in Figure 8. Again, the superior performance of FA compared to PSO can be observed by the sharper slope of the blue curve compared to the red curve. The maximum fitness value achieved for FA is –0.1456, whereas this value is –82.223. FA clearly outperforms PSO in terms of convergence rate and maximum fitness value reached within a limited number of iterations.

Due to the presence of random number parameters in FA and PSO, the results vary on each time the program is executed. However, in all cases, the FA converges faster than PSO. Since, PSO usually outperforms GA for antenna array problems, it is expected that FA will outperform GA as well. ABC usually requires over 100 iterations (with a population size of 30 to 40) to reach convergence in antenna array problems [7, 8]. So, it can be concluded that FA is a very fast converging algorithm.

Using the optimized position of the array elements, the normalized radiated power in the far-field zone of the antenna array is calculated for the first design problem. Equation (4) along with the fact that power varies at $1/r^2$ is

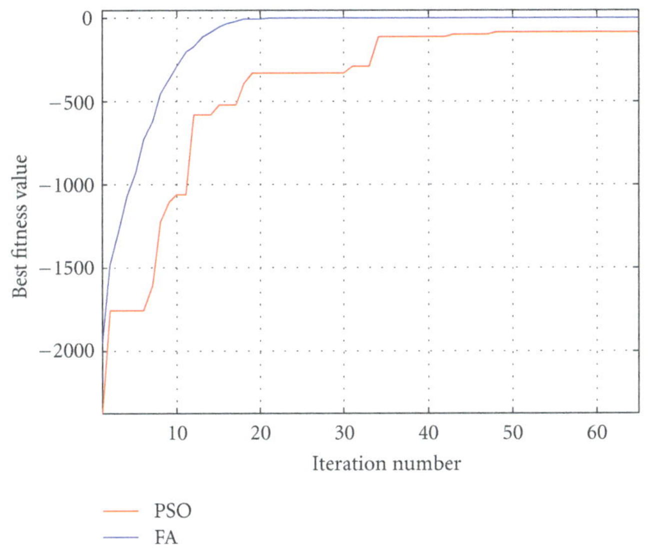

FIGURE 8: Fitness value versus iteration number for FA and PSO algorithm for the second design problem.

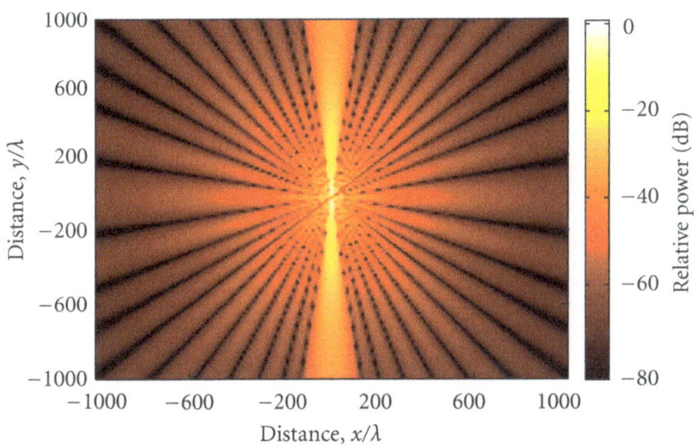

FIGURE 9: Relative power at far-field region in the xy plane.

used to calculate contour plots of normalized radiated power at xy plane and xz plane. The plots are shown in Figures 9 and 10.

As the array elements are located along the x axis, Figure 9 shows the broadside radiation characteristics of the array. The same results are also seen from Figure 3. The power distribution in the xz plane resembles the radiation pattern of a dipole antenna. This is expected, as the array elements are dipoles and the array factor does not contribute to the field distribution at xz plane ($\phi = 0^\circ$ plane).

5. Conclusion

Two nonuniformly spaced linear antenna arrays are designed for two distinct design requirements. The radiation characteristics of the designed arrays are numerically evaluated. The separations between the array elements are determined using firefly algorithm and PSO algorithm. It has been found that the design obtained from FA satisfies the predefined SLL,

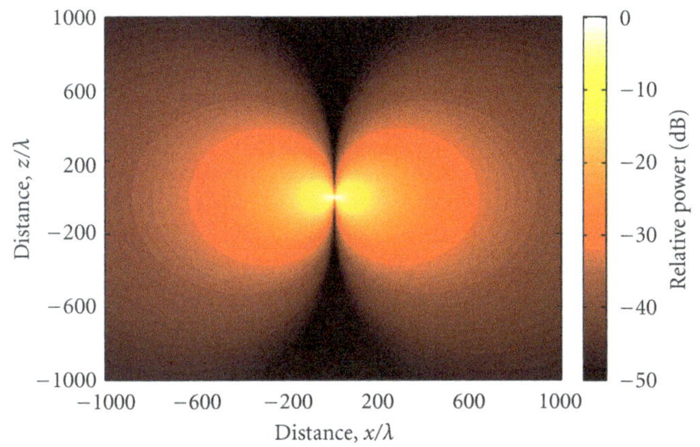

FIGURE 10: Relative power at far-field region in the xz plane.

beamwidth, and null requirements very well. The design obtained from PSO satisfies most of the design requirements but it is outperformed by FA in terms of convergence rate and obtained global best results. Also, a lower value of SLL is achieved using FA-based design method compared to other similar arrays designed by GA. Overall, the FA is found to be very suitable for NUSLA design compared to other existing optimization-algorithm-based methods.

References

[1] C. A. Balanis, *Antenna Theory Analysis and Design*, John Wiley & Sons, 3rd edition, 2005.

[2] C. A. Balanis, *Modern Antenna Handbook*, John Wiley & Sons, 2008.

[3] J. D. Kraus, R. J. Marhefka, and A. S. Khan, *Antennas for all Applications*, Tata McGraw-Hill, 3rd edition, 2007.

[4] R. C. Hansen, "Chapter 20: phased arrays," in *Antenna Engineering Handbook*, J. L. Volakis, Ed., McGraw-Hill, 4th edition, 2007.

[5] T. B. Chen, Y. B. Chen, Y. C. Jiao, and F. S. Zhang, "Synthesis of antenna array using particle swarm optimization," in *Proceedings of the Asia-Pacific Microwave Conference (APMC '05)*, vol. 3, p. 4, December 2005.

[6] A. Recioui, A. Azrar, H. Bentarzi, M. Dehmas, and M. Chalal, "Synthesis of linear arrays with sidelobe reduction constraint using genetic algorithm," *International Journal of Microwave and Optical Technology*, vol. 3, no. 5, pp. 524–530, 2008.

[7] M. A. Zaman, S. A. Mamun, M. Gaffar, S. M. Choudhury, and M. A. Matin, "Phased array synthesis using modified particle swarm optimization," *Journal of Engineering Science and Technology Review*, vol. 4, no. 1, pp. 68–73, 2011.

[8] M. A. Zaman, M. Gaffar, M. M. Alam, S. A. Mamun, and M. A. Matin, "Syntehsis of antenna arrays using artificial bee colony optimization algorithm," *International Journal of Microwave and Optical Technology*, vol. 6, no. 4, pp. 234–241, 2011.

[9] R. Harrington, "Sidelobe reduction by nonuniform element spacing," *IRE Transactions on Antennas and Propagation*, vol. 9, no. 2, pp. 187–192, 1961.

[10] F. Hodjat and S. A. Hovanessian, "Nonuniformly spaced linear and planar array antennas for sidelobe reduction," *IEEE Transactions on Antennas and Propagation*, vol. 26, no. 2, pp. 198–204, 1978.

[11] X. F. Ren, J. A. Azevedo, and A. M. Casimiro, "Synthesis of non-uniformly spaced arrays using the Fourier transform and window techniques," *IET Microwaves, Antennas and Propagation*, vol. 3, no. 8, pp. 1245–1253, 2009.

[12] D. G. Kurup, M. Himdi, and A. Rydberg, "Synthesis of uniform amplitude unequally spaced antenna arrays using the differential evolution algorithm," *IEEE Transactions on Antennas and Propagation*, vol. 51, no. 9, pp. 2210–2217, 2003.

[13] K. Chen, Z. He, and C. Han, "A modified real GA for the sparse linear array synthesis with multiple constraints," *IEEE Transactions on Antennas and Propagation*, vol. 54, no. 7, pp. 2169–2173, 2006.

[14] M. M. Khodier and C. G. Christodoulou, "Linear array geometry synthesis with minimum sidelobe level and null control using particle swarm optimization," *IEEE Transactions on Antennas and Propagation*, vol. 53, no. 8, pp. 2674–2679, 2005.

[15] E. Rajo-lglesias and Q. Quevedo-Teruel, "Linear array synthesis using an ant-colony-optimization-based algorithm," *IEEE Antennas and Propagation Magazine*, vol. 49, no. 2, pp. 70–79, 2007.

[16] H. Oraizi and M. Fallahpour, "Nonuniformly spaced linear array design for the specified beamwidth/sidelobe level or specified directivity/sidelobe level with coupling considerations," *Progress in Electromagnetic Research M*, vol. 4, pp. 185–209, 2008.

[17] X. S. Yang, "Firefly algorithms for multimodal optimization," in *Stochastic Algorithms: Foundations and Appplications* , O. Watanabe and T. Zeugmann, Eds., vol. 5792 of *SAGA 2009, Lecture Notes in Computer Science*, pp. 169–178, Springer, Berlin, Germany, 2009.

[18] X. S. Yang, "Firefly algorithm, stochastic test functions and design optimisation," *International Journal of Bio-Inspired Computing*, vol. 2, no. 2, pp. 78–84, 2010.

[19] B. Basu and G. K. Mahanti, "Firefly and artificial bees colony algorithm for synthesis of scanned and broadside linear array antenna," *Progress in Electromagnetic Research B*, vol. 32, pp. 169–190, 2011.

8

Systematic Design Methodology of a Wideband Multibit Continuous-Time Delta-Sigma Modulator

Awinash Anand,[1] Nischal Koirala,[1] Ramesh K. Pokharel,[2] Haruichi Kanaya,[1] and Keiji Yoshida[1]

[1] *Graduate School of Information Science and Electrical Engineering, Kyushu University, 744 Motooka, Fukuoka 819-0315, Japan*
[2] *E-JUST Center Kyushu University, 744 Motooka, Fukuoka 819-0315, Japan*

Correspondence should be addressed to Awinash Anand; awinash.anand@gmail.com

Academic Editor: Ahmed Allam

Systematic design of a low power, wideband and multi-bit continuous-time delta-sigma modulator (CTDSM) is presented. The design methodology is illustrated with a 640 MS/s, 20 MHz signal bandwidth 4th order 2-bit CTDMS implemented in 0.18 μm CMOS technology. The implemented design achieves a peak SNDR of 65.7 dB and a high dynamic range of 70 dB while consuming only 19.7 mW from 1.8 V supply. The design achieves a FoM of 0.31 pJ/conv. Direct path compensation is employed for one clock excess loop delay compensation. In the feedforward topology, capacitive summation using the last opamp eliminates extra summation opamp.

1. Introduction

Delta-sigma modulators embed low-resolution analog-to-digital converter in a feedback loop. The use of feedback and high oversampling pushes the quantization noise out of the band of interest and thereby provides a high in-band resolution. Delta-sigma modulator is well suitable for a high-resolution data conversion because a moderate accuracy of passive components is required. Recently, continuous-time delta-sigma modulator has brought tremendous attention because of its exceptional features such as inherent antialiasing filter (AAF), relaxed gain-bandwidth requirement on active elements resulting in a low-power consumption compared to its counterpart discrete-time delta-sigma modulator [1, 2]. Low-power consumption is the key for a CTDSM.

In [3], the design methodology for a multibit modulator with two-step quantizer is presented. However, the optimization of the peak SNR and the maximum stable amplitude is not taken into consideration. Also, excess loop delay compensation is for more than one clock, where, to achieve higher resolution, higher bit quantizer should be used. These all increase the design methodology complexity and are not simple to adopt for designers. To keep the design simple and the insight intact, we implement one-step quantizer with excess loop delay compensation for one clock. In [4], the optimal design methodology of a higher-order continuous-time wideband delta-sigma modulator is presented. However, this methodology requires summation amplifier and hence consumes higher power. In our approach, the summation amplifier is eliminated by using capacitive summation with last integrator's amplifier and this makes design simpler and saves significant power. Also, in [4] SNR and phase margin are optimized which could be replaced to simpler way to optimize the peak SNR and the maximum stable amplitude which are more obvious parameters.

Recent development in wireless communication standard demands a wideband and high-resolution data converters. To achieve a high SNR over a wideband, a higher clock rate, that is, a higher oversampling ratio (OSR), is desired. However, OSR is limited by the clock rate due to technology limitations and power consumption. Fortunately, the SNR

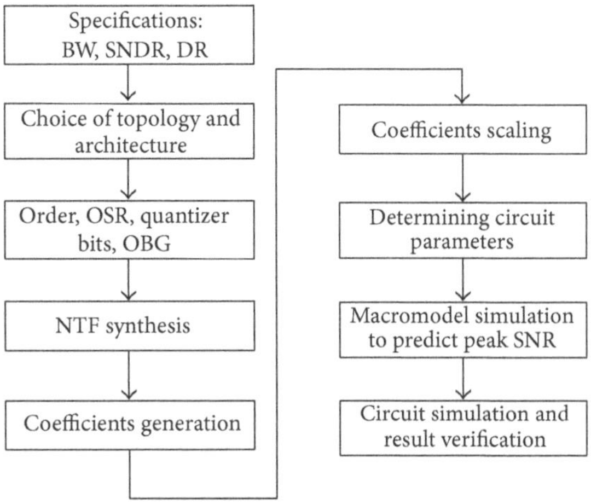

FIGURE 1: Flow chart of the design methodology.

degradation due to lower OSR can be compensated by a multi-bit quantizer.

We present a systematic methodology to design a wideband and high-resolution modulator at low power cost. To illustrate the methodology, we aim to design a continuous-time delta-sigma modulator which has signal bandwidth of 20 MHz and requires 10-11-bit resolution suitable for WLAN. Section 2 discusses the high-level synthesis. In Section 3, we present simple circuit implementation of the modulator. Section 4 presents the results and discussion and finally Section 5 concludes the paper.

2. High-Level Synthesis

In this section, we describe the design methodology in accordance with the flow chart in Figure 1 to synthesize a high-level wideband multi-bit continuous-time delta-sigma modulator in MATLAB to meet the specification for WLAN.

2.1. Choice of Topology and Architecture. A single-loop topology is preferred to a MASH topology to reduce the circuit complexity. To implement the loop filter, a feedforward (FF) topology is preferred to a feedback (FB). A FF topology has several advantages over a FB topology. Firstly, FF uses only one feedback DAC (without any compensation for excess loop delay (ELD)) in the main loop which results in smaller silicon area and better matching of coefficients. However, in the case of FB, multiple DACs equal to the order of the modulator are needed which increase the chip area and mismatch is a major concern [5]. Secondly, the integrating resistor in both the FF and the FB topologies is determined by the noise and distortion requirement. However, in a FF topology, the second and further resistors can be made larger. In a FF topology the first opamp is the fastest while in a FB topology, the first opamp is the slowest. Thus the capacitor size can be reduced in the second and higher integrators with increased resistor value which significantly reduces the silicon area [6]. Also, the necessity for scaling and also the requirements on integrator dynamics are much more relaxed which results in increase in power efficiency of FF topology compared to that of FB topology [1]. However, in general FF topology requires extra summation amplifier which could be eliminated by implementing capacitive as shown in Figure 4 and explained in Section 2.3 [7]. Thus a single-loop FF DSM is the most suitable choice for a high dynamic range and a low-power design.

2.2. Noise Transfer Function (NTF) Synthesis. Noise transfer function synthesis is critical for delta-sigma modulator design as it guides the overall performance and the stability of modulator. Before NTF can be synthesized, order of the modulator, oversampling ratio, quantizer's bit and out-of-band gain must be determined.

2.2.1. Oversampling Ratio (OSR). Among all these, oversampling is the most important driving factor as it is directed by the technology node and power consumption. In principle, increasing OSR by 2 times results in a 15 dB improvement in SNR. However, OSR or clock rate is limited by CMOS technology and the power consumption. To design a wideband modulator with 20 MHz signal bandwidth, an OSR of 16 results in a clock rate (f_s) of 640 MS/s which is high enough to design analog circuits in 0.18 um CMOS technology. Thus we need to design a comparator which can perform comparison at 640 MS/s and opamp which can have GBW higher than 640 MHz for integration to support sampling at 640 MS/s. Since these are pretty high-performance components, we limit the oversampling ratio to 16.

2.2.2. Modulator's Order. Higher-order modulator improves the SNR; however, it increases the circuit complexity and deteriorates the stability. Since we target a wide-signal band of 20 MHz, a higher-order modulator is essential and therefore we simulate the modulator for third, fourth, and fifth order. From simulation we find that a good choice of the modulator order is 4 for a wide bandwidth (20 MHz) and ideally produces a SNDR of 70 dB which is approximately 8 dB higher than the required 62 dB for 10-bit resolution. This 8 dB margin is kept to counter the loss due to circuit nonidealities. This is why 4th-order modulator is chosen for implementation.

2.2.3. Quantizer's Bit. A multi-bit quantizer has several advantages over a single-bit quantizer [1, 2] and compensates well the SNR limitation due to lower OSR. Firstly, a multibit quantizer reduces in-band quantization noise by 6 dB and allows more aggressive NTF with higher out-of-band gain (OBG) resulting in further significant drop in in-band quantization noise. Secondly, the amplitude of the noise in a multi-bit quantizer is much lower compared to that in a single-bit quantizer. Hence the slew rate requirement on the loop filter opamp is greatly relaxed to allow low-power opamp design. Thirdly, a multi-bit feedback DAC is less sensitive to clock jitter [8]. For low power, reduced circuit complexity, and to keep peak SNR well above 60 dB, a 2-bit quantizer is chosen.

2.2.4. Out-of-Band Gain (OBG). As a rule of thumb, the OBG for a single bit quantizer is 1.5 to ensure the stability

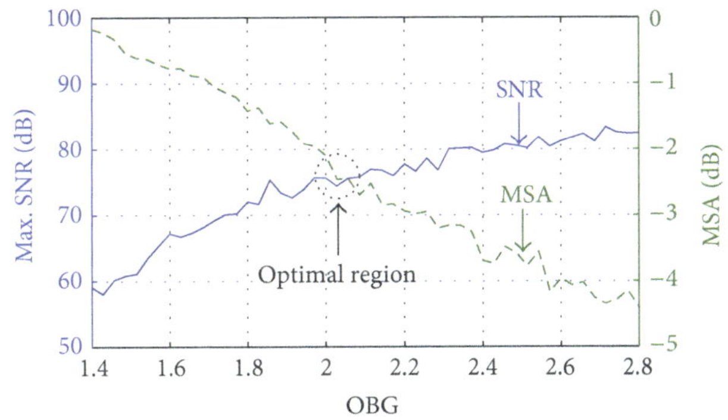

FIGURE 2: SNR and MSA versus the out-of-band gain (OBG) of the NTF for 4th-order 2-bit DSM.

[9]. However, in case of a multi-bit quantizer, the OBG can be increased to reduce the in-band noise and thereby improve the SNR. A 4th-order, 2-bit modulator is extensively simulated for various OBG to determine the maximum SNR and the maximum stable amplitude (MSA). Considering the tradeoff between the SNDR and the maximum stable amplitude (MSA), as depicted in Figure 2, the optimum OBG is chosen to be 2.

Now with all parameters in hand, the NTF is determined using the function *synthesizeNTF* from [10]. A 4th-order, 2-bit modulator with OSR of 16 results in a peak SNDR of 70 dB over a signal bandwidth of 20 MHz.

2.2.5. Excess Loop Delay (ELD) Compensation. The finite regenerative time of a flash converter and the digital logic delay time in the feedback add extra delay, called excess loop delay (ELD), in the loop and effectively increase the order of the modulator. For a modulator of order 2 or above, it needs to be compensated to ensure the stability and maintain a high SNR. One of the efficient methods to compensate ELD is coefficient tuning by adding a direct path between the DAC output and the flash input [1]. Though the compensation time could be any, from the circuit design and operation point of view, it is better to compensate for half a clock or integral multiple of half a clock. To use a single clock, one clock delay compensation is used which helps to relax the requirement on analog building blocks, opamp and comparator.

2.3. Coefficients Generation and RC Parameters. The function *synthesizeNTF* returns discrete-time (DT) coefficients of a modulator which must be translated into continuous-time (CT) coefficients. To reduce the clock jitter sensitivity, NRZ DAC pulse is preferred to other DAC pulse shapes. With NRZ DAC and one clock compensation for excess loop delay, the discrete-time coefficients are converted into the continuous time using the function *realizeNTF_ct* available in [10]. The obtained coefficients result in integrator's output which have much higher swing for modern low supply voltage like 1.8 V and direct implementation would result in large clipping and hence large distortion. Also, the output of one integrator is input to the next integrator and therefore large swing will demand high-input swing for opamp which costs high power.

TABLE 1: Scaled coefficients.

a_0	1.34
a_1	4.25
a_2	4.92
a_3	3.39
a_4	2.76
b_1	0.4
c_1	0.4
c_2	0.4
c_3	0.4
c_4	0.1
g_1	0.11
g_2	0.29

Therefore, the scaling is done to ensure that the output swings of all integrators are well below the maximum allowed voltage (in our case 1.8 V) such that they accommodate the saturation voltage of the output stages of opamps and they do not distort the signals. The resulting coefficients are tabulated in Table 1 for the modulator block diagram in Figure 3.

Figure 4 shows the block diagram of the loop filter. For simplicity, the diagram is shown single-ended; however, the actual circuit implementation is done fully differential. The fourth integrator is used to integrate with R_4C_4 and opamp and the same opamp is used to sum all the feedforward voltages with a_0C_4, a_1C_4, a_2C_4, and a_3C_4 along with C_4 [7]. The coefficients a_1, a_2, and a_3 are realized with the capacitive sum while the coefficient a_4 is embedded in the integration with R_4C_4. This helps to completely eliminate the summation opamp and thereby saves a significant amount of power. Delta-sigma

$$R_1 = \frac{V_{\text{in}}^2/2}{32kTf_B * 3*2^{2B-1}} \quad (1)$$

modulator is a thermal-noise-limited system and the resistor at the input of an active RC integrator contributes the majority of noise. So in a thermal-noise-limited modulator, the resistance value is calculated using (1) [1]. Here R_1 is the resistance at the input of the first integrator, V_{in} is the input signal voltage, k is Boltzmann constant, T is the temperature, f_B is the frequency bandwidth, and B is the effective number of bits. The determined coefficients are translated into "R" and "C" values with the thermal noise constraint as per (1) and keeping the capacitor values such that the feedforward capacitors values are not too large as it loads the last integrating opamp. The determined first resistance value is only 10.93 kΩ and the capacitance is 1.78 pF. The stability and performance of a continuous-time delta-sigma modulator are strongly dependent on process variation as it changes the coefficients drastically. To mitigate the effect, coefficient tuning is desirable. Since resistors are connected either between input and input to the first opamp or between output of one opamp and input of next opamp, it is imperative that tuning using capacitance will be much easier

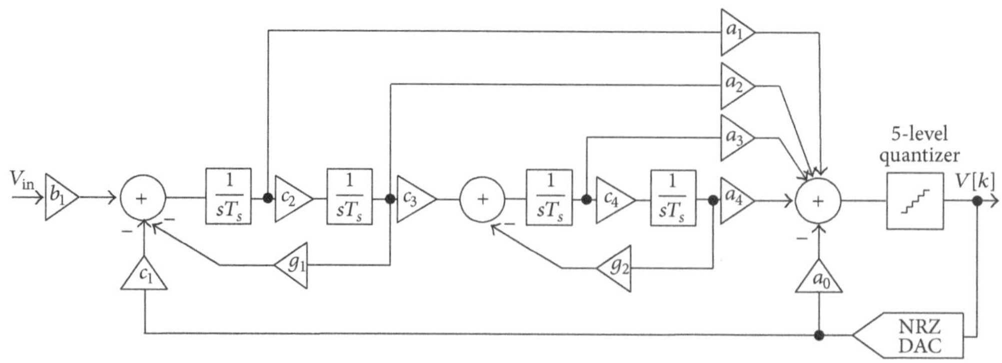

FIGURE 3: The block diagram of a 4th-order FF modulator with direct path for excess loop delay compensation.

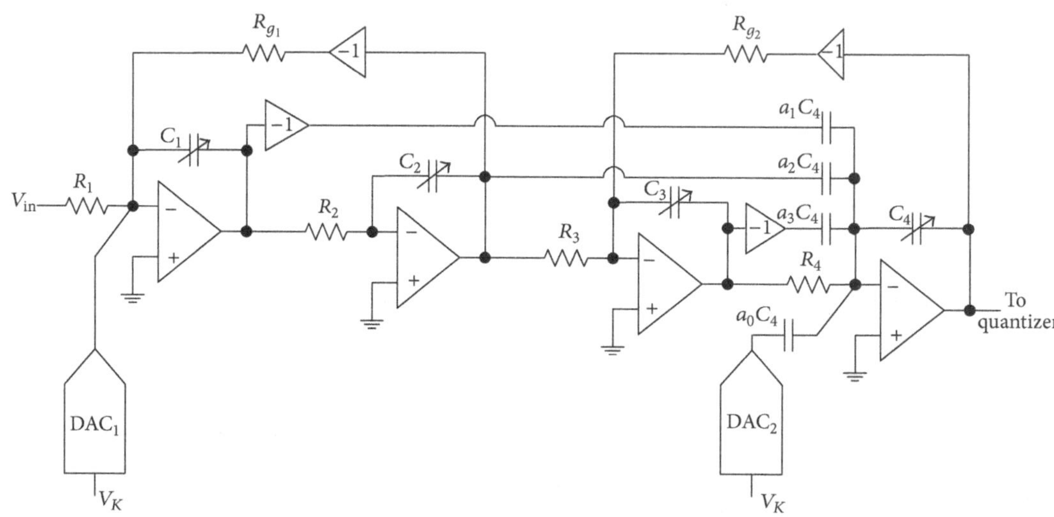

FIGURE 4: The block diagram of a 4th-order CIFF with direct path for excess loop delay compensation.

and effective to implement. Therefore, to combat process variation, capacitive tuning (C_1 to C_4) is implemented.

To predict the SNR, the behavioral simulation of the modulator is done with macro model of building blocks using the components from *analogLib* and *ahdlLib* of cadence. To include all the noises, thermal and circuit, transient noise is enabled while simulating the design. A 16384-point Hann window PSD predicts a SNDR of 69.7 dB for a tone at 1.0547 MHz.

3. Circuit Implementation

In this section, we describe the transistor level circuit designs of the building blocks used in the modulator.

3.1. Opamp. A generic two-stage miller-compensated opamp is used for a high-speed and a wide output swing. To mitigate input-referred flicker noise, long length input transistors are used. To keep the design simple and, power consumption low only one common mode feedback (CMFB) loop is used to maintain the output at V_{cm}. The opamp draws a total current of 2.2 mA, including the CMFB and biasing, from a supply of 1.8 V. The designed opamp has GBW of $1.56 f_s$.

3.2. Comparator. A preamp stage with a gain of 10 is used as input stage. A regenerative circuit follows the preamp stage and finally SR-latch is used to output the decision. Separate references for differential input are used to avoid the coupling between the two differential inputs. The comparator settles its output within 120 ps.

3.3. Feedback DAC. Feedback DAC is designed in two parts. First part is a d-flip-flop [11] which is used to retime the output of the quantizer. In the second part, a current steering DAC is used for fast response. This DFF and the quantizer effectively introduce a delay of one clock between the input of the flash converter and the output of the feedback DAC. The cascade current source in the DAC cell is used to achieve a high-output resistance. The output impedance of the current DAC is 70 kΩ.

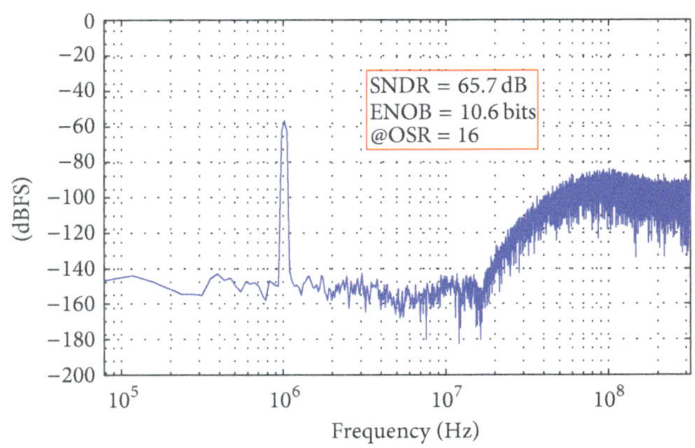

FIGURE 5: Output Spectrum of the modulator.

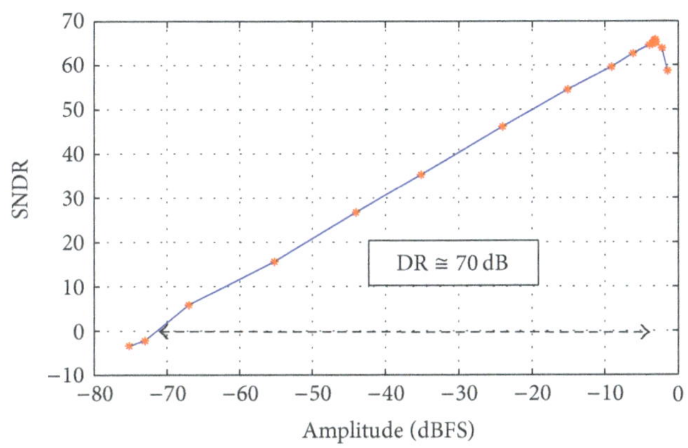

FIGURE 6: Amplitude versus SNDR of the modulator.

4. Results and Discussion

To illustrate the design methodology, a 4th-order 2-bit continuous-time delta-sigma modulator is designed in 0.18 μm CMOS technology. The implemented modulator is tested with a single tone at 1.0547 MHz. A 16384-point Hann window PSD is produced to ensure the sufficient accuracy. The resulted spectrum is shown in Figure 5. From Figure 5, it is determined that the peak SNDR is 65.7 dB over a bandwidth of 20 MHz. Figure 6 has the plot of SNR versus amplitude which gives a high dynamic range of 70 dB. The design consumes overall power of 19.7 mW to achieve a figure of merit (FoM) of 0.31 pJ/conv.

5. Conclusion

A systematic design methodology of a continuous-time delta-sigma modulator is described. A 640 MS/s, 20 MHz signal bandwidth 4th-order 2-bit continuous-time delta-sigma modulator is implemented in 0.18 μm CMOS technology to illustrate the design methodology. The CT coefficients are systematically computed compensating for excess loop delay. The designed modulator has a high SNDR of 65.7 dB and a high dynamic range of 70 dB for a signal band of 20 MHz. This modulator is well suited for WLAN applications. The modulator consumes 19.7 mW power from a 1.8 V supply to achieve FoM of 0.31 pJ/conv.

Acknowledgments

This work was partly supported by a grant of the Regional Innovation Cluster Program (Global Type 2nd Stage) from the Ministry of Education, Culture, Sports, Science and Technology (ME⊠T) and a Grant-in-Aid for Scientific Research (B) from JSPS, KAKENHI (Grant no. 23360159). This work was also partly supported by a grant of the Knowledge Cluster Initiative implemented by Ministry of Education, Culture, Sports, Science and Technology (ME⊠T), KAKENHI (B), and KIBAN (B) and the VLSI Design and Education Center (VDEC), The University of Tokyo in collaboration with CADENCE Corporation and Agilent Corporation.

References

[1] M. Ortmanns and F. Gerfers, *Continuous-Time Sigma-Delta A/D Conversion, Fundamentals, Performance Limits and Robust Implementations*, Springer, Berlin, Germany, 2006.

[2] J. A. Cherry and W. M. Snelgrove, *Continuous-Time Delta-Sigma Modulators for High Speed A/D Conversion Theory, Practice and Fundamental Performance Limits*, Kluwer Academic, New York, NY, USA, 2002.

[3] S. Balagopal, R. M. R. Koppula, and V. Saxena, "Systematic design of multi-bit continuous-time delta-sigma modulators using two-step quantizer," in *Proceedings of the IEEE 54th International Midwest Symposium on Circuits and Systems (MWSCAS '11)*, pp. 1–4, August 2011.

[4] Y. Ke, S. Radiom, H. R. Rezaee, G. Vandenbosch, J. Craninckx, and G. Gielen, "Optimal design methodology for high-order continuous-time wideband Delta-Sigma converters," in *Proceedings of the 14th IEEE International Conference on Electronics, Circuits and Systems (ICECS '07)*, pp. 743–746, Marrakech, Morocco, December 2007.

[5] R. Schreier and G. Temes, *Understanding Delta-Sigma Data Converters*, IEEE Press, Piscataway, NJ, USA, 2005.

[6] S. Pavan, N. Krishnapura, R. Pandarinathan, and P. Sankar, "A power optimized continuous-time ΔΣ ADC for audio applications," *IEEE Journal of Solid-State Circuits*, vol. 43, no. 2, pp. 351–360, 2008.

[7] S. Pavan, "Excess loop delay compensation in continuous-time delta-sigma modulators," *IEEE Transactions on Circuits and Systems II*, vol. 55, no. 11, pp. 1119–1123, 2008.

[8] J. A. Cherry and W. M. Snelgrove, "Clock jitter and quantizer metastability in continuous-time delta-sigma modulators," *IEEE Transactions on Circuits and Systems II*, vol. 46, no. 6, pp. 661–676, 1999.

[9] W. Lee, *A novel higher-order interpolative modulator topology for high resolution oversampling A/D converters [M.S. thesis]*, Massachusetts Institute of Technology, Cambridge, Mass, USA, 1987.

[10] R. Schreier, "The Delta-Sigma Toolbox Version 7. 1," http://www.mathworks.com/matlabcentral/fileexchange/19.

[11] M. J. Park, *A 4th order continuous-time ΔΣ ADC with VCO-based integrator and quantizer [Ph.D. thesis]*, Massachusetts Institute of Technology, Cambridge, Mass, USA, 2009.

Analysis and Design of Transformer-Based mm-Wave Transmit/Receive Switches

Ehsan Adabi[1] and Ali M. Niknejad[2]

[1] *Mobile and Wireless Group, Broadcom Corporation, Irvine, CA 92617, USA*
[2] *Berkeley Wireless Research Center (BWRC) and EECS Department, University of California, Berkeley, CA 94720-1770, USA*

Correspondence should be addressed to Ehsan Adabi, ehsan.adabi@gmail.com

Academic Editor: Paolo Colantonio

Transformer-based shunt single pole, double-throw (SPDT) switches are analyzed, and design equations are provided. A mm-wave transformer-based SPDT shunt switch prototype was designed and fabricated in 90 nm digital CMOS process. It has a minimum insertion loss of 3.4 dB at 50 GHz from the single pole to the ON-thru port and a leakage of 19 dB from the single pole to the OFF-thru port. The isolation is 13.7 dB between the two thru ports. Large signal measurements verify that the switch is capable of handling +14 dBm of input power at its 1 dB compression point. The fabricated SPDT switch has a minute active area size of $60\,\mu\text{m} \times 60\,\mu\text{m}$.

1. Introduction

Transmit/receives switches implemented in standard CMOS technologies are key mm-wave building blocks for demonstrating the ultimate goal of a mm-wave single chip solution. Previous works [1–5] have demonstrated the feasibility of many key mm-wave transceiver building blocks in standard digital CMOS processes. Operating at mm-wave frequencies has many advantages. One key advantage is due to small wavelengths, which allows antennas to be realized on chip or on the package, further reducing the cost of a system. Moreover, many antennas could be integrated with suitable phase shifters to create phased array systems that effectively increase the aperture size and directivity of transmit/receive antennas by a factor of N (number of antennas). Spatial filtering achieved by a phased antenna array system alleviates impairments such as delay spread and cochannel interference and further helps to extend the communication range and bandwidth. Without a transmit/receive (T/R) switch, two separate antennas should be employed for the receiver, and transmitter, which translates to half the transmit/receive antenna gain and aperture size for a given area. T/R switches with sufficient performance can allow a single antenna to be shared between the transmitter and receiver and thereby twice the number of antennas (antenna array gain) will be realizable out of the same die area. Hence T/R switches can save a great deal of area, and thus lower the total system cost, since antennas are relatively large, especially if implemented on chip. Therefore, designing CMOS T/R switches at mm-wave frequencies is an important goal toward the realization of a low-cost mm-wave system.

2. Antenna Array Reuse via T/R Switching

At mm-wave frequencies, antennas can be realized in the package or on chip due to the relatively small wavelength. Higher performance is realized from package antennas due to the smaller conductive and dielectric losses compared to a silicon substrate. Multilayer boards with high-quality metallization not only provide good design environments for the antenna array but they can also be employed to route signal, power, and ground lines and furthermore act as a heat sink to take the dissipated power off the chip.

Even when antennas are implemented on a package, having a smaller package makes the over-all solution more cost-effective and allows for easier adoption of the solution by WPAN applications such as smart phones and portable media players. But whenever the package size is a secondary issue, then removing T/R switches should result in a better overall performance, since there is no extra component after the PA to harm the TX linearity and output power nor is there a component before the LNA which degrades the receiver noise figure and consequently sensitivity. However in practical cases, for solutions without the T/R switch, the bigger size of the package means longer on-package routings to connect front ends residing on the chip to the antennas implemented on the package (Figure 1). The routing loss is as important as the T/R switch loss, since it directly affects the transmit power and receive sensitivity. Depending on the array size and on-package transmission line loss characteristics, the resultant routing loss can offset the benefit of not including a T/R switch.

To have a quantitative comparison, we consider the scenarios described in Figure 1. In Figure 1(b), one antenna array is shared between TX and RX via T/R switches (where each switch incorporates 3 dB of insertion loss), and in Figure 1(c) two separate arrays are used for RX and TX portions of the transceiver. To compare these two scenarios, we compare the maximum communication range achievable by each case while assuming nominal values of transmitter P_{out} = 10 dBm, receiver NF = 10 dB and a required SNR of 10 dB, for maintaining the communication link at the frequency band of 57–64 GHz.

As can be seen from Figure 2, for larger arrays (a 100 element array in this case) transmission line routing loss is more pronounced and unless very low loss routing is realizable on package, including the T/R switch will provide better overall performance. For smaller arrays (16 antenna elements in Figure 2) more routing loss is tolerable. However, for ultimate single-chip solutions where the antenna array is realized on chip, due to higher on-chip routing losses (~1 dB/mm), a solution with a shared array between TX and RX through a T/R switch looks more promising.

3. T/R Switching Configurations

3.1. Series versus Shunt Switching. By its nature, when the gate terminal of a MOS transistor is driven by a rail-to-rail voltage, it acts as a switch. In fact, this single transistor switch is sufficient for many digital and analog applications. When the switch is on, the transistor is in the triode region and a low resistive channel connects the MOS source and drain terminals. To the first order, the on resistance of a MOS switch (R_{ON}) is equal to $1/g_m$, where the transconductance is calculated at the edge of saturation. When the switch is off, there is no conducting channel between the source and drain terminals, and the resistive path that connects the source and drain nodes of the transistor exhibits a very large R_{OFF}. However, in the off mode there is a feedthrough path through parasitic capacitors between source and drain terminals. This path is through both C_{gd}-C_{gs} and C_{db}-C_{sb} networks. At low frequencies, when the transistor is operating at frequencies much lower than its f_t, the capacitive feedthrough path introduces a much higher off-mode reactance than the resistance of the channel in the on mode. But at mm-wave frequencies, when the device is operating close to the f_t limit, the impedance of the switch in the off state is comparable to when it is on. Figure 3 depicts the measured S_{21} of a 40 μm NMOS transistor acting as a switch in a 50 Ω environment. As can be seen in this figure, S_{21} for on and off states at 60 GHz is comparable.

There are techniques to diminish the feedthrough path by cascading multiple switches (decreasing the effective series capacitance) and adding a shunt switch in between two series switches with an inverted gate signal with respect to the series switches (shorting out the feedthrough signal). These techniques (Figures 4(a) and 4(b)) are mostly applicable to lower RF frequency systems, since at mm-wave frequencies having multiple transistors in the signal path introduces substantial insertion loss. Switches cannot be made too large due to parasitic capacitance limitations. In mm-wave regime, a switching structure with the least number of transistors is desirable to reduce parasitics. Furthermore, any transistors in series with the signal path can incur a noticeable loss, and a structure that removes series transistors is more suitable. This is the reason that in [6] an L-configuration with corresponding matching networks as depicted in Figure 4(c) is exploited with only one series transistor and two transistors in shunt with the signal path. An ultimate shunt SPDT switch at mm-wave frequencies has the capability to provide lower losses due to the elimination of transistors in series with the signal path.

The shunt SPDT switch configuration is demonstrated in Figure 5. Traditionally shunt switches were realized with two quarter-wave transmission line sections (Figure 5(a)) and shunt switches at each end. When a switch is closed, the corresponding thru port is grounded, and all the incident power reflects with negligible amount coupling through. The $\lambda/4$ line converts this low impedance of the short from the off-thru port to a high impedance (open circuit) at the common input port which results in directing all the input power toward the on port. The impedance of the on-port load is matched to the quarter-wave line characteristic impedance which is in turn matched to the input impedance. The off-state switch at the on-thru port has a very high shunt loading resistance and its parasitic capacitance is absorbed into the quarter wavelength transmission line design.

Even at mm-wave frequencies, $\lambda/4$ lines are rather bulky and occupy too much area to be implemented on chip. Absorption of parasitic capacitances added by the interfacing building blocks helps to reduce the transmission line size. The other approach to reduce the footprint of a transmission line is employing slow-wave structures by placing floating metal filaments underneath the line [12]. These filaments act as a shield preventing electric fields from penetrating down to the substrate and by confining the electric field to a small volume just underneath the transmission line, they increase the effective per unit length capacitance of the line, and as a result of that the wavenumber ($\beta = \omega\sqrt{LC}$) is increased at the price of having lower characteristic impedance ($Z_0 = \sqrt{L/C}$) which in turn requires higher power consumptions for interfacing blocks that drive such lines. Slow-wave techniques can enhance the capacitance of the line by up to four times which means that the length of the line can be halved for further area savings [13].

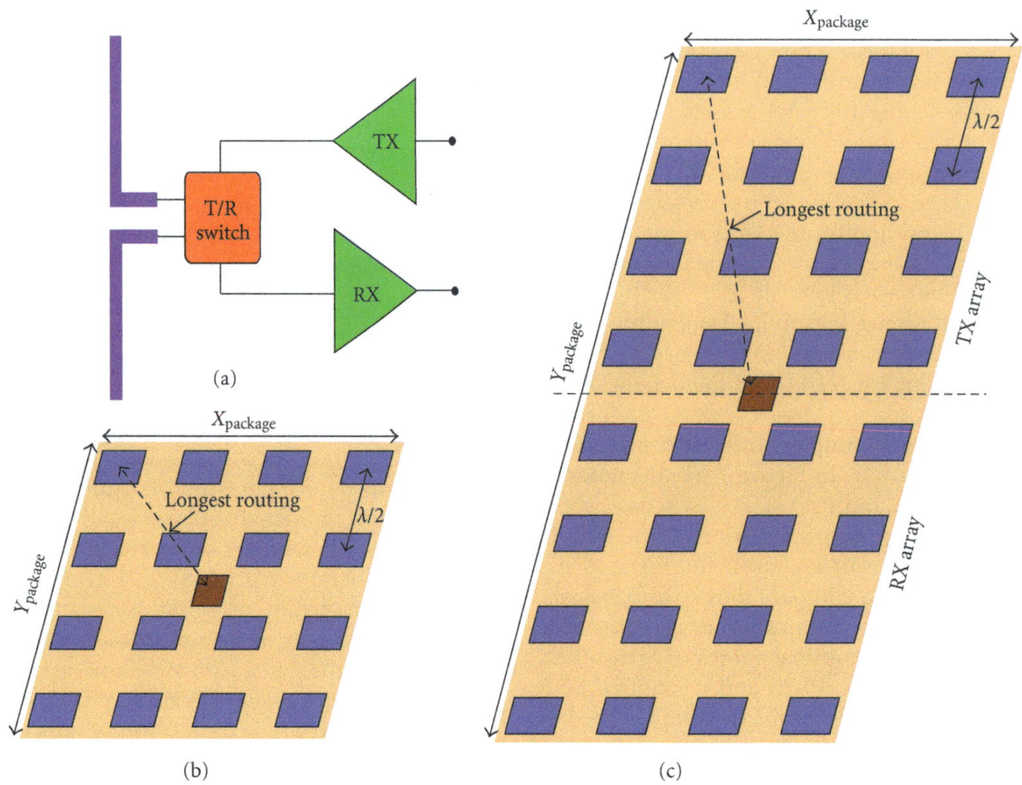

FIGURE 1: A T/R switch building block in a transceiver architecture (a). An antenna array which employs T/R switches (b) has half the footprint of the one without T/R switches (c).

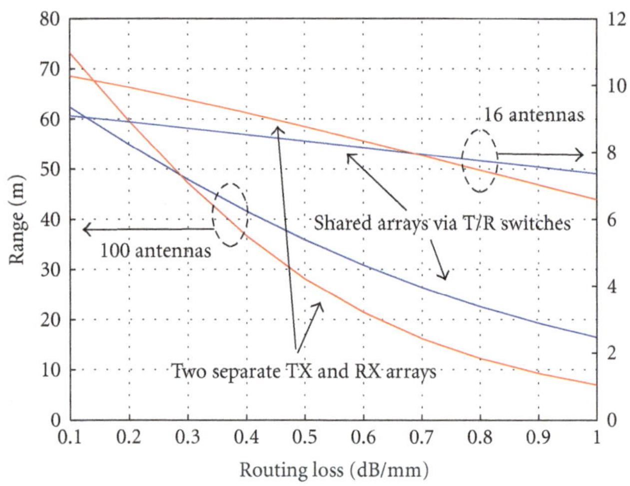

FIGURE 2: Graph of the maximum achievable communication range in two configurations of shared or separate arrays (for both small and large arrays). For larger arrays, the routing loss is more detrimental than the T/R switch loss and higher quality packaging should be employed (T/R switches were assumed to have 3 dB of insertion losses).

Even after absorbing the capacitance of the interfacing blocks and utilizing slow-wave techniques to reduce the transmission line footprint, they are still bulky and devising a lumped component counterpart for the shunt SPDT switch is highly valuable. In this work, we demonstrate and analyze a transformer based shunt switch employing a transformer and shunt NMOS switches as shown in Figure 5(b) [7].

3.2. The Transformer-Based Shunt Switching Structure. A transformer-based shunt switching structure is demonstrated in Figure 5(b). Comparing it with the traditional shunt SPDT structure, Figure 5(a) reveals that the two quarter-wavelength arms are replaced with a miniature transformer that saves a considerable amount of the die area. When a transistor is on, it introduces a low impedance at its port which quenches any voltage swing on it, and the induced voltage will majorly appear at the other end of the secondary winding. As shown in Figure 5(c), all parasitic capacitances in this structure are shunt capacitors to ground and do not provide a feedthrough path between the input and output terminals. Moreover, these capacitances will be resonated out with the equivalent inductance appearing at the transformer terminals at the desired frequency and will not be seen by load and source impedances.

Unlike the traditional distributed switch, this structure introduces a 180° phase difference between the two outputs. In a T/R switch, this 180° phase shift is of minor concern, since this corresponds to moving the transceiver a distance of $\lambda/2$, and most circuits that are sensitive to RF phase use carrier locking techniques. Shunt switches shown in Figure 5 are reflective type of switches which means that when the input is not connected to an output port, that port sees a

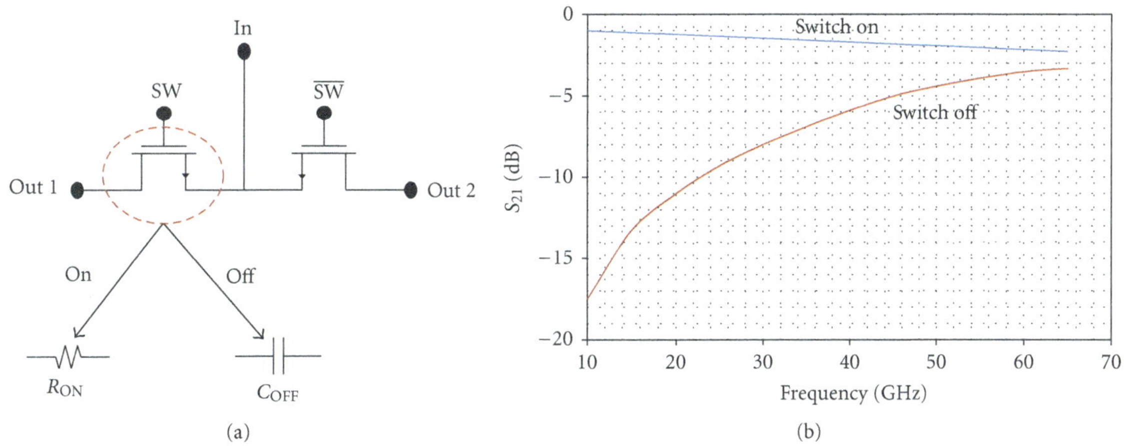

FIGURE 3: Measured S_{21} of an NMOS transistor acting as a series switch in a 50 Ω environment.

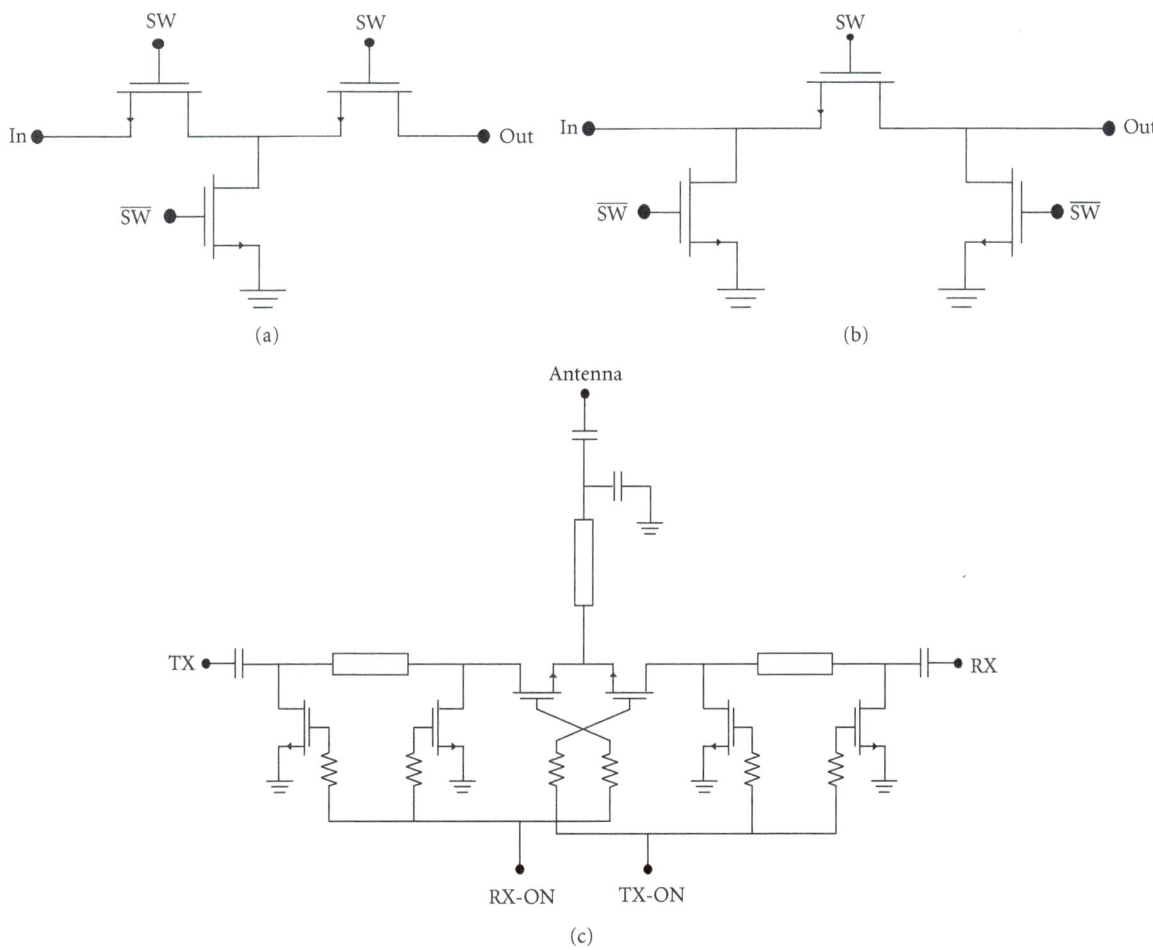

FIGURE 4: Combinations of series and shunt switches in a T (a) or π (b) or L-shape configuration (c). The technique in (c) [6] decreases the amount of off-state leakage at the price of degrading the on-state insertion loss.

short circuit and hence it is not matched. A mismatch at the input of LNA and output of PA can be problematic and cause oscillation. However, since these switches are used in a time division duplexing (TDD) scheme, whenever one output is disabled, the interfacing building block will be turned off in order to save some power dissipation. This eliminates the potential oscillation problems. Care must be taken to make sure that the recovery time of the interfacing building block to reach its steady state when it is turned on is within the overall system specification.

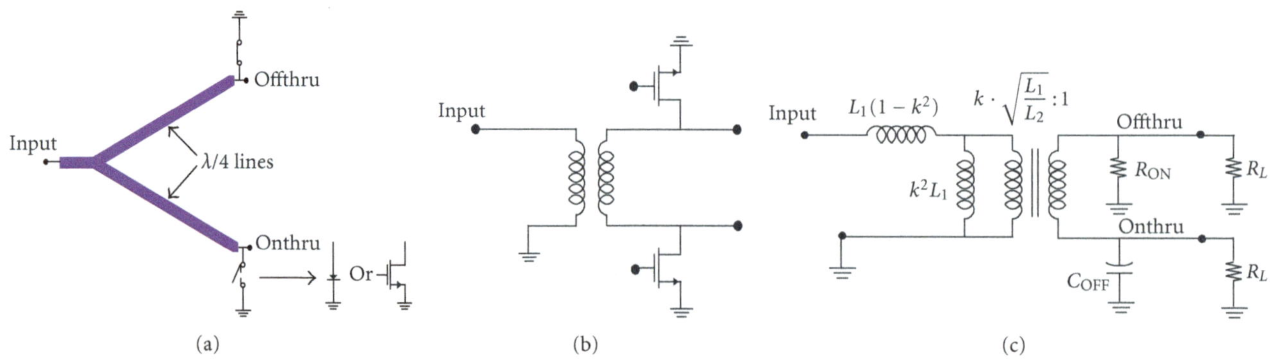

FIGURE 5: A traditional microwave shunt switch (a), the proposed miniature structure (b), and the equivalent circuit (c).

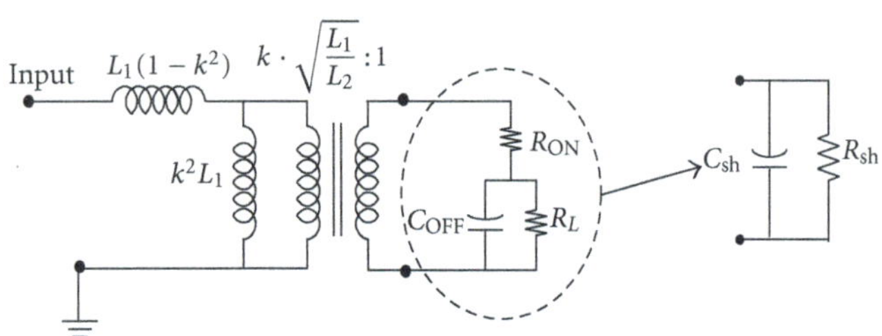

FIGURE 6: Simplified model of the SPDT network including loading effects.

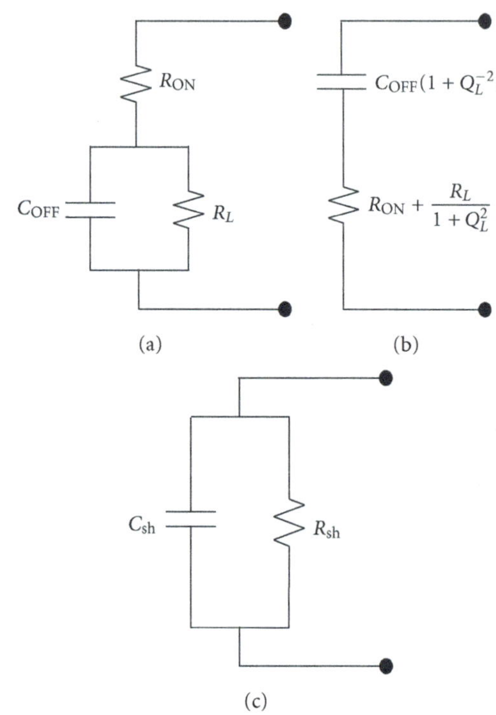

FIGURE 7: Calculating the shunt equivalent loading network.

4. Design Equations of a Transformer-Based T/R Switch

The equivalent circuit model depicted in Figure 5(c) will be used to drive design equations for the center frequency, insertion loss, leakage, and isolation of the switch. Since the on-resistance of the transistor is much smaller than the load resistance ($R_{ON} \ll R_L = Z0$) and for operating frequencies (ω) adequately less than the technology's transit frequency (ω_t), the on resistance is much lower than the shunt reactance of parasitic capacitances of the transistor (C_{drain}), then, we can assume that the loading at the off-thru port is only R_{ON} of the transistor. At the on-thru port, the transistor is off and the total loading at that port will be the load resistance(R_L) in parallel with the drain capacitance of the transistor at the off mode (C_{OFF}) as depicted in Figure 6.

4.1. Equivalent Shunt Loading. To derive the transfer function, the loading network depicted in Figure 7(a) is transformed into its equivalent shunt impedance (Figure 7(c)). Two quality factors are defined in order to accomplish this. The first one is the quality factor of the shunt network of R_L and C_{OFF} which is the loading network quality factor and can be written as

$$Q_L = R_L C_{OFF}\omega. \tag{1}$$

And the second one is the quality factor for the series network of R_{ON} and C_{OFF} which is the transistor-only quality factor and is determined by the choice of the technology. It improves by CMOS technology scaling and can be written as

$$Q_0 = \frac{1}{R_{ON}C_{OFF}\omega}. \tag{2}$$

In order to reach to the shunt equivalent network, first the intermediate series network depicted in Figure 7(b) is calculated as

$$R_S = R_{ON} + \frac{R_L}{1+Q_L^2}, \qquad C_S = C_{OFF}\left(1+Q_L^{-2}\right). \tag{3}$$

To transform this intermediate series network to the equivalent shunt network we should calculate its quality factor (Q_{int}) as follows

$$\frac{1}{Q_{int}} = \omega\left(R_{ON} + \frac{R_L}{1+Q_L^2}\right)C_{OFF}\left(1+Q_L^{-2}\right). \tag{4}$$

Expanding and regrouping the above equation yields to

$$\frac{1}{Q_{\text{int}}} = \frac{1}{Q_0} + \frac{Q_L(1+Q_L^{-2})}{1+Q_L^2}. \quad (5)$$

For large values of the load quality factor ($Q_L^2 \gg 1$) the intermediate quality factor can be simplified as:

$$\frac{1}{Q_{\text{int}}} = \frac{1}{Q_0} + \frac{1}{Q_L} \Longrightarrow Q_{\text{int}} = Q_0 \| Q_L. \quad (6)$$

Finally with the aid of the intermediate quality factor (Q_{int}), the series network depicted in Figure 7(b) can be transformed into its shunt equivalent network (Figure 7(c)). This shunt-loading impedance can be characterized as a resistance in parallel with a capacitance with values of

$$R_{\text{sh}} = R_{\text{ON}}\left(1+(Q_0\|Q_L)^2\right) + R_L\frac{1+(Q_L\|Q_0)^2}{1+Q_L^2}, \quad (7)$$

$$C_{\text{sh}} = C_{\text{OFF}}\frac{1+Q_L^{-2}}{1+(Q_L\|Q_0)^{-2}}. \quad (8)$$

This equivalent-loading network is used in the next section to calculate the center frequency.

4.2. Center Frequency and Matching. Since a lower desired loss requires a lower on resistance and consequently a larger transistor with more capacitive loading to operate at mm-wave frequencies, the smallest realizable inductance is preferred. So from now on we assume that the transformer is a 1 : 1 structure with self-inductance of L for each loop. As depicted in Figure 6, there is one remaining series structure. When this is transformed to its parallel equivalent network, the structure will resemble a parallel tank allowing the calculation of the center frequency. The series network of the source impedance (R_S) and the leakage inductance ($L(1-k^2)$) have a quality factor of $Q_S = L(1-k^2)\omega/R_S$. For a good transformer (tight coupling), the leakage inductance is small and the reactance associated with it at mm-wave frequencies is still much smaller than the source impedance ($Q_S \ll 1$). This small Q makes the equivalent shunt impedance of R_S stay at roughly the same value after the transformation ($R_S(1+Q_S^2) \sim R_S$), and the shunt equivalent of leakage inductance ($L(1-k^2)(1+Q_S^{-2})$) be much larger than the already existing shunt-magnetization inductance (k^2L) and be neglected in the parallel configuration. As the loading network ($R_{\text{sh}}\|C_{\text{sh}}$) is moved to the source side by multiplication by the factor k^2 (k being the coupling factor of the 1 : 1 transformer), the equivalent circuit looks like the one shown in Figure 8.

Neglecting the transformed shunt equivalent of the leakage inductance with respect to the magnetization inductance reduces the parallel tank depicted in Figure 8 to an inductance of k^2L in parallel with a capacitance of C_{sh}/k^2 with the resonance frequency of

$$\omega_0 \sim \frac{1}{\sqrt{(k^2L)\cdot(C_{\text{sh}}/k^2)}} = \frac{1}{\sqrt{LC_{\text{sh}}}}, \quad (9)$$

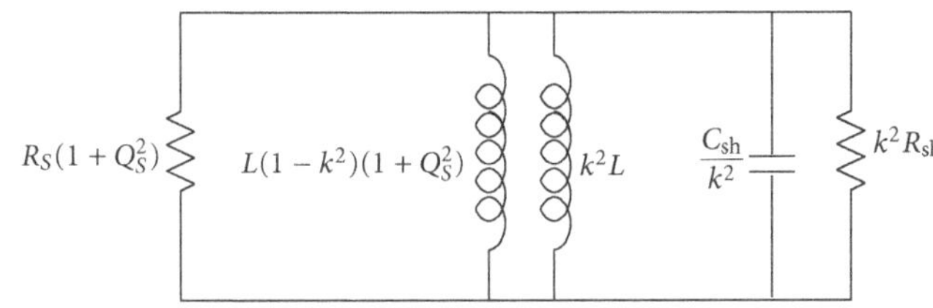

FIGURE 8: Parallel tank equivalent network of the switch for calculating the center frequency.

where L is the self-inductance of each loop of the transformer and C_{sh} was derived in (8). At the resonance frequency, the apparent load impedance at the source side is k^2R_{sh}, where R_{sh} is described in (7). For an ideal case where coupling factor is unity and the transistor-only quality factor (Q_0) is infinity, the load impedance is transformed to the source side intact and the match is perfect. However, for practical cases of $k \sim$ 0.7-0.9, $Q_L \sim$ 2-3, and $Q_0 \sim$ 5-7 getting a −10 dB of return loss (which is sufficient to meet the requirement for most communication systems) is fairly feasible.

4.3. Insertion Loss. The insertion loss due to the finite on resistance of the switch (R_{ON}) will be the ratio of the power delivered to the effective on-thru load impedance and the power transferred to the total effective impedance. Since we already have these two impedances in the series format (Figure 7(b)), the insertion loss is a resistance ratio and can be written as

$$\text{I.L.} = \frac{R_L\left(1+(Q_L\|Q_0)^2\right)/(1+Q_L^2)}{R_{\text{ON}}\left(1+(Q_0\|Q_L)^2\right)+R_L\left(1+(Q_L\|Q_0)^2\right)/(1+Q_L^2)}. \quad (10)$$

In practice, Q_0 is significantly larger than Q_L and as the design is moved to a more advanced technology in the future, Q_0 will be even higher with respect to Q_L and can be neglected in a parallel configuration ($Q_0\|Q_L \sim Q_L$), and hence the insertion loss can be simplified as

$$\text{I.L.} = \frac{RL}{RL + R_{\text{ON}}Q_L^2}. \quad (11)$$

The transformer is assumed to be lossless in (10) and (11), which means that the insertion loss of the transformer itself should be added to above equations to get the overall insertion loss of the switch. In practice, a typical transformer insertion loss is less than 1 dB at mm-wave frequencies.

4.4. Leakage. To calculate the leakage signal, we should consider that 1 − I.L. of the input power will be delivered to the network at the off-thru port. But since at that port R_L is in parallel with R_{ON}, most of the power will be dissipated in R_{ON} and only a portion of it $R_{\text{ON}}/(R_{\text{ON}}+R_L)$ will reach the off-thru output port. Therefore, the leakage can be formulated as

$$(1-\text{I.L.})\frac{R_{\text{ON}}}{R_{\text{ON}}+R_L} \sim \frac{R_{\text{ON}}}{R_L}\cdot\frac{R_{\text{ON}}Q_L^2}{R_L+R_{\text{ON}}Q_L^2}. \quad (12)$$

Again the transformer was assumed to be lossless here. Nonideal transformers help the leakage, number as the leakage signal will be attenuated by the insertion loss of the transformer as well.

4.5. Isolation. To calculate the isolation, the input signal is connected to the off-thru port, and the signal reaching to the on-thru port is calculated (or vice versa). The situation is depicted in Figure 9. The source impedance and the magnetization inductance can be moved to the secondary side as depicted in Figure 10. Assuming the shunt resistance of R_S/k^2 is large enough to be neglected in a parallel configuration with the reactance of $L\omega$ (which is not a very accurate assumption but can be thought of as an overestimation for the quality factor of the inductor and hence as an underestimation of the isolation value), the equivalent network can be simplified further which results in the inductance L being in series with the resistance $R_{ON} \| R_L$. Converting this series network to its shunt equivalent circuit (and taking into account that in this transformation the output voltage gets multiplied by the quality factor which is in fact the Q_L), we arrive at the parallel tank network described in Figure 10(b). At the center frequency L and C_{OFF} will resonate out each other and the equation for the isolation can be written as

$$\frac{1}{Q_0} \cdot \frac{Q_0^2 \cdot R_{ON}}{Q_0^2 \cdot R_{ON} + R_L} = \frac{Q_o R_{ON}}{R_L + Q_0^2 R_{ON}}. \quad (13)$$

Impact of the transistor on-resistance R_{ON} is noticeable again in the performance of the switch and as more advanced technologies are employed, better isolation is also achievable in the transformer-based T/R switches. These switches are reflective type of switches being used in TDD schemes, and as described earlier, the building block interfacing the off-thru port will be turned off, and hence low isolations are somewhat tolerable in such switches.

4.6. Effect of Interwinding Capacitances. The parasitic interwinding capacitances between the two loops of the transformer were not included in this analysis in order to have the order of the network low enough, and as a result of that to reach at some design equations that can be used as a guidance for obtaining close-to-optimum transistor and transformer sizes before plugging them in to the simulators for further optimizations.

However, if there are non-negligible feed-through capacitances between the two loops, the frequency response of the network would be down shifted, and there would be asymmetries between the frequency response from the single pole to the two different thru ports. As the polarity of the signal is inverted by choosing a different thru port, the voltage appearing on the feed-through capacitance changes and due to the Miller effect, it creates two different networks that the insertion loss through them can be different whether the operation is inverting or noninverting. The inverting configuration would have a higher insertion loss compared to the noninverting scenario [14].

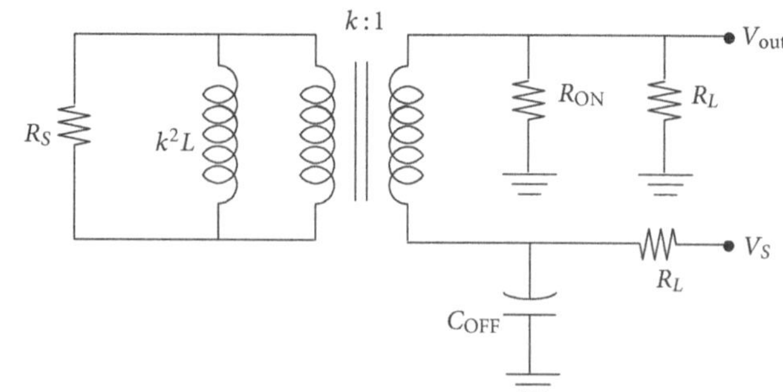

FIGURE 9: Equivalent circuit for calculating the isolation between the two thru ports.

5. Transformer-Based Switch Design Example in 90 nm CMOS Technology

After deriving design equations of a transformer-based shunt T/R switch presented in Section 4, a prototype is designed, fabricated, and tested to verify the validity of proposed equations. In upcoming subsections, MOS transistor performance in the switch mode is investigated, and an optimum transistor layout to be used in the T/R switch design is investigated. Chosen sizes and design parameters for the transformer and shunt transistors are presented and measurement results are shown at the end.

5.1. MOS Transistor's Performance in the Switching Mode. Layout concerns of a MOS transistor when it is used as a high frequency switch is different from when it is expected to function as a high frequency amplifier. When a MOS transistor is laid out to be used in a mm-wave amplifier, smaller finger widths are used to decrease the series resistance of the gate terminal and as a result of that, lower losses and higher f_{max} are achieved. To decrease the loss through the back-gate effect, bulk resistance should be either close to zero (the case with a solid substrate ring surrounding the device) or approaching infinity (which means very few contacts are placed relatively distant from the transistor or in case of having the deep n-well option, a bulk resonant network is employed). For high frequency amplifier design, the former option of having a well-defined substrate ring is preferred due to its compactness, more predictable modeling of the device, and less vulnerability to substrate noise and coupling issues. However, when a MOS transistor is laid out for switching purposes, especially transmit/receive switching, linearity is a major concern since any degradation to the output power will be extremely costly.

For MOS transistors acting as a shunt switch, the input signal is supposed to be directed to the output port that its corresponding shunt transistor is in the OFF mode (V_{GS} = 0). In case of having a small impedance at the gate, that terminal becomes an AC short which means that the gate voltage remains at the provided bias voltage throughout the operation. Therefore, for positive swings of the output voltage, the terminal connected to the output acts as the drain terminal of the transistor, while the AC fluctuation of the V_{GS} remains at zero and the transistor stays completely

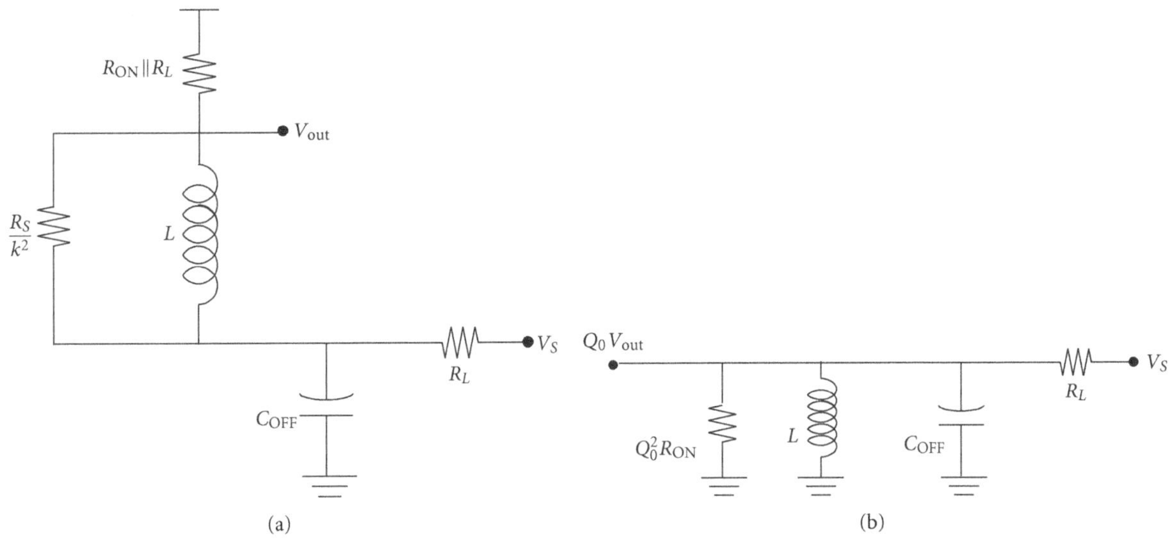

FIGURE 10: Source network is transferred to the load side (a) and then the structure is converted to a parallel tank configuration (b).

OFF at all times with no degradation to the output power linearity. However, in negative excursion of the swing the output node functions as the source of the transistor and if the output voltage goes below zero by a V_{TH}, the transistor will be turned on in the reverse direction which clamps the signal and produces severe distortion to the output voltage waveform (Figure 11).

On the other hand, if there is a large biasing impedance placed at the gate node, the gate terminal will be an AC floating node, and hence a feed through of output signal will appear at the gate terminal via parasitic paths provided by the transistor capacitances. As depicted in Figure 11, for positive swings of output signal that the output node becomes the drain of the transistor, the feedthrough signal appearing across gate-source terminal through the C_{GD}-C_{GS} network will be

$$V_{GS} = \frac{C_{GD}}{C_{GS} + C_{GD}} V_{out}. \quad (14)$$

For negative excursion of the output signal, the output node serves as the source of the transistor, and therefore, the voltage across gate-source terminal that modulates the channel can be written as

$$V_{GS} = \frac{C_{GS}}{C_{GS} + C_{GD}} V_{out} - V_{out} = \frac{-C_{GD}}{C_{GS} + C_{GD}} V_{out}. \quad (15)$$

Therefore, the gate-source voltage will be modulated by a fraction of the output voltage and the channel turns on when

$$|V_{out}| \geq \frac{C_{GS} + C_{GD}}{C_{GD}} V_{TH}, \quad (16)$$

which is a factor of $1 + (C_{GS}/C_{GD})$ larger than the turn-on voltage in the case of small biasing impedance being presented at the gate line ($|V_{out}| \geq V_{TH}$).

In conclusion, for the interest of linearity, large biasing resistors should be placed at the gate of MOS transistors intended for transmit/receive switching. Since that gate biasing resistance is in series with device's poly gate resistance and much larger than that, the value of intrinsic series gate resistance of the device loses significance, and therefore, larger finger widths are exploited in the layout to make the overall transistor structure more compact and the impact of parasitic inductances caused by long interconnections less effective.

The same argument applies to the bulk terminal through the back-gate effect which makes larger substrate resistances desirable. To do so one can use a deep n-well option and bias the bulk of the transistor through an inductor which makes the bulk node open at the operating frequency. However, having a triple-well transistor with a bulk resonant network is not appealing at mm-wave frequencies for the following reasons.

(1) The added deep n-well region increases parasitic capacitances of the MOS transistor and limits its operating frequency.

(2) Transformer-based switches at mm-wave frequencies are quite compact and adding two bulk resonating inductors (one for each shunt transistor) roughly triples the size of the SPDT structure. Ground shields should also be placed around inductors in order to diminish the inductive coupling among the loops.

The other method to increase the substrate resistance is by having fewer number of bulk contacts (with respect to a MOS transistor specialized for high frequency amplifier design) and place them at a relatively far distance. These few contacts will provide the correct bias voltage for the substrate. However, for latchup and substrate coupling purposes, it is still beneficial to have a well-defined ring at an outer area. The space between this ring and transistor's few bulk contacts can be filled with a p-well blocking layer in order to have a native substrate (as opposed to a highly doped surface) around the transistor to preserve the substrate resistance at a high desirable value (Figure 12).

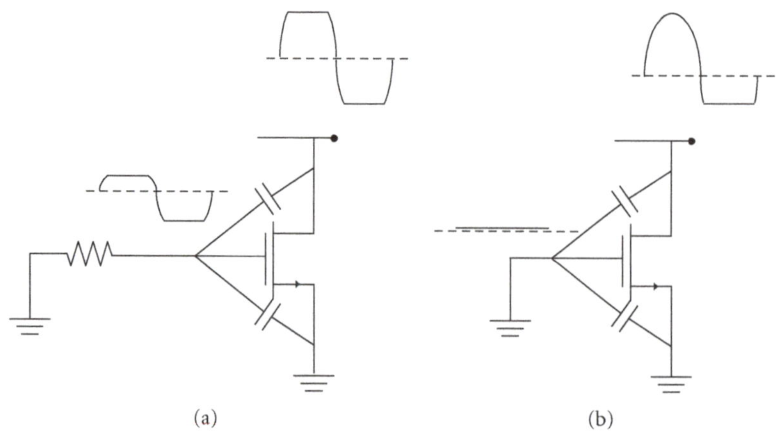

FIGURE 11: Higher biasing impedance at the gate not only improves the insertion loss but also results in less distortion to the output signal and a higher linearity number for the T/R switch.

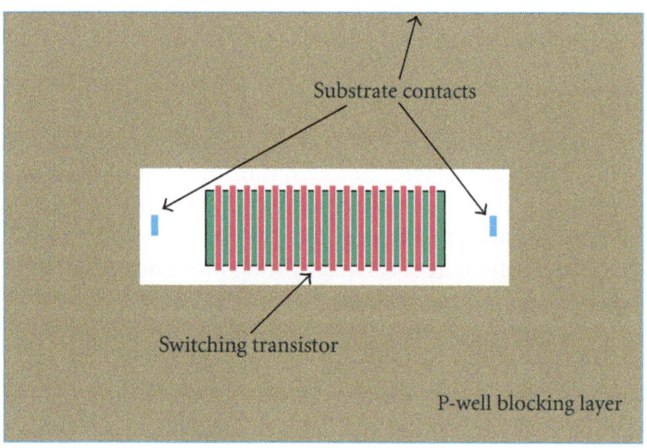

FIGURE 12: Layout example of a MOS transistor to be used as a switch.

Since this is a customized layout configuration and not included in standard libraries provided by the foundry, extraction tools will not be accurate in capturing all the device parasitic parameters. To have an accurate model of the transistor, a test structure is required to be fabricated and characterized so that a measured based custom model of that transistor is available for maximum accuracy of the design.

5.2. Prototype Design. In our design an overlay structure for a 1 : 1 transformer was used. Two equally thick top metal layers were employed. The diameter of the octagon loop is 42 μm and the width of the winding is $W = 8\,\mu$m. The two loops are identical in shape and have a self inductance of $L_1 \sim 90$ pH. Since the thickness and conductivity of the two top metal layers are equal, the quality factor numbers are similar for the loops ($Q \sim 12$). Although more parasitic capacitances is present at the bottom loop, its effect on the Q is not significant at 60 GHz. The coupling between the two loops is $k = 0.72$. These numbers for quality factors and mutual coupling at the operating frequency result in a minimum insertion loss of 0.85 dB.

The prototype uses 40 μm NMOS transistors in a 90 nm standard CMOS process. Channel lengths are set to the minimum for smallest R_{ON}. However, since higher R_{Gate} is beneficial for the switch insertion loss (less power will be dissipated in the gate network) and improves the linearity (by making the gate terminal a floating node), as previously discussed, finger widths that result in maximum $f_{\max}$ were not used. To further increase the impedance at the gate, a large biasing resistor of $R_G = 1\,\text{K}\Omega$ was inserted at the gate of each transistor in series. This large added R_G also alleviates the effect of any gate line inductance. Inductance at the gate line is detrimental to the linearity as it resonates out some part of parasitic gate capacitance to the ground (this is a capacitance that helps linearity as described in Section 5.1). This 1 KΩ of gate biasing resistance is high enough to have negligible effects on the insertion loss and linearity of the T/R switch. However it is low enough in order to provide a fast switching time—less than 250 pS—to turn the switch OFF and ON when 0 and 1.2 V are applied as gate switching voltages respectively. To eliminate the detrimental effect of the bulk network on linearity and insertion loss, a layout as described in Figure 12 with few close by bulk contacts is devised.

6. Prototype Measurement Results

A prototype SPDT T/R switch has been fabricated in a 90 nm CMOS process. The die photo is shown in Figure 13. As can be seen in Figure 13, employing a transformer and designing on a lumped component basis miniaturized the structure, and the active area of the switch is only $60 \times 60\,\mu\text{m}^2$. On-wafer measurement results are shown in Figures 14 and 15. The input and output GSG pads have been deembedded from the measurements using on-chip open and short deembedding structures of the pad. As depicted in Figure 14, the switch has its minimum insertion loss of 3.4 dB at 50 GHz, and its 3 dB bandwidth extends beyond the range of 40 GHz–60 GHz. In simulation, the T/R switch was designed to have a center frequency at the 60 GHz band with a 2.5 dB of insertion loss. However since a custom transistor

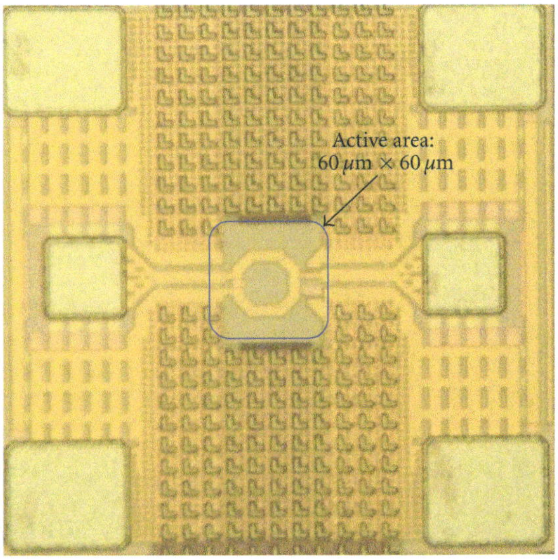

FIGURE 13: Die microphotograph of the miniaturized shunt switch employing a transformer.

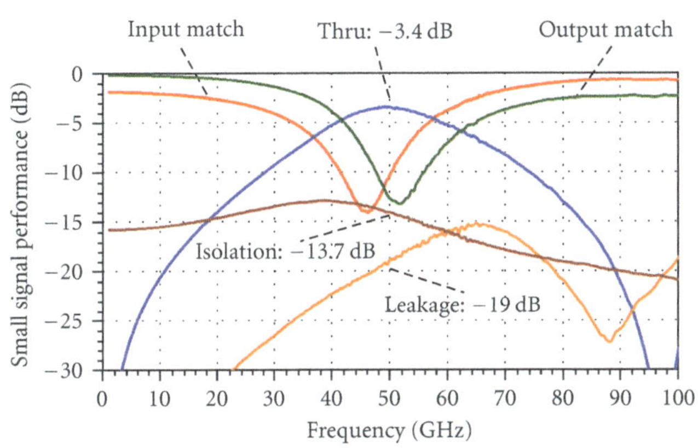

FIGURE 14: Measured insertion loss, leakage (input to the off-thru port), isolation (between the on-thru and off-thru ports), and return loss performance of the switch.

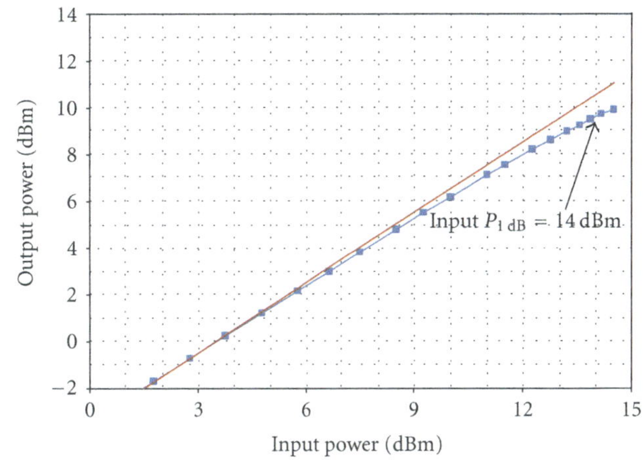

FIGURE 15: Large signal measurement ($P_{-1\,dB}$ = 14 dBm).

as described in Section 5.1 was used for switching, where its exact layout was not fabricated beforehand and characterized for the mm-wave library, a 0.9 dB difference in the insertion loss and a 10 GHz downshift in the frequency response can be observed.

When the input is connected to an on-thru output, there is 19 dB of leakage in the off-thru port. The isolation between two output ports is 13.7 dB. Input and output return loss numbers are better than −10 dB (Figure 14). The large-signal power measurement depicted in Figure 15 shows an input referred 1 dB compression point of +14 dBm, which is adequate for most CMOS mm-wave applications. Due to this high input power required and power handling limitations of the VNA, an external power amplifier module was used to bring up the power provided at the probe tips to a range that $P_{-1\,dB}$ curves could be captured. Table 1 summarizes the overall measured performance of the switch and compares them to recently published results for mm-wave T/R switches. This work has the best performance in terms of the die area. Linearity is comparable to other mm-wave switches, and the insertion loss will benefit from technology scaling in future demonstrations as suggested by [13].

This transformer-based shunt switch was implemented in a 90 nm MOS process with an f_t of ∼100 GHz. According to equations derived in Section 4, insertion loss, leakage, and isolation numbers get better as the R_{ON} of transistor improves. More advanced CMOS technologies with improved f_t numbers directly enhance the performance of the switch by providing more g_m at a constant transistor size (and hence a constant capacitive loading). This makes the future deployment of transformer-based shunt switches in mm-wave systems promising. In addition to transmit/receive (T/R) switching, single pole double throw (SPDT) switches have many other useful applications. Examples include modulators, stepped attenuators, and phase shifters.

7. Conclusion

A miniature lumped-element switch topology employing only a transformer and two shunt NMOS switches has been demonstrated. Shunt-only transistors make it more suitable for mm-wave frequencies. Design equations for the operating frequency, insertion loss, leakage, and isolation were derived in terms of the transformer inductance and coupling factor as well as transistors ON resistance and parasitic capacitances. A transformer-based single pole, double throw (SPDT) shunt switch prototype was designed and fabricated in 90 nm digital CMOS process. It has a minimum insertion loss of 3.4 dB at 50 GHz from the single pole to the ON-thru port and a leakage of 19 dB from the single pole to the OFF-thru port. The isolation is 13.7 dB between the two ports and the switch is capable of handling 14 dBm of power at its 1 dB compression point. The fabricated chip has a small active area of 60 μm × 60 μm.

Table 1: Comparison to recently published mm-wave T/R switches.

Process	Ref. This work/[7] 90 nm CMOS	[6] 130 nm CMOS	[8] 90 nm CMOS	[9] 90 nm CMOS	[10] 65 nm CMOS	[11] 90 nm CMOS
Frequency (GHz)	50	60	24	72	82	60
Insertion loss (dB)	3.4	4.5	3.5	2.7	4.2	1.5
Tx-RX isolation (dB)	13.7	24	16	27	25	25
Supply voltage (V)	1	1.2	1.2	1	1.2	1.2
$IP_{-1\,dB}$ (dBm)	14	4.1	28.7	15	N/A	13.5
Area (mm^2)	0.004	0.221	0.018	0.14	N/A	0.27
Switching time (nS)	0.25	0.4	10	N/A	N/A	3.1

Acknowledgments

This project was funded by C2S2. The authors acknowledge sponsors of Berkeley Wireless Research Center, the NSF Infrastructure Grant no. 0403427, and the foundry chip donation of STMicroelectronics. The authors would also like to thank Ashakn Borna for his help and support.

References

[1] C. H. Doan, S. Emami, A. M. Niknejad, and R. W. Brodersen, "Millimeter-wave CMOS design," *IEEE Journal of Solid-State Circuits*, vol. 40, no. 1, pp. 144–154, 2005.

[2] U. Pfeiffer, J. Gizyb, D. Liu et al., "A 60 GHz radio chipset fully-integrated in a low-cost packaging technology," in *Proceedings of IEEE 56th Electronic Components and Technology Conference*, pp. 1343–1346, June 2006.

[3] B. Razavi, "A 60 GHz direct-conversion CMOS receiver," in *Proceedings of IEEE International Solid-State Circuits Conference (ISSCC '05)*, vol. 1, pp. 400–606, February 2005.

[4] T. Yao, M. Gordon, K. Yau, M. T. Yang, and S. P. Voinigescu, "60 GHz PA and LNA in 90 nm RF-CMOS," in *Proceedings of IEEE Radio Frequency Integrated Circuits Symposium*, pp. 125–128, June 2006.

[5] B. Heydari, M. Bohsali, E. Adabi, and A. M. Niknejad, "Millimeter-wave devices and circuit blocks up to 104 GHz in 90 NM CMOS," *IEEE Journal of Solid-State Circuits*, vol. 42, no. 12, pp. 2893–2903, 2007.

[6] C. M. Ta, E. Skafidas, and R. J. Evans, "A 60 GHz CMOS transmit/receive switch," in *Proceedings of IEEE Radio Frequency Integrated Circuits Symposium (RFIC '07)*, pp. 725–728, June 2007.

[7] E. Adabi and A. M. Niknejad, "A mm-wave transformer based transmit/receive switch in 90 nm CMOS technology," in *Proceedings of the 39th European Microwave Conference (EuMIC '09) Digest*, pp. 389–392, October 2009.

[8] P. Park, D. H. Shin, J. J. Pekarik, M. Rodwell, and C. P. Yue, "A high-linearity, LC-tuned, 24 GHz T/R switch in 90 nm CMOS," in *Proceedings of IEEE Radio Frequency Integrated Circuits Symposium (RFIC '08)*, pp. 369–372, June 2008.

[9] S. F. Chao, H. Wang, C. Y. Su, and J. G. J. Chern, "A 50 to 94 GHz CMOS SPDT switch using traveling-wave concept," *IEEE Microwave and Wireless Components Letters*, vol. 17, no. 2, pp. 130–132, 2007.

[10] A. Tomkins, P. Garcia, and S. P. Voinigescu, "A passive W-band imager in 65 nm bulk CMOS," in *Proceedings of the Annual IEEE Compound Semiconductor Integrated Circuit Symposium (CSICS '09)*, pp. 1–4, October 2009.

[11] M. Uzunkol and G. Rebeiz, "A low-loss 50–70 GHz SPDT switch in 90 nm CMOS," *IEEE Journal of Solid-State Circuits*, vol. 45, no. 10, pp. 2003–2007, 2010.

[12] T. S. D. Cheung, "Shielded passive devices for silicon-based monolithic microwave and millimeter-wave integrated circuits," *IEEE Journal of Solid-State Circuits*, vol. 41, no. 5, pp. 1183–1200, 2006.

[13] G. Wang, W. Woods, H. Ding, and E. Mina, "Novel low-cost on-chip CPW slow-wave structure for compact RF components and mm-wave applications," in *Proceedings of the 58th Electronic Components and Technology Conference (ECTC '08)*, pp. 186–190, May 2008.

[14] J. R. Long, "Monolithic transformers for silicon RF IC design," *IEEE Journal of Solid-State Circuits*, vol. 35, no. 9, pp. 1368–1382, 2000.

10

Performance and Trends in Millimetre-Wave CMOS Oscillators for Emerging Wireless Applications

Marius Voicu,[1,2] Domenico Pepe,[1] and Domenico Zito[1,2]

[1] *Tyndall National Institute, Lee Maltings, Dyke Parade, Cork, Ireland*
[2] *Department of Electrical and Electronic Engineering, University College Cork, Cork, Ireland*

Correspondence should be addressed to Domenico Zito; domenico.zito@tyndall.ie

Academic Editor: Ramesh Pokharel

This paper reports the latest advances on millimeter-wave CMOS voltage-controlled oscillators (VCOs). Current state-of-the-art implementations are reviewed, and their performances are compared in terms of phase noise and figure of merit. Low power and low phase noise LC-VCO and ring oscillator designs are analyzed and discussed. Design and performance trends over the last decade are provided and discussed. The paper shows how for the higher range of millimeter-waves (>60 GHz) the performances of ring oscillators become comparable with those of LC-VCOs.

1. Introduction

In the last few years several standards have been, or have been planned to soon be, released, regarding millimetre-waves (mm-waves, i.e., 30–300 GHz) systems for emerging wireless applications. Some of the most attractive applications are 60 GHz unlicensed wireless data communication [1], 77-GHz automotive radars [2], and 94 GHz passive imaging [3]. Key enabler for high-volume and low-cost mass market implementation of these systems is the significant improvement of device performance in the latest CMOS technology nodes (i.e., 130 nm and smaller), which offer a great potential for the realization of millimeter-waves wireless transceivers on a single chip.

One of the most important building blocks in a wireless transceiver is the frequency synthesizer. Performance of the voltage controlled oscillator (VCO) dictates the performance of the frequency synthesizer and thus of the whole communication system.

The aim of the present paper is to provide a review of the state-of-the-art (SoA) of millimeter-wave (mm-wave, 30–300 GHz) VCOs in CMOS technology in order to identify the trends over the last decade and derive some useful observations regarding the past and possible future evolution of design and performance. In particular, the paper reports a comparison of performances among SoA design solutions and highlights the achievements and trends in terms of phase noise (PN) and figure of merit (FOM).

The present paper is organized as follows. Section 2 provides an overview of two of the most widespread VCO topologies, LC-tank, and ring oscillators and recalls briefly their main causes responsible for the phase noise. In Section 3, SoA millimeter-wave CMOS LC-VCO and ring oscillator design solutions are reported, and their performances are discussed and compared. In Section 4, the conclusions are drawn.

2. CMOS VCOs

The most widespread CMOS VCO topologies at mm-wave frequencies are LC-tank and ring oscillators. Section 2.1 provides a brief overview of LC-VCOs and their PN contributions. Section 3.1 provides a brief review of ring oscillators and their PN contributions.

2.1. LC-Tank VCOs. LC-VCOs consist of a resonant circuit (LC-tank) and an amplifier that provides adequate gain to compensate the losses of the resonant circuit. The amplifier can be a single transistor in one of the known configurations (common-source, common-gate, or source follower) or the widespread cross-coupled differential pair (see Figure 1(a)).

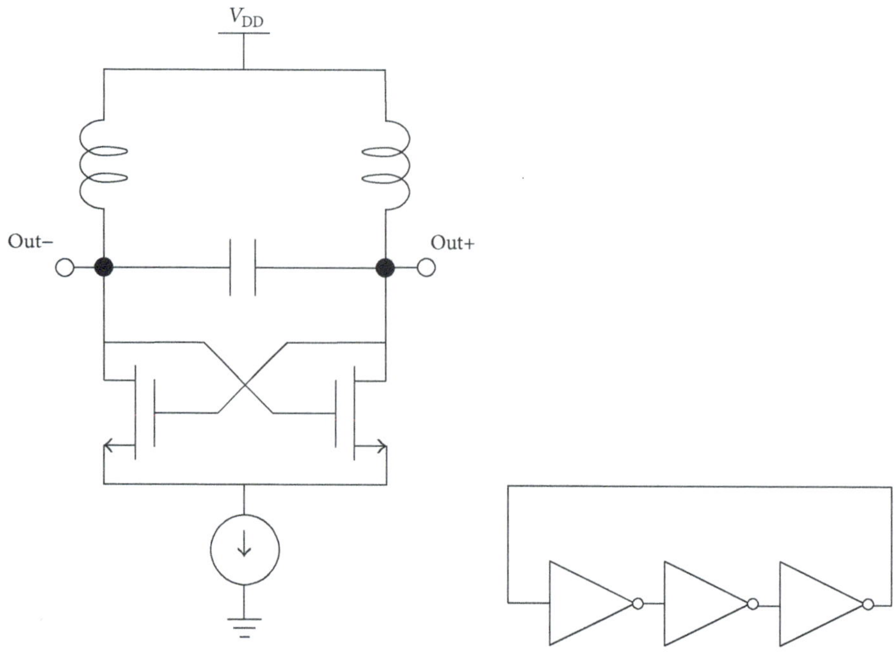

FIGURE 1: (a) LC-VCO; (b) ring oscillator.

The main causes of PN in LC-VCO are due to the losses in the resonator and the amplifier noise. For instance, in the case of cross-coupled differential pair LC-VCOs, they are (i) resonator thermal noise (due to the loss conductance in the resonator), (ii) tail current noise (the switching action of the differential pair translates noise up and down in frequency, and so the noise enters the resonator), and (ii) differential pair noise (due to the finite switching time of the pair) [4].

2.2. Ring Oscillators. Ring oscillators (ROs) are composed of a cascade of inverting amplifiers, and the output of the last element is fed back to the input of the first (see Figure 1(b)). These inverter stages can be implemented by differential amplifiers, CMOS inverters, or even LC-VCOs.

The main causes of PN in ring oscillators are (i) the thermal noise (due to MOSFET drain-source channel resistance and load resistors) and (ii) flicker noise (in CMOS inverter-based ROs, the pull-up and pull-down currents contain flicker noise which varies slowly over many transitions, while, in differential ROs, the flicker noise in the tail current modulates the propagation delay of the stages) [5].

3. State-of-the-Art of mm-Wave CMOS VCOs

In this Section, a review of SoA mm-wave CMOS LC-VCO and RO design solutions is provided, and their performances are discussed and compared. In Section 3.1, three SoA mm-waves CMOS LC-VCO implementations, operating at 30, 60, and 140 GHz, respectively, are reported. In Section 3.2, two SoA mm-wave RO designs (the first implemented at 50 and 60 GHz, the second at 104 and 121 GHz) are reported. In Section 3.3, performance trends over the last decade of

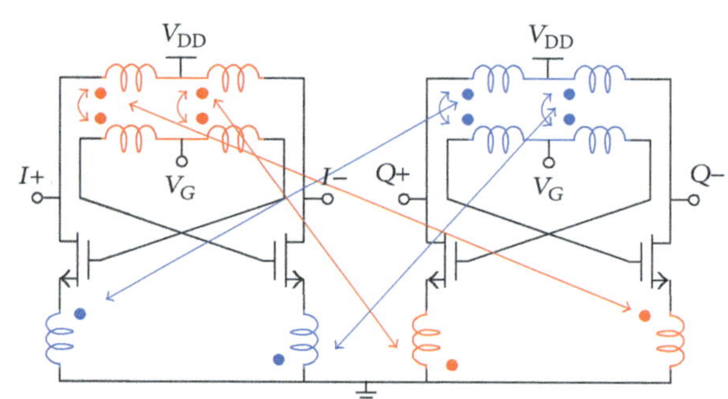

FIGURE 2: Schematic of the VCO presented in [6]. The two trifilar transformers of I and Q VCOs are highlighted in red and blue. Spirals couplings are represented by the thin arrowed lines. I and Q are VCO outputs.

mm-wave CMOS LC-VCOs and ROs are provided and discussed. The performances of the SoA VCOs are summarized in Tables 1 and 2, and their figure of merit ((FOM) see (1)) are evaluated:

$$\mathrm{FOM} = PN - 20\mathrm{Log}\left(\frac{f_0}{\Delta f}\right) + 10\mathrm{Log}\left(\frac{P_{DC}}{1\,\mathrm{mW}}\right), \quad (1)$$

where f_0 is the oscillation frequency, Δf is the offset at which the PN is evaluated, and P_C is the power consumption.

3.1. SoA mm-Wave CMOS LC-VCOs. In [6] a 30 GHz quadrature VCO (QVCO) implemented in 0.13 μm CMOS technology is presented. The circuit schematic is shown in Figure 2. It exploits the use of a trifilar (1 : 1 : 1) transformer with a high quality factor (i.e., with respect to spiral inductors)

Table 1: SoA mm-wave CMOS LC-VCOs: summary of performance.

Reference	Tech. (nm)	f_0 (GHz)	V_{DD} (V)	P_C (mW)	Phase noise (dBc/Hz)	FOM (dBc/Hz)
[6]	130	30.3	0.6	7.8	−114 @ 1 MHz	−196
[7]	90	57.6	0.6	7.2	−102 @ 1 MHz	−188
[9]	90	139.8	1.2	9.6	−75 @ 1 MHz	−178

Table 2: SoA mm-wave CMOS ROs: summary of performance.

Reference	Tech. (nm)	f_0 (GHz)	V_{DD} (V)	P_C (mW)	Phase noise (dBc/Hz)	FOM (dBc/Hz)
[10]	130	50.3	0.8	35	−104 @ 1 MHz	−186.4
[10]	130	58.5	0.8	34	−95 @ 1 MHz	−180.6
[11]	130	104	1.5	28	−93.3 @ 1 MHz	−180
[11]	130	121	1.3	21	−88 @ 1 MHz	−176.5

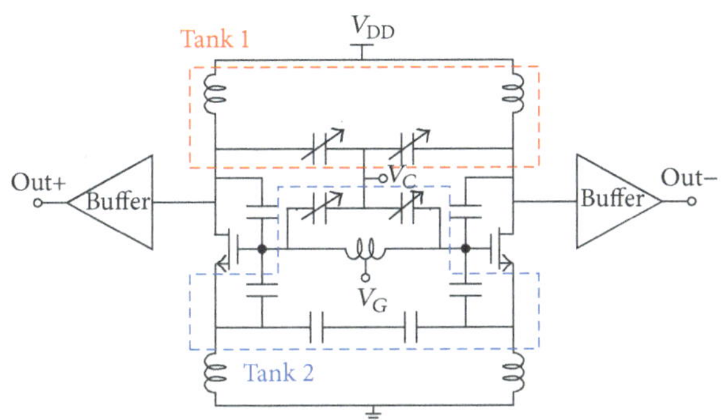

Figure 3: Schematic of the 60 GHz Colpitts LC-VCO presented in [7].

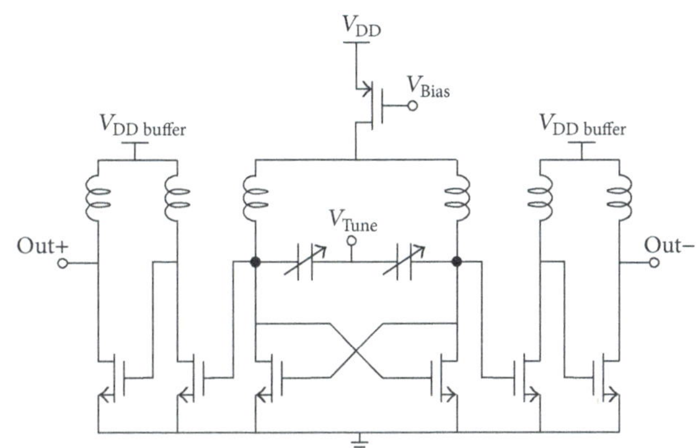

Figure 4: Schematic of the 140-GHz LC-VCO presented in [9].

in order to improve the PN performance. In fact, with respect to inductors, transformers can provide higher quality factors due to the mutual coupling between the spirals. The trifilar transformer couples two series cascaded cross-coupled VCO structures. The transformer couples in-phase and in-quadrature drain and source spirals, allowing for a reduction of device noise, parasitic capacitances, and power consumption. The PN is −114 dBc/Hz @ 1 MHz from the carrier frequency of 30.3 GHz. The power consumption amounts to 7.8 mW from a 0.6 V supply voltage.

In [7], a 60 GHz Colpitts LC-VCO implemented in 90 nm CMOS technology is presented. The circuit schematic is shown in Figure 3. Although Colpitts oscillators have good PN performances, they suffer from the Miller capacitance effects, which cause an increase in the parasitic gate-drain capacitance of the MOSFET transistors. This issue is solved by combining a conventional Colpitts oscillator and a tuned-input tuned-output (TITO) oscillator [8]. In this way, start-up issues of the Colpitts oscillator have been solved, and phase noise performance improved (thanks to an extra LC-tank for noise filtering). The circuit consumes 7.2 mW from a 0.6 V supply voltage. The PN is −102 dBc/Hz @ 1 MHz offset from the carrier (57.6 GHz). The tuning range is 5.3 GHz (from 55.8 to 61.1 GHz).

In [9] a 140 GHz cross-coupled LC-VCO implemented in 90 nm CMOS technology by UMC is presented. The circuit schematic is shown in Figure 4. A low parasitic cross-coupled transistor layout is developed in order to achieve a high fundamental frequency. The VCO core has been biased through a p-MOSFET in order to reduce the flicker noise contribution to the overall close-in PN. Moreover, to minimize the load capacitance connected to the LC-tank, a two-stage tapered buffer has been used to drive the 50 Ω load. The VCO core consumes 9.6 mW from a 1.2 V voltage supply. The buffers consume 7.2 mW. The PN amounts to −75 dBc/Hz @ 1 MHz offset from the carrier frequency of 139.8 GHz.

Table 1 summarizes the main characteristics and performances of the aforementioned LC-VCOs.

3.2. SoA mm-Wave CMOS Ring Oscillators. In [10], 50 GHz and 60 GHz ring oscillators implemented in 0.13 μm CMOS are presented. The block diagram is shown in Figure 5(a). An interpolative-phase-tuning (IPT) technique is used to tune frequency of multiphase mm-wave LC-based ROs without using varactors (see Figure 5(b)). In order to vary the output frequency, the delay of each stage of the ROs is varied by means of tunable phase shifters. A fixed phase shift is used to introduce a delayed current i_1 via M_3 and M_4; i_1 is interpolated with the undelayed current provided by M_1 and M_2. The phase shift can be tuned from 0 to β by controlling the biasing dc current I_0 and I_1. Two output

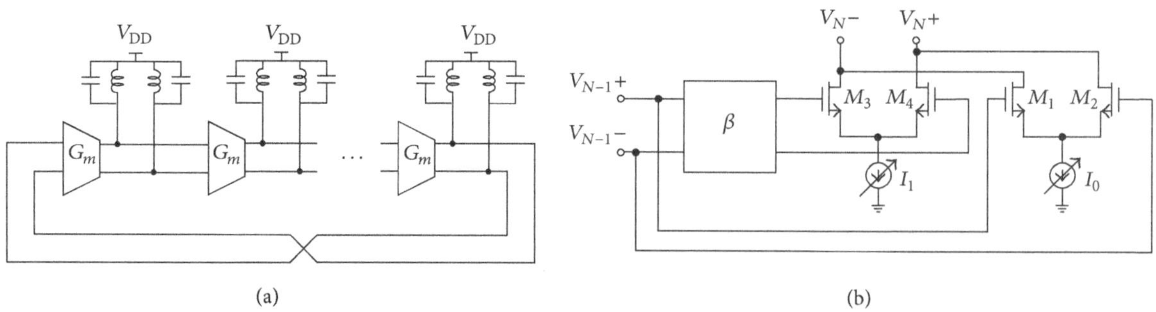

FIGURE 5: (a) Block diagram of the 50 and 60 GHz ring oscillators presented in [10]. (b) Schematic of Gm block of Figure 4(a): β is a fixed phase shifter implemented by a LC-based differential stage.

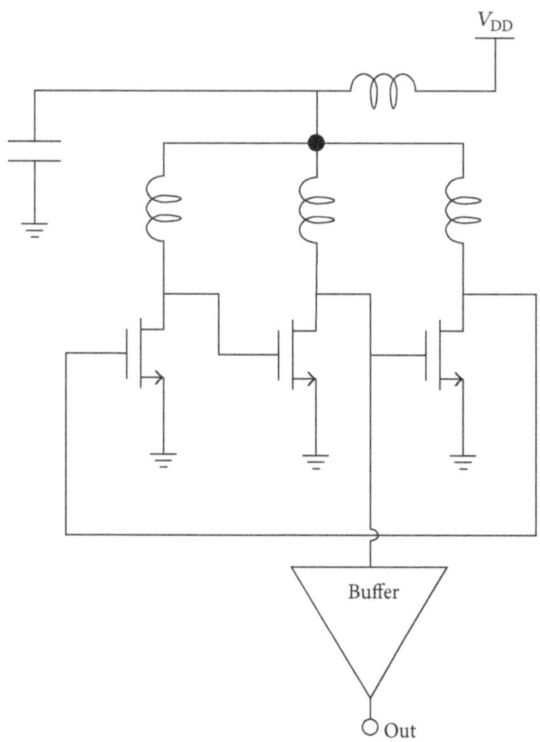

FIGURE 6: Schematic of the 104 and 121 GHz ring oscillators presented in [11].

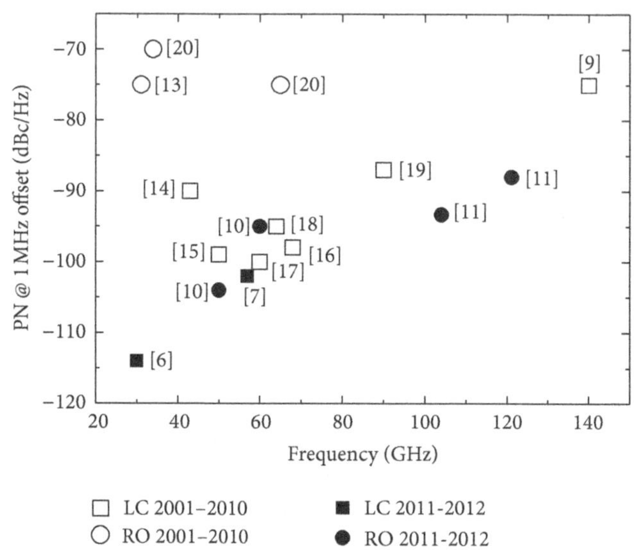

FIGURE 7: Oscillator phase noise performances versus oscillation frequency [6, 7, 9–11, 13–20].

current-controlled oscillators (CCOs), with 4 and 8 phases, are implemented using this technique. The 8-phase CCO can be tuned from 48.6 to 52 GHz, and it consumes from 32 to 48 mW from a 0.8 V voltage supply. The 4-phase CCO can be tuned from 56 to 61.3 GHz and consumes from 30 to 37 mW. The PN of the 8-phase CCO amounts to −104 dBc/Hz at 1 MHz offset from the carrier (50.3). The PN of the 4-phase CCO is −95 dBc/Hz @ 1 MHz offset from the carrier (58.5 GHz).

In [11] two fundamental three-stage ROs implemented in a 0.13 μm CMOS process and oscillating at 104 GHz and 121 GHz, respectively, are presented. The circuit schematic is shown in Figure 6. A new design methodology for designing high-frequency oscillators has been developed. This method finds the best topology to achieve frequencies close to the maximum frequency of oscillation of the transistors. It is based on the activity condition of the transistors. A device is called active at a certain frequency, if it can generate power in the form of a single sinusoidal signal at that frequency [12]. This method determines also the maximum frequency of oscillation for a fixed circuit topology. Each stage of the implemented ROs is implemented using a double gate transistor with a substrate contact ring around the transistor and an inductive load. The measured peak output powers of the two oscillators are −3.5 dBm and −2.7 dBm at 121 GHz and 104 GHz, respectively. The DC power consumptions, including the output buffer, is 21 mW from a 1.28 V supply and 28 mW from a 1.48 V supply for the 121 GHz and 104 GHz oscillators, respectively. The PN at 1 MHz offset frequency is −88 dBc/Hz and −93.3 dBc/Hz for the 121 GHz and 104 GHz oscillators, respectively.

The main figures of merit and performances of the aforementioned ring oscillators are presented in Table 2.

3.3. Performance Trends in mm-Wave CMOS VCOs. PN versus oscillation frequency of SoA mm-wave CMOS LC-VCOs and ROs published in the last 11 years are shown in

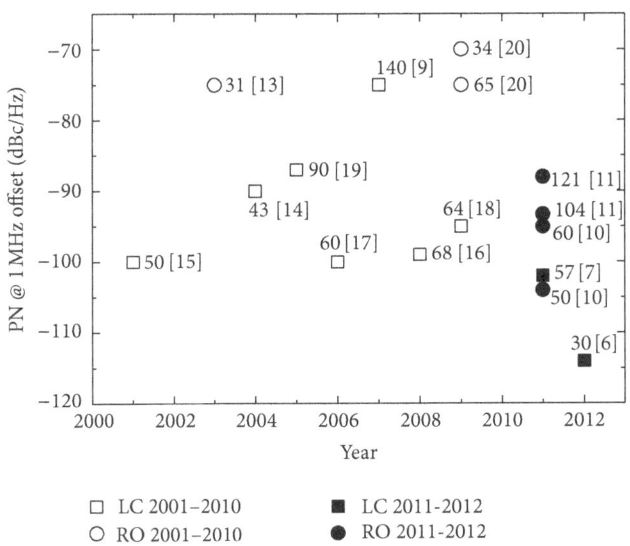

Figure 8: Oscillator phase noise performances versus publication year [6, 7, 9–11, 13–20].

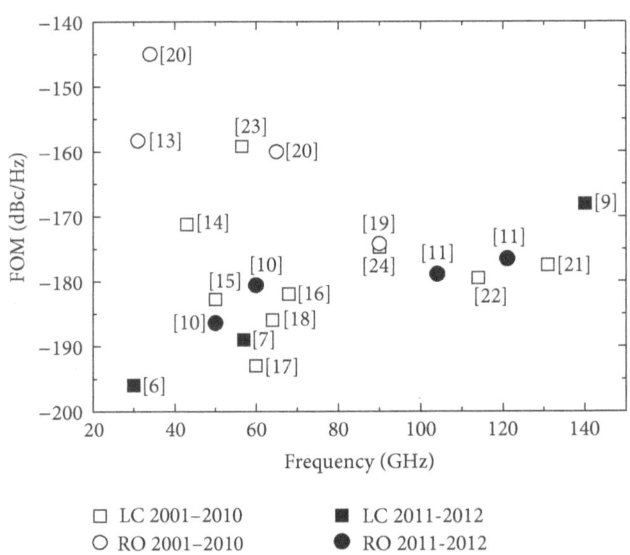

Figure 9: Oscillator FOM versus oscillation frequency [6, 7, 9–11, 13–24].

Figure 7. It can be observed that PN performances of ring oscillators are becoming closer to those of LC-VCOs while moving towards very high frequencies.

Figure 8 shows PN versus publication year. It can be noted that in, the last couple of years, while in the low mm-wave range (30 GHz) the PN of LC-VCOs is still better (−114 dBc/Hz @ 1 MHz offset from 30 GHz [6]), at very high frequencies PN of RO becomes comparable to that of LC-VCOs, achieving a PN of −88 dBc/Hz @ 1 MHz at 121 GHz [11].

The FOM achieved by the state-of-the-art mm-wave CMOS LC-VCOs and ROs published in the last 11 years are shown in Figure 9. Also in this case, as in Figure 7 relative to PN, it can be noted that in the lower part of the mm-wave range (below 60 GHz) LC-VCOs attain overall better FOM

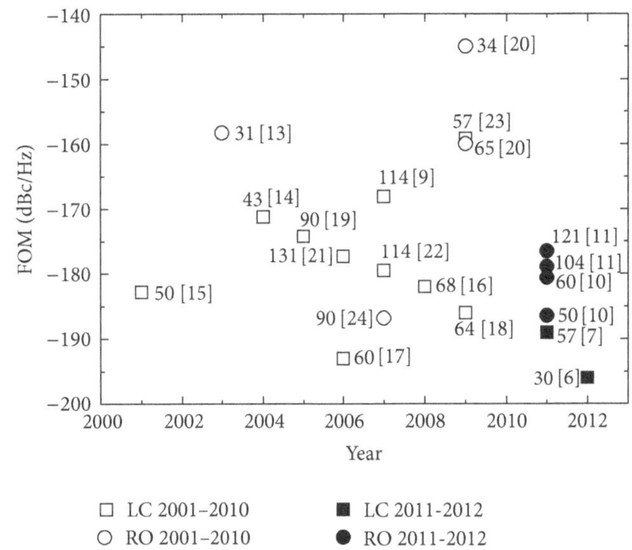

Figure 10: Oscillator FOM versus publication year [6, 7, 9–11, 13–24].

than ROs, but for very high frequencies FOM of ROs became comparable to that of LC-VCOs.

Figure 10 shows FOM versus publication year. It can be noted that the trend in the last couple of years is that the FOM of LC-VCOs is still superior, but it is achieved for lower frequencies [6, 7] than the ROs in [10, 11]. In fact, the solutions in [10, 11] achieve FOM comparable to those of previous implementations of LC-VCOs at lower frequencies.

4. Conclusions

A review of the state-of-the-art of millimeter-wave CMOS VCOs has been presented. State-of-the-art LC-VCOs and ring oscillators have been presented and discussed, and their performances have been compared. The trends for VCO design and performance over the last decade have been traced and discussed.

From these evaluations it appears that while moving in the higher part of the mm-wave spectrum (>60 GHz) phase noise and FOM performance of ring oscillators tend to become closer, and even comparable, to those of LC-VCOs, which are dominant at lower frequencies. Thus, ring oscillators appear to be a strong candidate for the implementation of CMOS VCOs operating at the higher region of the mm-wave frequency spectrum.

References

[1] D. Pepe and D. Zito, "System-level simulations investigating the system-on-chip implementation of 60-GHz transceivers for wireless uncompressed HD video," in *Applications of MATLAB in Science and Engineering*, pp. 181–196, InTech Publisher, 2011.

[2] M. Schneider, "Automotive radar—status and trends," in *Proceedings of the German Microwave Conference (GeMiC '05)*, Ulm, Germany, April 2005.

[3] L. O. Mereni, D. Pepe, F. Alimenti, and D. Zito, "Feasibility and challenges of a 95-GHz SoC radiometer for W-band passive

imaging," in *Proceedings of the IET Irish Signals and Systems Conference*, pp. 28–29, Maynooth, Ireland, June 2012.

[4] J. J. Rael and A. A. Abidi, "Physical processes of phase noise in differential LC oscillators," in *Proceedings of the 22nd Annual Custom Integrated Circuits Conference (CICC '00)*, pp. 569–572, May 2000.

[5] A. A. Abidi, "Phase noise and jitter in CMOS ring oscillators," *IEEE Journal of Solid-State Circuits*, vol. 41, no. 8, pp. 1803–1816, 2006.

[6] J. S. Syu, H. L. Lu, and C. Meng, "A 0.6 V 30 GHz CMOS quadrature VCO using microwave 1 : 1 : 1 trifilar transformer," *IEEE Microwave and Wireless Components Letters*, vol. 22, no. 2, 2012.

[7] L. Li, P. Reynaert, and M. Steyaert, "A colpitts LC VCO with Miller-capacitance gm enhancing and phase noise reduction techniques," in *Proceedings of the 37th European Solid-State Circuits Conference, (ESSCIRC '11)*, pp. 491–494, Helsinki, Finland.

[8] S. Shekhar, J. S. Walling, S. Aniruddhan, and D. J. Allstot, "CMOS VCO and LNA using tuned-input tuned-output circuits," *IEEE Journal of Solid-State Circuits*, vol. 43, no. 5, pp. 1177–1186, 2008.

[9] C. Cao, E. Seok, and K. O. Kenneth, "Millimeter-wave CMOS voltage-controlled oscillators," in *Proceedings of the IEEE Radio and Wireless Symposium, (RWS '07)*, pp. 185–188, Long Beach, Calif, USA, January 2007.

[10] S. Rong and H. C. Luong, "Design and analysis of varactor-less interpolative-phase-tuning millimeter-wave LC oscillators with multiphase outputs," *IEEE Journal of Solid-State Circuits*, vol. 46, no. 8, pp. 1810–1819, 2011.

[11] O. Momeni and E. Afshari, "High power terahertz and millimeter-wave oscillator design: a systematic approach," *IEEE Journal of Solid-State Circuits*, vol. 46, no. 3, pp. 583–597, 2011.

[12] R. Spence, *Linear Active Networks*, Wiley-Interscience, New York, NY, USA, 1970.

[13] J. O. Plouchart, J. Kim, N. Zamdmer et al., "A 31 GHz CML ring VCO with 5.4 ps delay in a 0.12-/spl mu/m SOI CMOS technology," in *Proceedings of the 29th European Solid-State Circuits Conference, (ESSCIRC '03)*, pp. 357–3360, Estoril, Portugal, September 2003.

[14] A. P. Van der Wel, S. L. J. Gierkink, R. C. Frye, V. Boccuzzi, and B. Nauta, "A robust 43-GHz VCO in CMOS for OC-768 SONET applications," *IEEE Journal of Solid-State Circuits*, vol. 39, no. 7, pp. 1159–1163, 2004.

[15] H. M. Wang, "A 50 GHz VCO in 0.25 μm CMOS," in *Proceedings of the IEEE International Solid-State Circuits Conference (ISSCC '01)*, pp. 372–3373, San Francisco, Calif, USA, February 2001.

[16] H.-K. Chen, H. J. Chen, D. C. Chang, Y. Z. Juang, and S. S. Lu, "A 0.6 V, 4.32 mW, 68 GHz low phase-noise VCO with intrinsic-tuned technique in 0.13 μm CMOS," *IEEE Microwave and Wireless Components Letters*, vol. 18, no. 7, pp. 467–469, 2008.

[17] D. Huang, W. Hant, N. Y. Wang et al., "A 60GHz CMOS VCO using on-chip resonator with embedded artificial dielectric for size, loss and noise reduction," in *Proceedings of the IEEE International Solid-State Circuits Conference, (ISSCC '06)*, pp. 1218–1227, San Francisco, Calif, USA, February 2006.

[18] L. Li, P. Raynaert, and M. Steyaert, "A low power mm-wave oscillator using power matching techniques," in *Proceedings of the IEEE Radio Frequency Integrated Circuits Symposium (RFIC '09)*, pp. 469–472, Boston, Mass, USA, 2009.

[19] C. Cao and O. K. Kenneth, "A 90-GHz voltage-controlled oscillator with a 2.2-GHz tuning range in a 130-nm CMOS technology," in *Proceedings of the Symposium on VLSI Circuits*, pp. 242–2243, Kyoto, Japan, June 2005.

[20] C. C. Chen, C. C. Li, Bo-Jr Huang, K. Y. Lin, H. W. Tsao, and H. Wang, "Ring-based triple-push VCO with wide continous tuning ranges," *IEEE Transactions on Microwave Theory and Techniques*, vol. 57, no. 9, pp. 2173–2183, 2009.

[21] P. C. Huang, R. C. Liu, H. Y. Chang et al., "A 131-GHz push-push VCO in 90-nm CMOS technology," in *Proceedings of the IEEE Radio Frequency Integrated Circuits (RFIC '05)*, pp. 613–616, June 2005.

[22] P.-C. Huang, M. D. Tsai, G. D. Vendelin, H. Wang, C. H. Chen, and C. S. Chang, "A low-power 114-GHz push-push CMOS VCO using LC source degeneration," *IEEE Journal of Solid-State Circuits*, vol. 42, no. 6, pp. 1230–1239, 2007.

[23] C. W. Tsou, C. C. Chen, and Y. S. Lin, "A 57-GHz CMOS VCO with 185.3% tuning-range enhancement using tunable LC source-degeneration," in *Proceedings of the International Symposium on VLSI Design, Automation and Test, (VLSI-DAT '09)*, pp. 146–149, Hsinchu, Taiwan, 2009.

[24] Z.-M. Tsai, C.-S. Lin, C. F. Huang, J. G. J. Chern, and H. Wang, "A fundamental 90-GHz CMOS VCO using new ring-coupled quad," *IEEE Microwave and Wireless Components Letters*, vol. 17, no. 3, pp. 226–228, 2007.

11

CMOS Ultra-Wideband Low Noise Amplifier Design

K. Yousef,[1] H. Jia,[2] R. Pokharel,[3] A. Allam,[1] M. Ragab,[1] H. Kanaya,[3] and K. Yoshida[3]

[1] *Electronics and Communications Engineering Department, Egypt-Japan University of Science and Technology, New Borg Al-Arab, 21934 Alexandria, Egypt*
[2] *E-JUST Center, Kyushu University, Nishi-ku, Fukuoka 819-0395, Japan*
[3] *Graduate School of ISSE, Kyushu University, Nishi-ku, Fukuoka 819-0395, Japan*

Correspondence should be addressed to K. Yousef; khalil.yousef@ejust.edu.eg

Academic Editor: Mohammad S. Hashmi

This paper presents the design of ultra-wideband low noise amplifier (UWB LNA). The proposed UWB LNA whose bandwidth extends from 2.5 GHz to 16 GHz is designed using a symmetric 3D RF integrated inductor. This UWB LNA has a gain of 11 ± 1.0 dB and a NF less than 3.3 dB. Good input and output impedance matching and good isolation are achieved over the operating frequency band. The proposed UWB LNA is driven from a 1.8 V supply. The UWB LNA is designed and simulated in standard TSMC 0.18 μm CMOS technology process.

1. Introduction

CMOS technology is one of the most prevailing technologies used for the implementation of radio frequency integrated circuits (RFICs) due to its reduced cost and its compatibility with silicon-based system on chip [1]. The use of ultra-wideband (UWB) frequency range (3.1–10.6 GHz) for commercial applications was approved in February 2002 by the Federal Communications Commission. Low cost, reduced power consumption, and transmission of data at high rates are the advantages of UWB technology. UWB technology has many applications such as wireless sensor and personal area networks, ground penetrating radars, and medical applications [2].

Low noise amplifier is considered the backbone of the UWB front-end RF receiver. It is responsible for signal reception and amplification over the UWB frequency range. LNA has many desired design specifications such as low and flat noise figure, high and flat power gain, good input and output wide impedance matching, high reverse isolation, and reduced DC power consumption [1, 3].

Nowadays one of the most suitable configurations suggested for LNA implementation is current reuse cascaded amplifier. This LNA configuration can attain low DC power consumption, high flattened gain, minimized NF, and excellent reverse isolation while achieving wide input and output impedance matching [1–3].

Radio frequency integrated inductors play a significant role in radio frequency integrated circuits (RFICs) implementation. Design, development, and performance improvement of RF integrated inductors represent a challenging work. Achieving high integration level and cost minimization of RFICs are obstructed because of the difficulties facing the RF integrated inductors designers which are related to obtaining high quality factors [4–6].

In this paper, the implementation of LNAs using 3D integrated inductors will be investigated. A symmetric 3D structure is proposed as a new structure of integrated inductors for RFICs.

This paper discusses the design procedure of current reuse cascaded UWB LNA and its bandwidth expansion. In addition, the employment of suggested symmetric 3D RF integrated inductor will be demonstrated. This paper is organized as follows. Section 2 introduces the suggested UWB LNA circuit. Section 3 gives simulation results and discussion. Conclusion is driven in Section 4.

2. Circuit Description

As shown in Figure 1, the proposed UWB LNA is a current reuse cascaded core based on a common source topology with a shunt resistive feedback technique implemented over the input stage.

This current reuse cascaded amplifier achieved good wideband input impedance matching through the use of source degeneration input matching technique. Figure 2 shows the small signal equivalent circuit of this LNA input stage. The input port of this UWB LNA is desired to match source impedance R_s at resonance frequency ω_o. This matching circuit bandwidth is defined through the quality factors of source degeneration and gain-peaking inductors (L_s and L_g) where the input impedance is given by

$$\begin{aligned} Z_{\text{in}} &= j\omega\left(L_s + L_g\right) + \frac{1}{j\omega C_{\text{gs}}} + \omega_T L_s \\ &= j\omega\left(L_s + L_g\right) + \frac{1}{j\omega C_{\text{gs}}} + R_s, \end{aligned} \quad (1)$$

where Z_{in} is the UWB LNA input impedance and ω_T is the current-gain cut-off frequency, where $\omega_T = g_m/C_{\text{gs}}$ and g_m and C_{gs} are the input stage transconductance and gate-source capacitance, respectively. V_s represents the RF signal source. R_s is the output impedance of V_s.

Although the shunt resistive feedback loop leads to LNA noise performance degradation [7], it is widely used in recently proposed LNAs due to its superior wideband characteristics. Shunt capacitive-resistive feedback technique is employed to widen the input-matching bandwidth and increase the LNA stability.

Shunt-peaked amplifiers are known to have wide gain bandwidth and high low frequency power gain [8]. To have a high flattened gain of the proposed UWB LNA, shunt-peaking technique is used. In addition the gate-peaking technique is used to enhance the LNA gain at high frequencies. Besides the shunt- and gate-peaking techniques, the shunt resistive feedback loop is used in gain flattening [2, 8]. The LNA approximate gain is given by

$$\begin{aligned} A &\cong \frac{V_{\text{out}}}{V_s} \\ &\cong \frac{g_{m1} g_{m2} \left[R_L // \left(R_{d2} + SL_{d2}\right)\right] \left[SL_{d1}\right]}{2 \cdot SC_{\text{gs1}} \left[S\left(L_{s1} + L_{g1}\right) + 1/SC_{\text{gs1}}\right]}. \end{aligned} \quad (2)$$

Ultra-wideband applications require good noise performance in addition to high and flat gain. Low noise design techniques which are suitable for narrowband applications cannot be used for wideband applications. Main contribution of cascaded matched stages noise figure is due to first stage [9]. The reduction of noise figure of input stage will lead to the reduction of the overall noise figure of the proposed design. Optimization and control of factors affecting the NF will improve this UWB LNA noise performance. An equivalent circuit of the input stage for noise factor calculation is shown in Figure 3 [1].

An estimated value of the noise figure (NF ⊠ 10 $\log_{10} f$) of this topology is given in [1] where f is the noise factor of the UWB LNA. The noise factor f can be given by

$$\begin{aligned} f = 1 &+ \frac{R_g + R_{\text{lg}} + R_{\text{ss}} + R_{\text{ls}}}{R_s} + \frac{\delta\alpha\omega^2 C_{\text{gs1}}^2 R_s}{5 g_{m1}} \\ &+ \frac{R_{\text{FB}}\left(\left(L_{g1} + L_{s1}\right) C_{\text{gs1}}\right)^2}{R_s\left(g_{m1} R_{\text{FB}} - 1\right)^2} \\ &\cdot \left| s^2 + s\left(\frac{\omega_{\text{o,rfbn}}}{Q_{\text{rfbn}}}\right) + \omega_{\text{o,rfbn}}^2 \right|^2 \\ &+ \frac{\gamma g_{m1}\left(R_{\text{FB}} + R_s\right)^2 \left(\left(L_{g1} + L_{s1}\right) C_{\text{gs1}}\right)^2}{\alpha R_s \left(g_{m1} R_{\text{FB}} - 1\right)^2} \\ &\cdot \left| s^2 + s\left(\frac{\omega_{\text{o,dn}}}{Q_{\text{dn}}}\right) + \omega_{\text{o,dn}}^2 \right|^2, \end{aligned} \quad (3)$$

$$f = 1 + \frac{R_g + R_{\text{lg}} + R_{\text{ss}} + R_{\text{ls}}}{R_s} + f_{\text{gn}} + f_{\text{rfbn}} + f_{\text{dn}}, \quad (4)$$

where

$$\begin{aligned} \omega_{\text{o,rfbn}} &= \sqrt{\frac{1 + g_{m1} R_s}{\left(L_{g1} + L_{s1}\right) C_{\text{gs1}}}}, \\ Q_{\text{rfbn}} &= \frac{1}{R_s + \omega_{T1} L_{s1}} \cdot \sqrt{\frac{\left(1 + g_{m1} R_s\right)\left(L_{g1} + L_{s1}\right)}{C_{\text{gs1}}}}, \\ \omega_{\text{o,rfbn}} &= \sqrt{\frac{1}{\left(L_{g1} + L_{s1}\right) C_{\text{gs1}}}}, \\ Q_{\text{dn}} &= \frac{1}{\left(R_s \parallel R_{\text{FB}}\right) + \omega_{T1} L_{s1}} \cdot \sqrt{\frac{\left(L_{g1} + L_{s1}\right)}{C_{\text{gs1}}}}, \end{aligned} \quad (5)$$

where f_{gn}, f_{dn}, and f_{rfbn} are gate, drain, and feedback resistor noise factors, respectively and α, δ, and γ are constants equal to 0.85, 4.1, and 2.21, respectively.

It is clear from (4) that, to reduce the noise figure, high quality factors of L_{s1} and L_{g1} are desired. It can also be noted that the noise factor is inversely proportional to feedback resistor R_f. In other words, weak feedback topology decreases the noise factor value while strong feedback implementation degrades the noise performance of the suggested UWB LNA.

In addition, the noise factor formula given by (4) states that the noise figure is also inversely proportional to the transconductance of the input stage (g_{m1}). This goes along with the known fact that noise performance trades off with power consumption.

For output matching, the series resonance of the shunt peaking technique is used to match the proposed UWB LNA to the load impedance R_L while the series drain resistance R_{d2} is used to extend the output matching bandwidth.

This proposed UWB LNA (LNA1) has an operating bandwidth of 3.1–10.6 GHz. The proposed LNA2 whose schematic

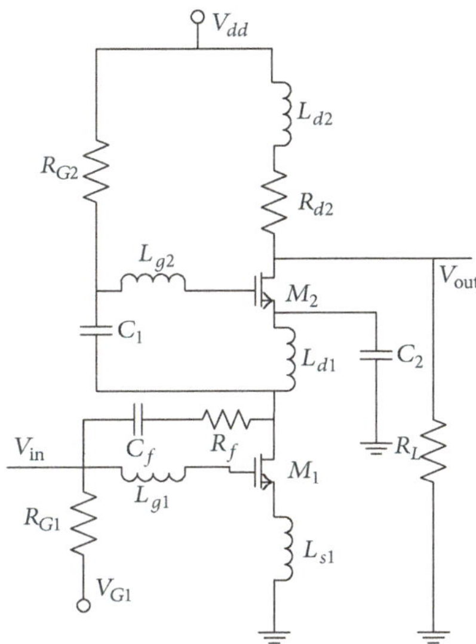

FIGURE 1: Current reuse UWB LNA (LNA1).

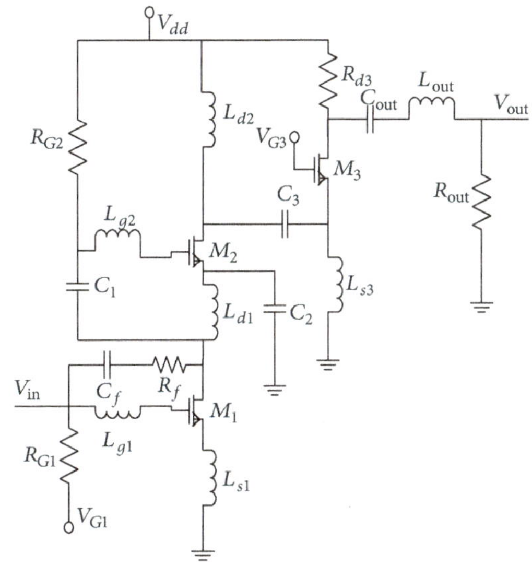

FIGURE 4: Schematic circuit of LNA2.

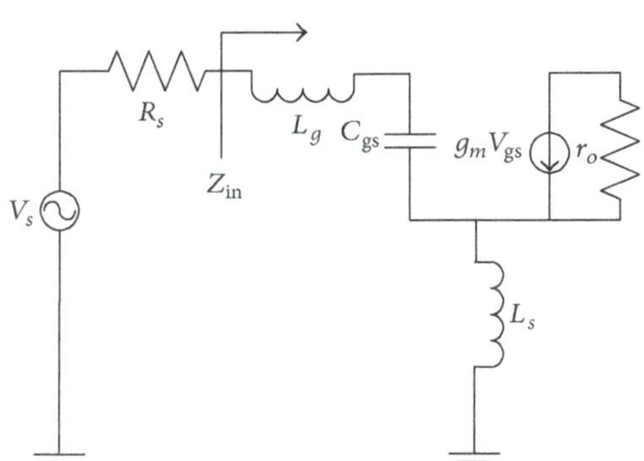

FIGURE 2: Input stage small signal equivalent circuit.

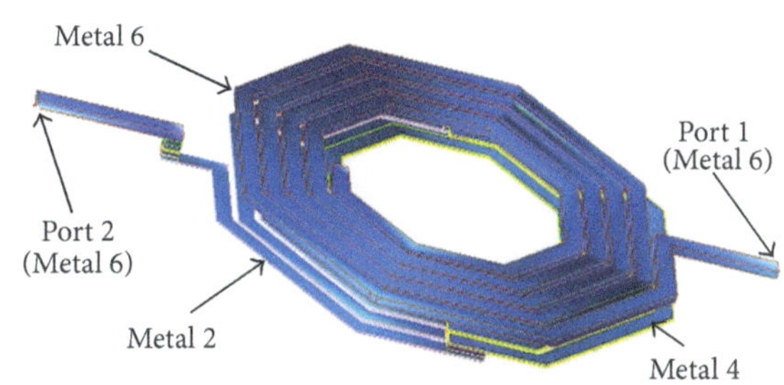

FIGURE 5: 3D view of the symmetric 3D proposed structure.

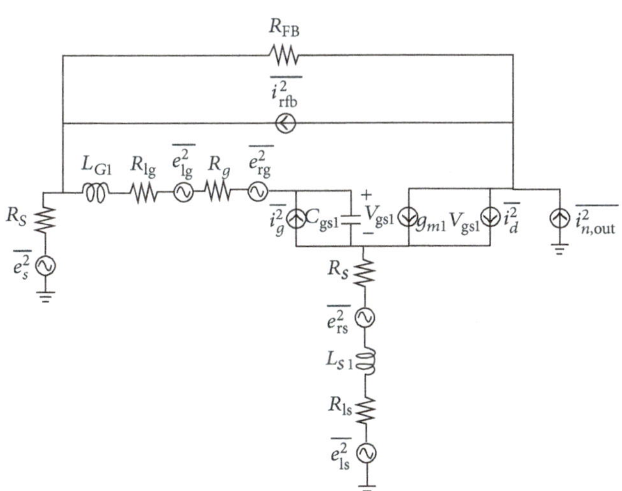

FIGURE 3: Equivalent circuit of the fisrt stage for noise calculation [1].

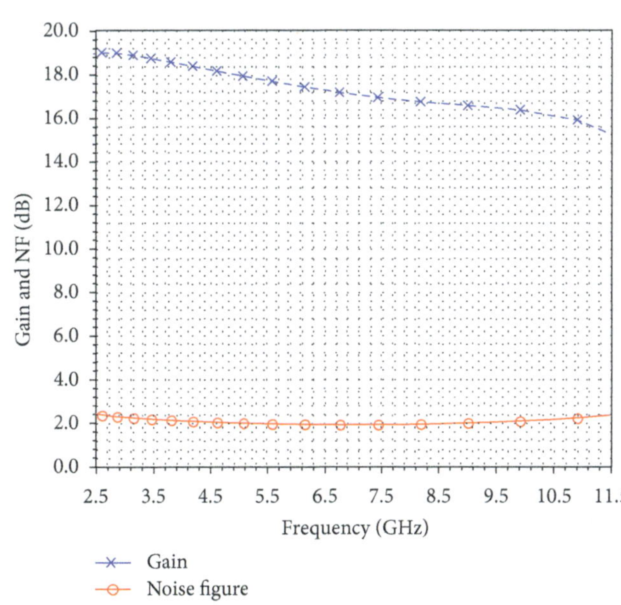

FIGURE 6: S_{21} (dB) and NF (dB) of LNA1.

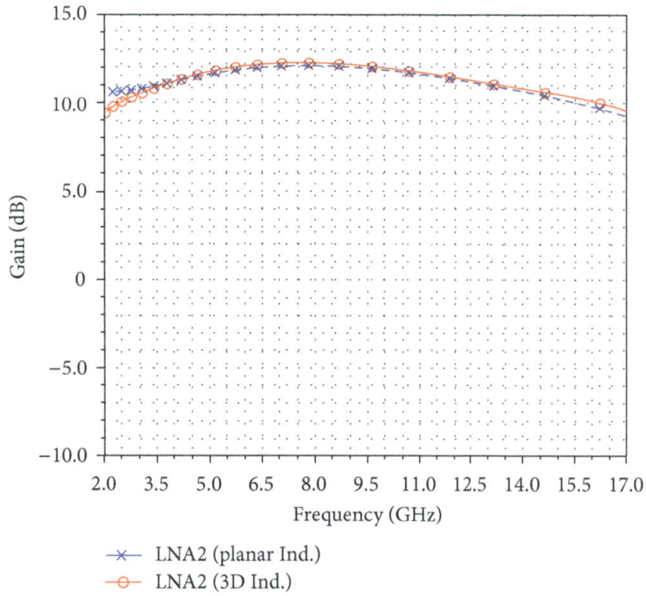

FIGURE 7: S_{21} (dB) of LNA2.

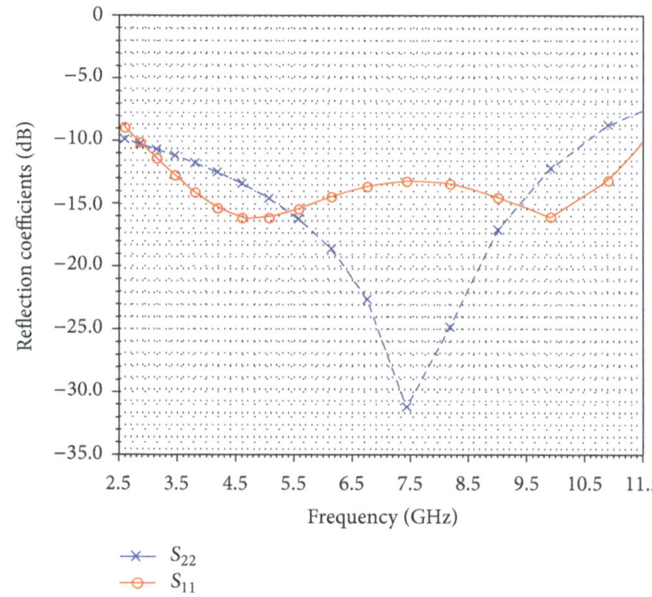

FIGURE 9: S_{11} (dB) and S_{22} (dB) of LNA1.

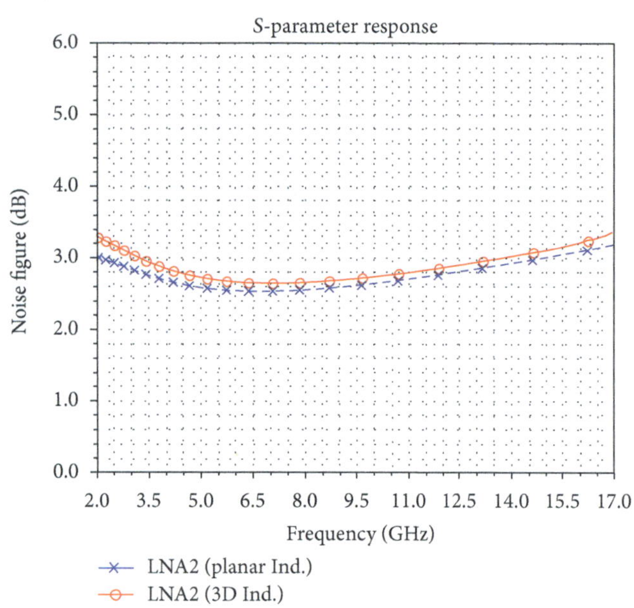

FIGURE 8: NF (dB) of LNA2.

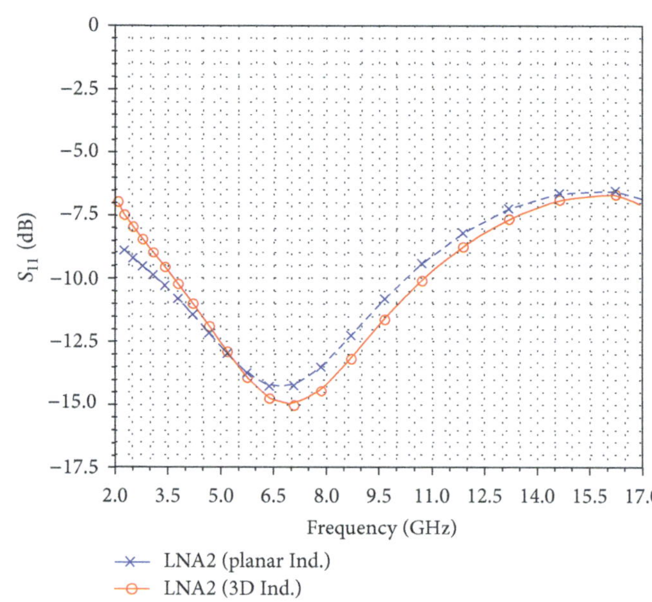

FIGURE 10: S_{11} (dB) of LNA2.

circuit is shown in Figure 4 is an extended version of LNA1. It has a wider operating band of frequency which extends from 2.5 GHz to 16 GHz.

Input impedance match has a special importance and consideration especially in wideband sensitive circuits design. Input impedance matching bandwidth is broadened by the use of a weaker shunt capacitive-resistive feedback loop which mainly leads to quality factor reduction of the input matching circuit. Weakness of shunt feedback strength not only reduces the input reflection coefficient over this wide bandwidth but it also reduces the input side injected thermal noise which decreases the proposed LNA2 noise figure indicating the enhanced noise performance of the suggested design.

Shunt-peaking technique increases the low frequency gain and hence decreases the gain flatness while having a wide operating bandwidth. In spite of shunt-peaking drawbacks, it mainly facilitates LNA output impedance to load matching. LNA2 bandwidth extension and gain flatness over its operating band of frequency are achieved through the removal of shunt peaking. Moreover the control of gate peaking is used to enhance the current reuse amplifier core gain.

For wideband output impedance matching, a unity common gate (CG) matching topology in addition to series

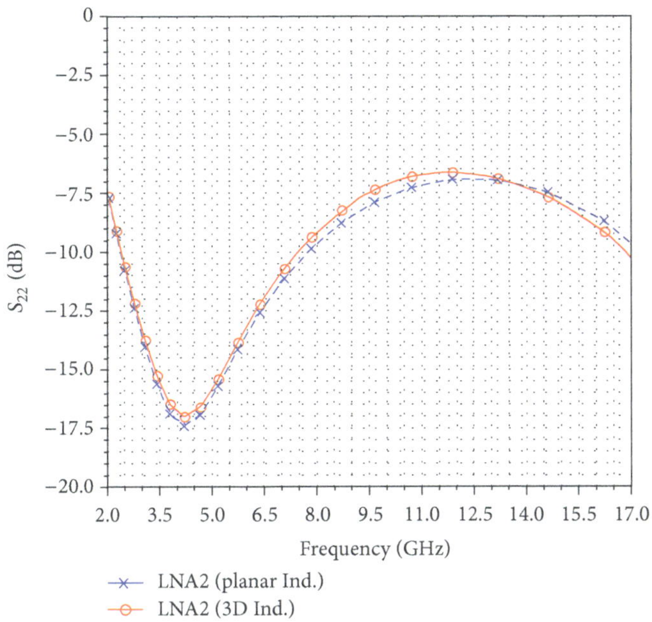

FIGURE 11: S_{22} (dB) of LNA2.

resonance circuit consisting of capacitor C_{out} and inductor L_{out} is used to match the LNA2 output impedance to its load (succeeding RF stage). The resistive termination R_{out} is used to control the load-output impedance match bandwidth.

A planar RF on-chip spiral inductor (L_{d1}) having an inductance of 14.5 nH and a maximum quality factor of 8.0 is needed as a load of the input CS stage to improve the current reuse stages matching. This RF integrated inductor occupies an area of 428 μm × 425 μm which represents a considerable part of the UWB LNA total die area.

One of the well-known difficulties facing the development of RFICs is inductors large area relative to other passive and active components. This area problem becomes more severe with the recent intensive shrinking of active devices and competitive reduction of fabrication cost [10].

Inductors quality factor (Q) reduction is another limiting factor of RFICs performance enhancement. The reduction of inductor Q factor is due to ohmic and substrate losses. Ohmic losses can be decreased by using a high conductive metal for inductor implementation. On the other hand placing a high resistive layer underneath the inductor can minimize the substrate losses. Lately optimized 3D structures and implementations of RF integrated inductors are suggested to overcome all of these limitations and improve the RF integrated inductors performance [4, 5].

For LNA2 circuit area reduction and RF inductor characteristics improvement, a symmetric 3D structure for RF integrated inductor implementation is suggested to replace the planar RF integrated inductor (L_{d1}). Similar to the design of planar RF inductor, 3D metallic structure layout should be drawn on a substrate to design and test a 3D integrated inductor [11]. 3D RF inductors structures are mainly consisting of serially connected different metal layers spirals having the same current flow direction. This 3D structure inductance is dependent on these different spirals inductances and the positive mutual coupling they have [11].

For 1P6 M CMOS technology which has six different metal layers, the proposed symmetric 3D RF integrated inductor has a complete spiral inductor on the highest metal layer ($M6$) . Half of the lower spiral is implemented using fourth metal layer ($M4$) to increase its inductance value due to the increased mutual coupling. The second metal layer ($M2$) which is distant from the top metal layer is employed to implement the lower spiral other half to reduce the parasitic components of that 3D metal structure and increase its quality factor. The suggested symmetric 3D inductor has an inductance of 14.5 nH, a quality factor of 8.5, and an area of 185 μm × 165 μm. 80% of planar inductor area is saved through this symmetric 3D structure while achieving the same inductance value and higher quality factor. Figure 5 shows a 3D view of the proposed symmetric RF integrated inductor.

3. Simulation Results and Discussion

The proposed UWB LNA (LNA1 and LNA2) circuits are designed in TSMC CMOS 0.18 μm technology process using Agilent Advanced Design System (ADS). Electromagnetic simulation is verified by the post-layout simulation results which are obtained using the Cadence design environment. The suggested symmetric 3D structure is designed and tested using Momentum simulation software and verified using Cadence design environment. The LNAs simulation results are given below.

3.1. Power Gain and Noise Figure. LNA1 has a gain of 17 ± 1.5 dB as shown in Figure 6. It also has a noise figure less than 2.3 dB over its operating band of frequency (3.1–10.6 GHz).

S_{21}(dB) of LNA2 is higher than 10 dB with a maximum value of 12 dB over the desired band of frequency (2.5–16 GHz). This high and flat gain is due to the use of inductive gain-peaking technique in addition to the control of the unity gain current cut-off frequencies of LNA2. Figure 7 shows that the proposed LNA2 employing the symmetric 3D RF integrated inductor achieves a gain of 11 ± 1.0 dB.

The proposed UWB LNA2 has an enhanced LNA noise performance. LNA2 NF ranges from 2.5 dB to 3.3 dB over the operating bandwidth (2.5–16 GHz). This NF reduction is accomplished due to the optimization of the LNA noise factor given by (4) and the use of weak shunt capacitive-resistive feedback implemented over the input stage. LNA2 achieves a NF less than 3.3 dB over the operating band of frequency as shown in Figure 8.

3.2. Input and Output Impedance Matching. LNA1 input and output ports have good matching conditions to its source and load, respectively. Simulation results of input and output reflection coefficients of LNA1 are shown in Figure 9. LNA1 has S_{11} and S_{22} less than −11 dB and −10 dB, respectively, over the UWB range of frequencies.

The proposed UWB LNA2 achieves good input impedance matching as shown in Figure 10. Good impedance

Table 1: Proposed UWB LNA performance summery in comparison to recently published UWB LNAs.

Reference	BW (GHz)	Gain (dB)	NF (dB)	S_{11} (dB)	S_{22} (dB)
This work (LNA2)*	**2.5~16**	**11 ± 1.0**	**<3.3**	**<−7**	**<−7.25**
This work (LNA1)*	**3.1~10.6**	**17 ± 1.5**	**<2.3**	**<−11**	**<−10**
LNA-1 [1]	1.7~5.9	11.2 ± 2.3	<4.7	<−11.8	<−12.7
LNA-2 [1]	1.5~11.7	12.2 ± 0.6	<4.8	<−8.6	<−10
[2]	3~10.6	15	<4.4	<−7	NA
[12]	3.1~10.6	10.8 ± 1.7	<6	<−10	<−9.3
[13]	1~5	12.7 ± 0.2	<3.5	<−8	NA

*Post-layout simulation results.

match between LNA2 and its source is obtained using the series-resonant input matching technique. The input return loss (S_{11}) is less than −7.0 dB over this wide range of frequency (2.5–16 GHz).

Figure 11 shows that better output impedance matching is obtained using the planar integrated inductor while simulating LNA2. Good output impedance matching of LNA2 over its operating band of frequency (2.5–16 GHz) is accomplished due to the optimization of the CG output matching stage with the aid of the output LC resonant circuit. R_{out} termination is used to widen the matching bandwidth. The output return loss (S_{22}) shown in Figure 11 is less than −7.25 dB for LNA2 using the planar inductor while it is less than −6.0 dB for LNA2 employing the proposed 3D inductor over the desired frequency band (2.5–16 GHz).

3.3. DC Power, Reverse Isolation, and Stability. LNA1 and LNA2 consume DC power of 12.8 mW and 20 mW, respectively, from a 1.8 V power source. The increased DC consumption of LNA2 is due to having enough driving bias for the CG output match stage.

Both of the proposed UWB LNA1 and LNA2 have a reverse isolation factor (S_{12}) less than −28 dB over each design bandwidth. The proposed UWB LNAs (LNA1 and LNA2) are unconditionally stable over their bandwidths.

Table 1 shows a summary of the proposed UWB LNAs performance in comparison to other recently published UWB LNAs implemented in 0.18 μm CMOS technology.

4. Conclusion

In this paper, two different UWB LNAs were presented. LNA1 has high gain, minimized noise figure, and good impedance match over the UWB range of frequencies. LNA2 has a wide range of operating frequency (2.5 GHz–16 GHz). UWB LNA2 consists of a current reuse cascaded amplifier with shunt resistive feedback followed by a CG output stage with resistive termination. LNA2 input stage use series-resonant impedance matching technique and employs a symmetric 3D RF integrated inductor as a load. The post-layout simulation results of LNA1 and LNA2 demonstrate the performance improvement achieved through theses designs. The next step is to implement these UWB LNAs to have a comparison between post-layout simulation results and measured results.

References

[1] Y. S. Lin, C. Z. Chen, H. Y. Yang et al., "Analysis and design of a CMOS UWB LNA with dual-RLC-branch wideband input matching network," *IEEE Transactions on Microwave Theory and Techniques*, vol. 58, no. 2, pp. 287–296, 2010.

[2] A. I. A. Galal, R. K. Pokharel, H. Kanay, and K. Yoshida, "Ultra-wideband low noise amplifier with shunt resistive feedback in 0.18 μm CMOS process," in *Proceedings of the 10th Topical Meeting on Silicon Monolithic Integrated Circuits in RF Systems (SiRF '10)*, pp. 33–36, January 2010.

[3] K. Yousef, H. Jia, R. Pokharel, A. Allam, M. Ragab, and K. Yoshida, "A 2–16 GHz CMOS current reuse cascaded ultra-wideband low noise amplifier," in *Proceedings of the Saudi International Electronics, Communications and Photonics Conference (SIECPC '11)*, April 2011.

[4] K. Yousef, H. Jia, R. Pokharel, A. Allam, M. Ragab, and K. Yoshida, "Design of 3D Integrated Inductors for RFICs," in *Proceeding of 2012 Japan Egypt Conference on Electronics, Communications and Computers (JECECC '12)*, pp. 22–25.

[5] ⊠. N. Wang, ⊠. L. Zhao, Y. Zhou, ⊠. H. Dai, and B. C. Cai, "Fabrication and performance of novel RF spiral inductors on silicon," *Microelectronics Journal*, vol. 36, no. 8, pp. 737–740, 2005.

[6] A. M. Niknejad and R. G. Meyer, "Analysis, design, and optimization of spiral inductors and transformers for Si RF IC's," *IEEE Journal of Solid-State Circuits*, vol. 33, no. 10, pp. 1470–1481, 1998.

[7] T. H. Lee, *The Design of CMOS Radio-Frequency Integrated Circuits*, Cambridge University Press, 2nd edition, 2004.

[8] S. S. Mohan, M. Del Mar Hershenson, S. P. Boyd, and T. H. Lee, "Bandwidth extension in CMOS with optimized on-chip inductors," *IEEE Journal of Solid-State Circuits*, vol. 35, no. 3, pp. 346–355, 2000.

[9] H. T. Friis, "Noise figure of radio receivers," *Proceedings of the IRE*, vol. 32, no. 7, pp. 419–422, 1944.

[10] H. Y. Tsui and J. Lau, "Experimental results and die area efficient self-shielded on-chip vertical solenoid inductors for multi-GHz CMOS RFIC," in *Proceedings of the IEEE Radio Frequency Integrated Circuits (RFIC) Symposium*, pp. 243–246, June 2003.

[11] H. Garcia, S. Khemchndai, R. Puildo, A. Lturri, and J. Pino, "A Wideband active feedback LNA with a Modified 3D inductor," *Microwave and Optical Technology Letters*, vol. 52, pp. 1561–1567, 2010.

[12] P. Sun, Sh. Liao, H. Lin, Ch. Yang, and Y. Hsiao, "Design of 3.1 to 10.6 GHz ultra-wideband low noise amplifier with current reuse techniques and low power consumption," in *Proceedings of the Progress In Electromagnetics Research Symposium*, pp. 901–905, Beijing, China, March 2009.

[13] A. I. A. Galal, R. K. Pokharel, H. Kanaya, and K. Yoshida, "1-5GHz wideband low noise amplifier using active inductor," in *2010 IEEE International Conference on Ultra-Wideband, ICUWB2010*, pp. 193–196, chn, September 2010.

12

A Quasi-Yagi Antenna Backed by a Jerusalem Cross Frequency Selective Surface

Sergio E. Melais,[1] David Cure,[2] and Thomas M. Weller[2]

[1] *Tampa Microwave, Tampa, FL 33626, USA*
[2] *Department of Electrical Engineering, University of South Florida, Tampa, FL 33620, USA*

Correspondence should be addressed to Sergio E. Melais; semelais@gmail.com

Academic Editor: Walter De Raedt

A quasi-Yagi antenna is developed to operate at 2.4 GHz (ISM band) presenting a low profile and off-axis radiation when packaged over a metal ground plane. The off-axis radiation is realized by incorporating a Jerusalem cross frequency selective surface (JC-FSS) as the ground plane for the antenna. A JC-FSS is preferred because of its frequency stability in the operating band for a large angular spectrum ($\approx 70^\circ$) of TE- and TM-polarized incident waves. In this research, the substrate of the antenna flush-mounted on top of the FSS is added to the JC-FSS model and allows for a smaller cell grid. The prepared quasi-Yagi antenna over the JC-FSS offered 260 MHz of functional bandwidth and 54° of beam tilt towards the end-fire direction. To the best of the authors' knowledge this is the first instance that these two structures are combined for off-axis radiation. Additionally, to support the preferred use of the JC-FSS, the quasi-Yagi is backed by a square patch (SP) FSS for comparison purposes.

1. Introduction

The work presented in this paper introduces a quasi-Yagi antenna over a metal reflector with off-axis radiation at 2.4 GHz (ISM band). The main application of interest for this antenna is on sensor nodes comprising a wireless sensor network inside a multipath rich environment such as an aircraft fuselage. In practice these antennas would be mounted atop the metal package of a sensor node and used to communicate preferentially toward the front or rear of the aircraft. The packaging of the antenna over a metal ground plane presents a challenge, however, as this configuration results in undesired phase reflections ($1e^{j180^\circ}$) and image currents from the ground canceling the current of the antenna, degrading its operational bandwidth, and tilting the beam away from the end-fire direction. A previous solution was proposed in [1] which consisted of displacing the metal reflector from the antenna by a suitable distance ($0.19\lambda_g = 7.5$ mm). Though off-axis radiation of 40° was achieved, the arrangement resulted in an inherently high profile.

The configuration proposed herein introduces a new alternative to [1] as it packages the quasi-Yagi antenna over a high impedance surface (HIS) or electromagnetic band-gap (EBG) structure. The HIS eliminates the out-of-phase reflections generated at the ground from radiating to the antenna and supports the radiation of leaky TE waves in the frequency region of high impedance. The preferred HIS configuration used here is the Jerusalem cross frequency selective surface (JC-FSS) from [2], because of its compact size, numerous parameters for tuning, and its stability over the band gap for a large angular ($<70^\circ$) spectrum of TE and TM incident waves. The unique feature in [2] is the addition of the antenna substrate layer into the JC-FSS model, allowing for a smaller FSS grid.

The use of an HIS to back existing antenna technologies presents advantages such as preserving a considerable fraction of the antenna bandwidth while reducing profile, improving gain, reducing back radiation, and increasing radiation efficiency. Work described in [3] for an Archimedean spiral antenna over a square patch EBG (versus the conventional 0.25λ grounded cavity) offered an overall antenna/EBG profile reduction >69% and preserved the bandwidth by 71% without losing directive gain. Another example, developed in [4], shows a wearable dual-band (2.4 and 5.1 GHz) coplanar patch antenna combined with a square patch EBG made of common clothing fabrics. The EBG used here improved

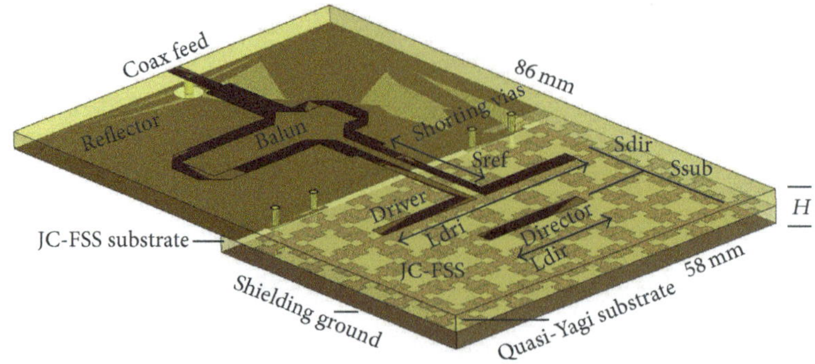

FIGURE 1: Quasi-Yagi antenna over Jerusalem cross frequency selective surface. The beam tilt is measured along the theta direction, aligned with the feed line.

antenna gain (3 dB) and reduced back radiation into the body (>10 dB). An equivalent design in [5] provides the same characteristics (high gain and low back radiation), but it used a nonconformal material and the EBG arrangement consisted of a Jerusalem cross-slot array. An example published in [6] shows a flared dipole with quasi end-fire radiation at 3 GHz over an HIS consisting of an array of hexagonal grids (thumbtacks). The amount of angular beam tilting towards the end-fire direction depends on the impedance of the HIS, which can be tuned in real time by a variety of methods (e.g., electronically, using varactor diodes [7]). Finally, work presented in [8] illustrates a broadband diamond dipole antenna over a Jerusalem cross frequency selective surface. The bandwidth for the combined structure extends from 5 to 11 GHz with high gain (>6 dB) from 3.7 to 6.8 GHz and a nondisturbed (without unwanted nulls) directive pattern up to 7 GHz.

As suggested above, the majority of existing antenna/EBG configurations present radiation in the broadside direction. The design prepared here combines for the first time a quasi-Yagi antenna with a JC-FSS over a metal ground with radiation directed toward the end-fire direction. The combined structure has an overall profile of 5 mm ($0.125\lambda_g$), functional bandwidth extending from 2.29 to 2.55 GHz, and 54° of beam tilt towards the end-fire direction. In comparison to [1] the new design provides 33% reduced profile and 14° of additional beam tilting. Also, in comparison with a conventional quasi-Yagi antenna of the same substrate thickness (5 mm) but lacking the HIS layer, the proposed design offers wider bandwidth (2.24–2.46 GHz versus 2.35 GHz) and 27° more beam tilting.

The following sections present the antenna design, the derivation of the JC-FSS model with the antenna substrate, and comparisons between simulated and measured results for the combined structure. Also, the performance of the quasi-Yagi antenna when backed by a square patch FSS is shown to highlight the advantages of the JC-FSS.

2. Quasi-Yagi Antenna

2.1. Design Characteristics. A quasi-Yagi antenna, such as that in [9], consists of an array of dipoles printed on a substrate and fed by a microstrip to coplanar strip-line (CPS) transition. The transition is used as a transformer to connect the unbalanced microstrip input line to the balanced (CPS) antenna feed line. In addition, the ground plane from the microstrip transition is used as the reflector element for the array, eliminating the need for a reflecting dipole and resulting in a more compact length ($<\lambda_0/2$), along with direct compatibility with microstrip circuitry. Further advantages inherent in the design are mechanical support and planar transmission line compatibility due to the presence of a substrate. The use of a high permittivity substrate means that the antenna will be extremely compact in terms of free space wavelength (λ_0). In regard to frequency of operation and radiation pattern, quasi-Yagi antennas are broadband (≈50%) and radiate in the end-fire direction.

TABLE 1: Optimized dimensions of antenna elements.

Antenna Element (dimensions in mm.)	[1]	QY-JCFSS
Length of CPS (Sref)	18.33	20.5
Length of driver (Ldri)	34	35.5
Length of director (Ldir)	18	18
Separation of driver and director (Sdir)	16	11.16
Separation of director and end of substrate (Ssub)	10.83	16
Height of substrate layer (H)	7.5	5

The new feature developed in this research, Figure 1, consists of shielding the antenna with a metal ground. To overcome out-of-phase reflection ($1e^{j180^\circ}$) from the metal ground and surface currents from shorting the director and driver dipoles, a JC-FSS is implemented. The ground planes of the FSS and the truncated microstrip are connected to the same potential through shorting vias. The distance separating the antenna and the FSS was determined based on antenna bandwidth requirements and commercially available substrate thickness options. The effect of this distance is accounted for as a superstrate during the derivation of the closed form equations for the JC-FSS model.

The dimensions for the antenna elements in Figure 1 are optimized from [1] for end-fire radiation and to account for the added JC-FSS through simulations in Ansoft HFSS. The substrate material is RT/Duroid 6010 LM ($\varepsilon_r = 10.2$). The overall size of the antenna is 58 mm × 86 mm. The optimized dimensions for the quasi-Yagi antenna are listed in Table 1.

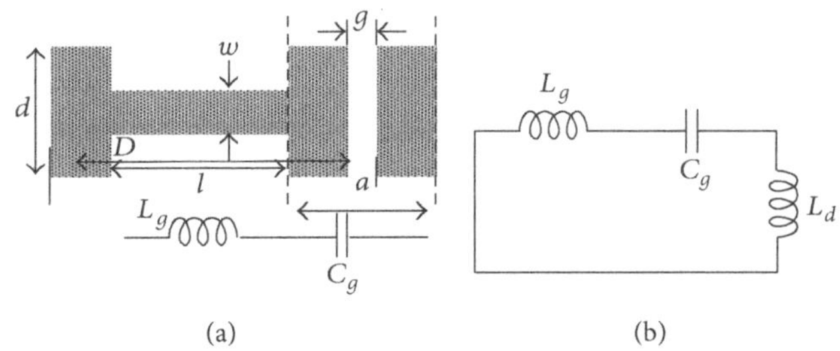

FIGURE 2: (a) Equivalent circuit for the self-resonant grid. (b) Series combination of the self resonant grid with the dielectric slab impedance.

The most significant adjustments are the reduction of the driver to director separation (30%) and the decrease of the overall substrate profile (33%).

2.2. Jerusalem Cross Frequency Selective Surface. The JC-FSS implemented here was previously developed by the authors in [2]. The design offers in-phase reflection ($1e^{j0^\circ}$) for an operational band extending from 2.39 to 2.5 GHz at normal incidence. In addition, at the center frequency (2.45 GHz) the JC-FSS offers frequency stability for a large angular spectrum (>70°) for both TE- and TM-polarized incident waves. As previously stated, the main feature from the FSS in [2] is the addition of the antenna substrate into the JC-FSS model, which decreases the center frequency of the high impedance band from 3.2 GHz to 2.45 GHz for the same FSS dimensions.

2.2.1. Surface Waves on a Metal Surface versus a Textured Surface. The properties of surface waves on a metal surface versus a textured surface are compared herein, to explain the use of the latter in the quasi-Yagi antenna. If the radiating element is placed near the ground plane, it will generate currents that propagate along the metal sheet. Any break or discontinuity (e.g., the edge of board) on the flat surface will promote radiation from that location. The result is a destructive interference which cancels the radiation from the antenna and decreases the radiation efficiency. By adding a special texture to a metal surface, it is possible to suppress surface currents over a range of frequencies (band gap). As discussed in [10], the electromagnetic properties of the structure can be described by a single value, the surface impedance, if the period of the textured surface is much smaller than the wavelength in the dielectric media (λ_d). A smooth or flat conductive sheet has low surface impedance, while a textured surface can be engineered to have high surface impedance.

The fields radiated by the quasi-Yagi antenna over the FSS are a combination of those produced by the antenna elements themselves and those that exist due to the presence of the FSS. For a quasi-Yagi antenna without a ground or underlying structure, the radiated fields are TE with respect to the substrate surface and the direction of propagation (end fire). If a conventional ground was placed beneath the antenna, TM surface wave propagation is possible; however, these waves are unlikely to be excited by the antenna elements due to their orientation. The main problem with this configuration is field cancelation due to image currents. With the textured FSS layer beneath the antenna both TE and TM wave propagations are possible. However, TE wave excitation is dominant because of the orientation of the antenna elements. When the surface impedance is large, these TE waves are leaky and radiate readily, causing the overall radiation pattern to tilt away from broadside. The low cross-polarization levels achieved with the antenna presented herein support the conclusion that TM surface wave radiation is not significant.

2.2.2. Derivation of the JC-FSS Model. The JC-FSS of this work is effectively modeled by a parallel resonant LC circuit following the condition that the grid period is smaller than the wavelength ($D \ll \lambda_g$). The LC model consists of the parallel combination of the self-resonant grid impedance (Z_g), which represents a strip, with the grounded dielectric slab impedance (Z_d). Figure 2(a) shows that Z_g can be expanded into the series combination of the narrow strip impedance and the edge impedance between end loading strips. The narrow strip impedance is mostly inductive (L_g) and is derived from Telegrapher's equations or from the stepped impedance equations from [11]. The equation for the grid inductance is written as

$$L_g = \frac{Z_o \beta \ell}{\omega}. \tag{1}$$

The impedance between end loading plates is mostly capacitive (C_g) and is a result of the charge builtup between plates [10]. This capacitance is given by

$$C_g = \frac{2d}{\pi} \varepsilon_o \varepsilon_{\text{eff}} \cosh^{-1}\left(\frac{a}{g}\right), \tag{2}$$

where ε_{eff} is the effective permittivity including the superstrate layer, d is the length of an end loading plate, g is the gap between crosses, and a is the period between adjacent capacitive plates.

As illustrated in Figure 2(b) Z_d, is mostly inductive (L_d) and is derived from the TEM transmission line equation for a dielectric slab backed by a perfect electric conductor [12]. In (3) k is the wavenumber $\omega\sqrt{\mu_o \varepsilon_o}\sqrt{\varepsilon_r}$, h is the dielectric height, and η_o is the intrinsic impedance in free space:

$$L_d = \frac{\eta_o}{\varepsilon_r} \frac{\tan(kh)}{\omega}. \tag{3}$$

From the parallel LC circuit in Figure 2(b) the equivalent surface impedance is calculated by

$$Z_s = j\omega L_d \frac{1 - \omega^2 L_g C_g}{1 - \omega^2 \left(L_g + L_d\right) \cdot C_g}. \quad (4)$$

The resonant frequency can then be derived by equating the denominator from (4) to zero which results in

$$f_r = \frac{1}{2\pi\sqrt{\left(L_g + L_d\right) \cdot C_g}}. \quad (5)$$

The bandwidth is obtained by dividing the equivalent impedance of the JC-FSS by η_o and following the criteria in [12, 13] that the phase of the reflection coefficient should fall between $\pm 0.25\pi$.

The dimensions for the JC-FSS are listed in [2]. Both dielectric layers have a high relative permittivity ($\varepsilon_r = 10.2$) offering better angular stability and smaller dimensions at low resonant frequencies. Each dielectric layer is relatively thick (2.5 mm), increasing the inductance (L_d) of the equivalent surface impedance (4). The use of a small gap ($g = 0.32$ mm) between crosses leads to larger edge capacitance (C_g) in the surface impedance.

2.2.3. Impact of Superstrate on JC-FSS Model. The most attractive trait from the JC-FSS in [2], compared to others [13], is that the antenna substrate (or superstrate) is included into the JC-FSS model. Here, the impinging wave is excited on top of this layer, (Figure 3). The equivalent reflection coefficient (Γ_{IN}) for the structure at oblique incidence for TE- and TM-polarized waves is the result of the combination of the reflected waves from the two dielectric layers and ground. Also, the addition of the superstrate varies the effective permittivity (ε_{eff}) between layers from 7.2 to 9.7, which is considered during the derivation of the grid capacitance in (2).

The reflection coefficient for TE and TM waves at the JC-FSS/superstrate boundary is calculated by combining (4) and $\eta_1 = \eta_o/\sqrt{10.2}$ with

$$\Gamma_s^{TE} = \frac{Z_s \cdot \cos\left(\theta^1\right) - \eta_1 \cdot \cos(\theta)}{Z_s \cdot \cos\left(\theta^1\right) + \eta_1 \cdot \cos(\theta)},$$
$$\Gamma_s^{TM} = \frac{Z_s \cdot \cos(\theta) - \eta_1 \cdot \cos\left(\theta^1\right)}{Z_s \cdot \cos(\theta) + \eta_1 \cdot \cos\left(\theta^1\right)}, \quad (6)$$

where the incident (θ^1) and refracted (θ) angles have minimum effect on the phase of the TE and TM reflected wave ($\angle\Gamma_s^{\circ}$) for the frequency range in which the JC-FSS exhibits high impedance, resulting in angular stability.

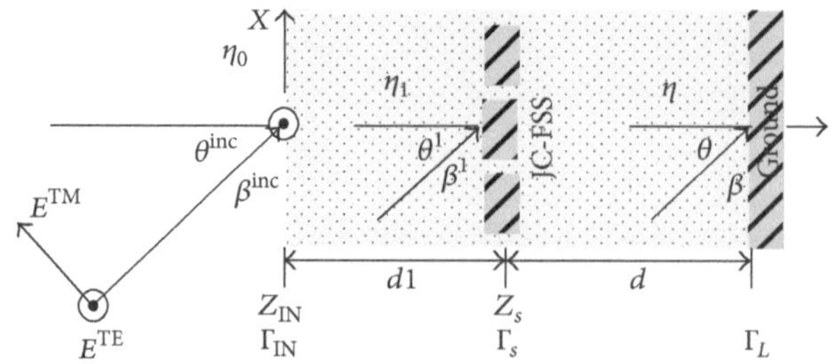

FIGURE 3: TE- and TM-polarized waves incident on the superstrate layer covering the FSS.

Next, the reflection coefficient at the free-space/superstrate boundary for TE and TM waves is derived. First the input impedance is found for the wave reflected from the FSS:

$$Z_{IN}^{TE} = \eta_1 \cdot \frac{1 + \Gamma_s^{TE} e^{-j \cdot 2 \cdot \beta^1 d1}}{1 - \Gamma_s^{TE} e^{-j \cdot 2 \cdot \beta^1 d1}},$$
$$Z_{IN}^{TM} = \eta_1 \cdot \frac{1 + \Gamma_s^{TM} e^{-j \cdot 2 \cdot \beta^1 d1}}{1 - \Gamma_s^{TM} e^{-j \cdot 2 \cdot \beta^1 d1}}, \quad (7)$$

and the results are included in

$$\Gamma_{IN}^{TE} = \frac{Z_{IN}^{TE} \cdot \cos\left(\theta^{inc}\right) - \eta_0 \cdot \cos\left(\theta^1\right)}{Z_{IN}^{TE} \cdot \cos\left(\theta^{inc}\right) + \eta_0 \cdot \cos\left(\theta^1\right)},$$
$$\Gamma_{IN}^{TM} = \frac{Z_{IN}^{TM} \cdot \cos\left(\theta^1\right) - \eta_0 \cdot \cos\left(\theta^{inc}\right)}{Z_{IN}^{TM} \cdot \cos\left(\theta^1\right) + \eta_0 \cdot \cos\left(\theta^{inc}\right)}. \quad (8)$$

The relation between the angle of incidence and refraction at the boundary of each layer is determined from Snell's law of refraction in [14]. This equation demonstrates that the addition of the superstrate layer reduces the angle of incidence for the wave impinging on the JC-FSS. For example, a travelling wave with an angle of incidence of 60° at the superstrate surface has an angle of refraction of 15°. This angle of refraction will be the angle of incidence for the FSS.

A comparison is performed on the design of a JC-FSS including the superstrate versus a design without the superstrate. For simplicity the evaluation is carried out for a wave with normal incidence. At normal incidence Γ_{IN} is independent of the polarization of the incident wave since the *E* and *H* fields are both tangential to the boundary. Simulation results in Figure 4 demonstrate that the superstrate shifts the center frequency down to the desired band from 3.2 to 2.45 GHz (750 MHz). This analysis supports the importance of accounting for any additional layer covering the JC-FSS during the FSS closed-form modeling to prevent undesired frequency shifts. In addition, an overall cell size reduction has been achieved by considering the extra layer; an equivalent cell size for 3 GHz, with no superstrate, is used at 2.4 GHz with superstrate.

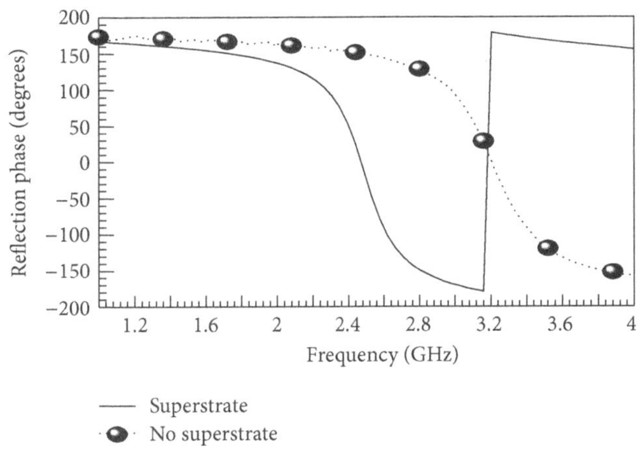

FIGURE 4: Reflection phase (deg.) versus frequency at normal incidence for a JC-FSS with and without a superstrate.

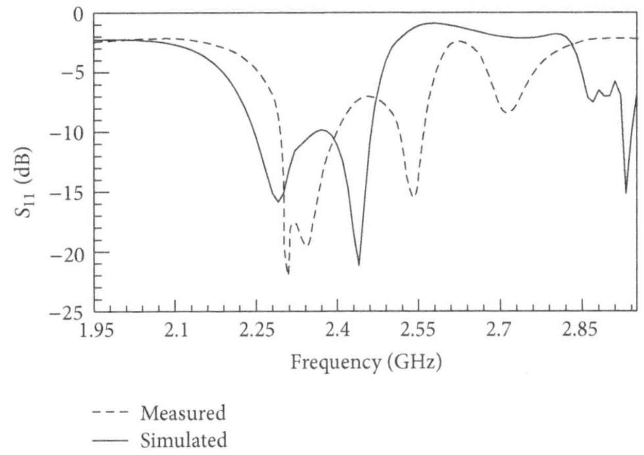

FIGURE 5: Simulated versus measured reflection coefficient for the quasi-Yagi antenna over the JC-FSS. The simulated return loss bandwidth extends from 2.24 to 2.46 GHz while the measured bandwidth extends from 2.29 to 2.55 GHz.

3. Simulation and Measurement Results

A comparison between measured and simulated data on return loss and radiation pattern is presented in this section for the quasi-Yagi antenna backed by the JC-FSS. Additionally, these same results are compared against data for a conventional grounded quasi-Yagi antenna of the same substrate height (5 mm) and the quasi-Yagi design presented in [1]. Finally, a comparison is made between the JC-FSS design and one using a square patch FSS (SP-FSS).

The simulated and measured reflection coefficients for the quasi-Yagi antenna over the JC-FSS are compared in Figure 5. The simulated data demonstrate an operational bandwidth from 2.24 to 2.46 GHz. However, the measured frequency band is shifted up in frequency, and the response exhibits undesired reflections in the 2.4 to 2.5 GHz frequency range. These effects are a result of the sensitivity of the JC-FSS to small air gaps from the adhesive used to attach the antenna substrate to the JC-FSS substrate. This explanation is confirmed through simulations, (Figure 6), where the bond line is approximated by a 1.5 mil air gap between layers. Furthermore, a small air gap will also affect ε_{eff}, disturbing the impedances of the grid design.

The results for the simulated and measured normalized H-plane patterns at 2.45 GHz are illustrated in Figures 7 and 8, respectively. The simulated and measured results demonstrate a copolarized H-plane (H-CPOL) pattern with beam tilt of 45° and 54°, respectively, towards the end-fire direction. At 45° the simulated beam peak is 1 and 3 dB larger than at the end-fire ($\theta = 90°$) and broadside ($\theta = 0°$) directions. Correspondingly the measured beam peak at 54° is 2 and 3 dB larger than at $\theta = 90°$ and 0°.

The simulated and measured H-plane cross-polarization (H-⊠POL) levels are −22.6 and −13 dB. The drastic increase in the measured H-⊠POL level is attributed to measurement set-up tolerances, sensitivity of the JC-FSS to large angles of incidence, and to the air gap resulting from the adhesive. If the resonant frequency of the high impedance band moves up enough such that the frequency of interest (2.45 GHz) falls below the band gap, then the surface impedance is inductive and TM surface waves from the ground radiate readily thereby increasing the ⊠-POL levels [10].

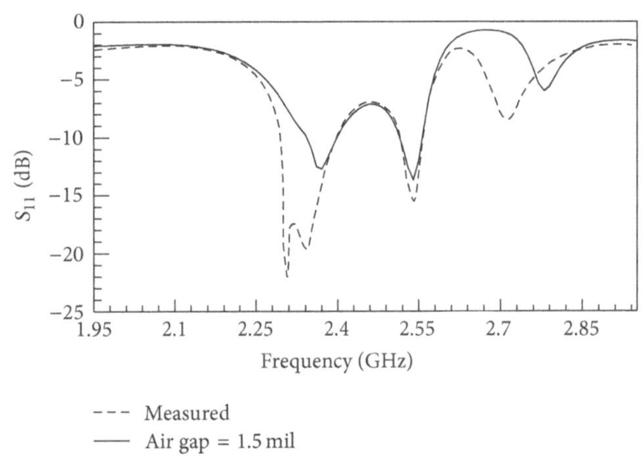

FIGURE 6: Simulation results including a 1.5 mil air gap between substrate layers versus measurement data for the quasi-Yagi antenna over the JC-FSS, to attest the frequency shift on the measured design.

Simulations were performed on a quasi-Yagi antenna printed on a 5 mm thick grounded dielectric slab in order to assess the impact of including the JC-FSS layer. The comparison presented in Figure 9 shows a drastic improvement of 220 MHz in the return loss bandwidth for the design including the JC-FSS versus the design over the 5 mm thick grounded slab. A similar evaluation is shown in Figure 10 comparing the copolarization (C-POL) level for the H-plane pattern versus frequency (2.4, 2.45, and 2.5 GHz). For the quasi-Yagi design backed by the JC-FSS, the H-plane C-POL pattern provides an additional 27° of beam tilt towards the end-fire direction. Furthermore, in the end-fire direction ($\theta = 90°$) the gain increases by 4.5 dB at 2.4 GHz, 5.03 dB at 2.45 GHz, and 5.68 dB at 2.5 GHz relative to the design on the grounded slab.

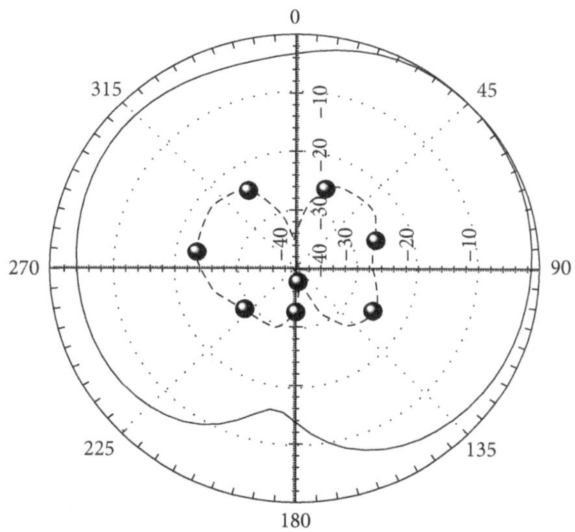

FIGURE 7: Simulation data of the normalized H-plane pattern at 2.45 GHz for the quasi-Yagi antenna over the JC-FSS. The beam peak angle shows a maximum tilt of 45°. The pattern is normalized to the H-CPOL value.

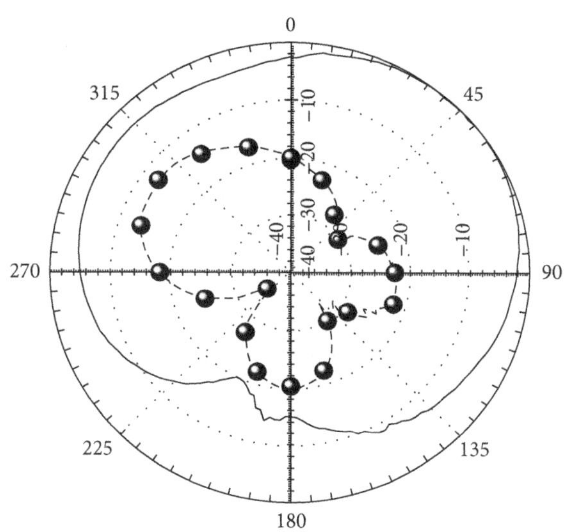

FIGURE 8: Measurement data for normalized H-plane pattern at 2.45 GHz. The maximum beam peak is recorded at 54°. The pattern is normalized to the H-CPOL value.

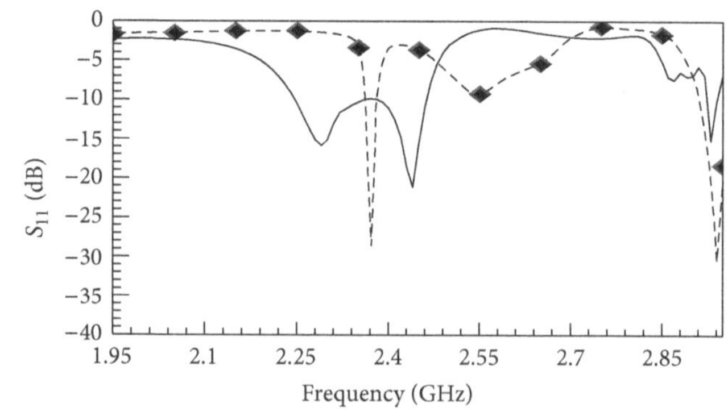

FIGURE 9: Simulation comparing the return loss bandwidth for a quasi-Yagi antenna resting on top of a grounded dielectric slab versus the design on top of the JC-FSS. The simulated bandwidth for the grounded slab is very narrow (2.35 GHz) while the bandwidth of the design that includes the JC-FSS extends from 2.24 to 2.46 GHz.

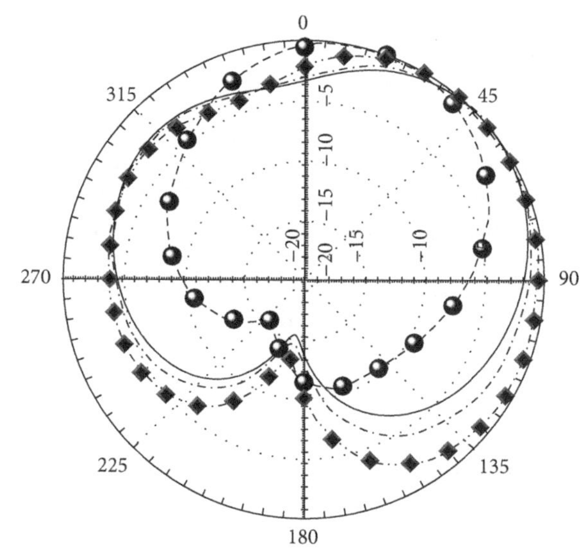

FIGURE 10: Copolarization of the H-plane pattern at 2.4, 2.45, and 2.5 GHz for the quasi-Yagi design on top of a grounded dielectric slab versus the quasi-Yagi design on top of the JC-FSS.

Previous work in [1] presented a quasi-Yagi antenna packaged over a grounded dielectric slab (7.5 mm thick) with an operational bandwidth from 2.36 to 2.55 GHz and off-axis radiation of 40°. In comparison to [1], the antenna backed by the JC-FSS presents a wider bandwidth extending from 2.29 to 2.55 GHz, 14° of additional beam tilt towards the end-fire direction and an overall profile reduction of 33%.

3.1. Quasi-Yagi Backed by an SP-FSS. In this section the SP-FSS derived in [15] is realized as the HIS structure for the quasi-Yagi antenna. The objective is to assess the dependence of the quasi-Yagi on the chosen HIS to promote principal beam tilting towards the off-axis direction. The proposed SP-FSS consists of a periodic cell with length and width of 3 mm and a gap of 0.1 mm between adjacent cells. In addition, as suggested by [16], the size of the ground plane beneath

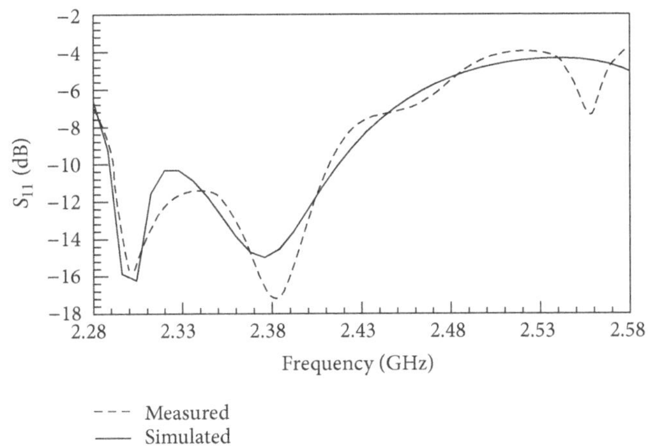

FIGURE 11: Simulated versus measured return loss for the quasi-Yagi antenna over the SP-FSS. The simulated and measured return loss bandwidth extends from 2.28 to 2.43 GHz

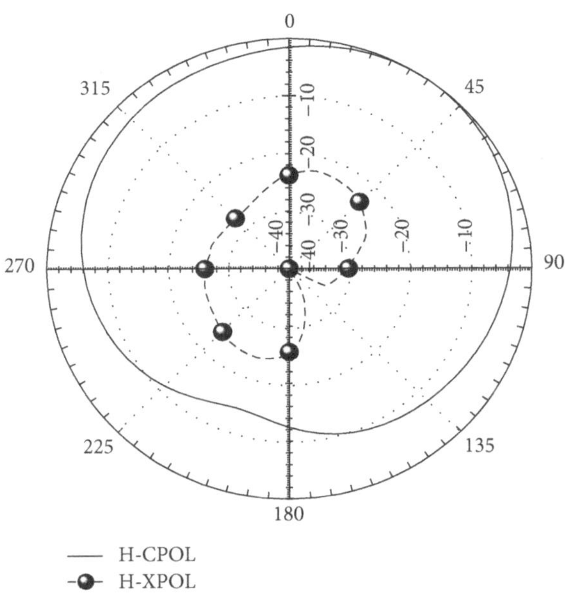

FIGURE 12: Simulation data for the normalized H-plane pattern at 2.38 GHz for the quasi-Yagi antenna over the SP-FSS. The beam peak angle shows a maximum tilt of 38°. The pattern is normalized to the H-CPOL value.

the director/driver dipoles was kept unchanged from JC-FSS design to avoid the introduction of unwanted resonances in the return loss response.

The SP-FSS design was fabricated and assembled using the same process used for the JC-FSS design, and the measured performance was compared to HFSS simulation data. Figure 11 demonstrates close agreement between the simulated and measured return loss with an operational bandwidth from 2.28 to 2.43 GHz. Figures 12 and 13 show the simulated and measured normalized H-plane patterns at 2.38 GHz. The simulated and measured results demonstrate a copolarized H-plane (H-CPOL) pattern with beam tilt of 38° and 35°, respectively, towards the off-axis direction. At 38° the

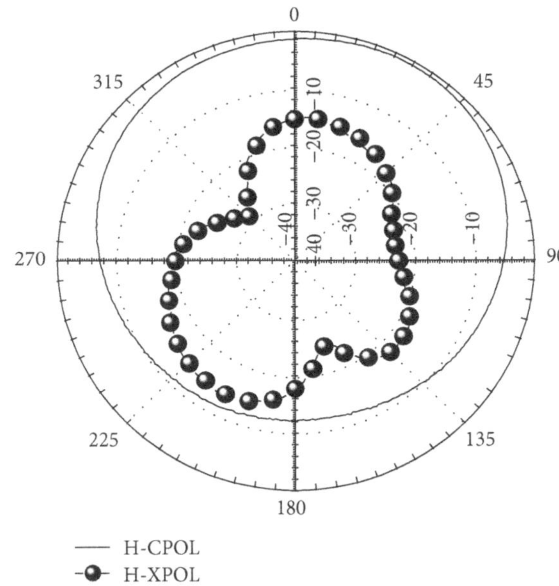

FIGURE 13: Measurement data for normalized H-plane pattern at 2.38 GHz. The maximum beam peak is recorded at 35°. The pattern is normalized to the H-CPOL value.

simulated beam peak is 1.5 and 3.5 dB larger than at the end-fire ($\theta = 90°$) and broadside ($\theta = 0°$) directions. Correspondingly the measured beam peak at 35° is 1.3 and 4.2 dB larger than at $\theta = 90°$ and 0°. The simulated and measured H-plane cross-polarization (H-⊠POL) levels are −25 and −14 dB.

The presented evaluation of the JC-FSS versus the SP-FSS has demonstrated that the JC-FSS provides additional beam tilting of 19° towards the off-axis direction. This is the result of the inherent angular stability and high inductance from the JC grid which makes it a preferable shielding candidate for antennas with off-axis radiation.

4. Conclusion

A new design for a quasi-Yagi antenna backed by a metal ground and with end-fire-like radiation has been proposed. The design consisted of packaging the antenna over a JC-FSS. This is the first time that these two structures are combined for end-fire operation. The results on return loss show an operational bandwidth from 2.29 to 2.55 GHz. The H-plane pattern showed beam tilt of 54° towards the end-fire direction.

In comparison to a design of same substrate height (5 mm) but without the JC-FSS, the proposed design offers 220 MHz more bandwidth and 27° of extra beam tilting in the end-fire direction. Furthermore, when compared to the option previously proposed in [1] with the quasi-Yagi antenna placed over a thick grounded slab, the proposed design offers a profile reduction of 33% and 14° of additional beam tilt in the end-fire direction. Additionally, the presence of the superstrate above the JC-FSS reduces the physical size of the unit cells by 23%. Finally, the comparison carried out on the quasi-Yagi backed by a SP-FSS has demonstrated that selecting an FSS with inherent angular stability for oblique angles of incidence is preferred for antennas with radiation patterns towards the end-fire direction.

Acknowledgments

The authors wish to express appreciation to Ansoft Corporation and Rogers Corporation for providing software and materials necessary for this work. This work was supported in part by Goodrich Lighting Systems in Oldsmar, Florida.

References

[1] S. E. Melais and T. M. Weller, "A Quasi Yagi antenna backed by a metal reflector," *IEEE Transactions on Antennas and Propagation*, vol. 56, no. 12, pp. 3868–3872, 2008.

[2] S. E. Melais and T. M. Weller, "A multilayer Jerusalem cross frequency selective surface," in *Proceedings of the IEEE 10th Annual Wireless and Microwave Technology Conference (WAMICON '09)*, April 2009.

[3] J. M. Bell and M. F. Iskander, "A low-profile archimedean spiral antenna using an EBG ground plane," *IEEE Antennas and Wireless Propagation Letters*, vol. 3, no. 1, pp. 223–226, 2004.

[4] S. Zhu and R. Langley, "Dual-band wearable antennas over EBG substrate," *Electronics Letters*, vol. 43, no. 3, pp. 141–143, 2007.

[5] Q. Luo, H. M. Salgado, A. M. Moura, and J. R. Pereira, "Dual-band antenna design using an EBG artificial magnetic conductor ground plane," in *Proceedings of the Loughborough Antennas and Propagation Conference (LAPC '08)*, pp. 217–220, Loughborough, UK, March 2008.

[6] J. J. Lee, R. J. Broas, and S. Livingston, "Flush-mounted antennas on Hi-Z ground planes," in *Proceedings of the IEEE Antennas and Propagation Society International Symposium*, vol. 3, pp. 764–767, June 2002.

[7] D. F. Sievenpiper, J. H. Schaffner, H. Jae Song, R. Y. Loo, and G. Tangonan, "Two-Dimensional Beam Steering Using an Electrically Tunable Impedance Surface," *IEEE Transactions on Antennas and Propagation*, vol. 51, no. 10I, pp. 2713–2722, 2003.

[8] L. Akhoondzadeh-Asl, P. S. Hall, J. Nourinia, and C. Ghobadi, "Influence of angular stability of EBG structures on low profile dipole antenna performance," in *Proceedings of the IEEE International Workshop on Antenna Technology (IWAT '06)*, pp. 253–256, March 2006.

[9] N. Kaneda, Y. Qian, and T. Itoh, "A novel Yagi-Uda dipole array fed by a microstrip-to-CPS transition," in *Proceedings of the Asia-Pacific Microwave Conference*, pp. 1413–1416, Yokohama, Japan, December 1998.

[10] D. Sievenpiper, *High-impedance electromagnetic surfaces [Ph.D. dissertation]*, Electrical Engineering Department University of California, Los Angeles, Los Angeles, Calif, USA, 1999.

[11] D. M. Pozar, *Microwave Engineering*, Wiley, New York, NY, USA, 3rd edition, 2005.

[12] M. Hosseini and M. Hakkak, "Characteristics estimation for jerusalem cross-based artificial magnetic conductors," *IEEE Antennas and Wireless Propagation Letters*, vol. 7, pp. 58–61, 2008.

[13] M. Hosseini, A. Pirhadi, and M. Hakkak, "Design of an AMC with little sensitivity to angle of incidence using an optimized Jerusalem cross FSS," in *Proceedings of the IEEE International Workshop on Antenna Technology (IWAT '06)*, pp. 245–248, New York, NY, USA, March 2006.

[14] C. A. Balanis, *Advanced Engineering Electromagnetics*, John Wiley ⊠ Sons, 1989.

[15] D. Cure, T. Weller, and F. A. Miranda, "A comparison between Jerusalem Cross and Square Patch Frequency Selective Surfaces for low profile antenna applications," in *Proceedings of the IEEE International Conference on Electromagnetics in Advanced Applications (ICEAA '11)*, pp. 1019–11022, Torino, Italy, September 2011.

[16] F. Costa, O. Luukkonen, C. R. Simovski, A. Monorchio, P. de Maagt, and S. A. Tetryakov, "TE surface wave resonances on high-impedance surface based antennas: analysis and modeling," *IEEE Transactions on Antennas and Propagation*, vol. 59, no. 10, pp. 3588–3596, 2011.

13

Complex Permittivity Measurements of Textiles and Leather in a Free Space: An Angular-Invariant Approach

B. Kapilevich, B. Litvak, M. Anisimov, D. Hardon, and Y. Pinhasi

Department Electrical and Electronics Engineering, Ariel University Center of Samaria, Ariel 40700, Israel

Correspondence should be addressed to B. Kapilevich, boriskap@ariel.ac.il

Academic Editor: Andrea Randazzo

The paper describes the complex permittivity measurements of textiles and leathers in a free space at 330 GHz. The destructive role of the Rayleigh scattering effect is considered and the angular-invariant limit for an incidence angle has been found out experimentally within 25–30 degrees. If incidence angle exceeds this critical parameter, the uncertainty caused by the Rayleigh scattering is drastically increased preventing accurate measurements of the real and imaginary parts of a bulky material. The phenomenon must be taken into consideration in predicting shielding effectiveness of materials covering hidden object in concealed threat detection with millimeter-wave radar systems.

1. Introduction

Submillimeter and THz waves demonstrate a reasonable penetration depth in certain common materials, such as fabric, plastic, and wood. Therefore, they are good candidates for detecting various hidden objects such as concealed weapons, drags, and explosives. In the last decade, there has been a growing demand in mm, sub-mm, and THz imaging systems for homeland security applications [1, 2]. In particular, efficiency of these systems depends on shielding effectiveness of materials screening the detected object [3].

Complex permittivity $\varepsilon(f)$ is widely used for characterization of dielectric materials. It is frequency-dependent quantity, resulting in attenuation and phase shifts of electromagnetic wave propagating in media. In a slab structure, the permittivity can be acquired from the transmission and reflection coefficients. Knowledge of the dielectric properties of these materials is important in predicting overall performance of mm-wave imagers because a penetration depth is governed by these characteristics. There are a lot of microwave methods suggested for measuring complex permittivity [4], but free-space approach is more suitable in sub-mm range since the sample size has reasonable dimensions in order to avoid undesirable edge diffraction effect. However, the drawback of most free-space measuring systems is the ripples (or parasitic resonances) in the measured data caused by multiple reflections from sample surfaces and antennas that cannot be accurately predicted in the theoretical models. As a result, various smoothing algorithms based on heuristic criteria must be used to overcome degrading measurement accuracy [5, 6].

An isotropic slab in air [7] is the basic theoretical model employed for retrieving the complex permittivity from the measured data. However, approximation of textiles having typical woven fabric nature and leather materials with nonregular surface's shaping by uniform dielectric slab is incorrect. Indeed, texture and surface roughness maybe comparable with wavelength and responsible for the Rayleigh scattering that may drastically degrade the measurement accuracy. One to overcome the problem is the development of special data smoothing algorithms [8] needed to avoid undesirable rapid oscillations caused by the parasitic standing wave interference between R_x and T_x antennas as well as other factors. As a result, the originally measured transmittance cannot be used for validation of the best-fitting permittivity without additional specific processing.

Assuming that textiles and leathers are isotropic materials, they must have invariant values of complex permittivity regardless of an incidence angle. The deviations of real and imaginary parts of the dielectric constant can be considered as a measure of a nonuniformity of the thickness and surface's shape of the dielectric slab approximating a real sample. This parameter has statistical nature and should be characterized by standard deviation of the measured material characteristics in angular domain. In the present study, the matched THz power meter (MPM) has been used in the transmittance measurements at 330 GHz resulting in a minimization of destructive role of the parasitic standing wave interference [9]. High directivity Gaussian beam horn-lens antenna was employed in order to suppress undesirable spillover effect during variations of the sample's position.

2. Theoretical Basis of the Model

Assume that dielectric slab of thickness d is illuminated by plane wave with incident angle θ_0. The field reflection r and transmission t coefficients being basic physical parameters measured in free-space method can be presented in general form via proper Fresnel's formulas [10] as nonlinear functions of d, θ_0, the frequency f, and the relative complex dielectric constant $\varepsilon_r' - j\varepsilon_r''$:

$$t = \frac{\left(1-\rho_0^2\right) e^{-jk_z(f)d}}{1-\rho_0^2 e^{-j2k_z(f)d}} = \frac{\left(1-\rho_0^2\right) e^{-(\alpha+j\beta)\cos(\theta)d}}{1-\rho_0^2 e^{-2(\alpha+j\beta)\cos(\theta)d}},$$
$$r = \rho_0 \cdot \frac{1-e^{-j2k_z(f)d}}{1-\rho_0^2 e^{-j2k_z(f)d}} = \rho_0 \cdot \frac{1-e^{-2(\alpha+j\beta)\cos(\theta)d}}{1-\rho_0^2 e^{-2(\alpha+j\beta)\cos(\theta)d}}, \tag{1}$$

where

$$\rho_0 = \begin{cases} \dfrac{\sqrt{\varepsilon_0}\cos(\theta_0) - \sqrt{\varepsilon_r}\cos(\theta)}{\sqrt{\varepsilon_0}\cos(\theta_0) + \sqrt{\varepsilon_r}\cos(\theta)} & \text{TE wave} \\ \\ \dfrac{\sqrt{\varepsilon_0}\cos(\theta) - \sqrt{\varepsilon_r}\cos(\theta_0)}{\sqrt{\varepsilon_0}\cos(\theta) + \sqrt{\varepsilon_r}\cos(\theta_0)} & \text{TM wave.} \end{cases} \tag{2}$$

According to the generalized Snell's law, we can write the following for lossy dielectric medium:

$$tg(\theta) = \frac{\sin(\theta_0)}{\operatorname{Re}\left\{\sqrt{\underbrace{(\varepsilon' - j\varepsilon'')}_{\varepsilon_r} - \sin^2(\theta_0)}\right\}}. \tag{3}$$

If the experimental setup is equipped by a vector network analyzer (VNA), the magnitude and phase of r and t can be measured. Otherwise the power transmission and reflection coefficients are available only

$$|t|^2 = \frac{\left(1-\rho_0^2\right)^2 e^{-2\alpha\cos(\theta)d}}{\left[1-\rho_0^2 e^{-2\alpha\cos(\theta)d}\right]^2 + 4\rho_0^2 e^{-2\alpha\cos(\theta)d}\sin^2\left[\beta\cos(\theta)d\right]},$$
$$|r|^2 = \rho_0^2 \cdot \frac{\left[1-e^{-2\alpha\cos(\theta)d}\right]^2 + 4e^{-2\alpha\cos(\theta)d}\sin^2\left[\beta\cos(\theta)d\right]}{\left[1-\rho_0^2 e^{-2\alpha\cos(\theta)d}\right]^2 + 4\rho_0^2 e^{-2\alpha\cos(\theta)d}\sin^2\left[\beta\cos(\theta)d\right]}. \tag{4}$$

The real and imaginary parts of the dielectric constant as a function of incidence angle were evaluated from the measured transmittance using the numerical root-finding algorithm applied to the system of nonlinear equations:

$$\left|t(f,d,\theta_1,\varepsilon_r',\varepsilon_r'')\right|^2 = T_{p1},$$
$$\left|t(f,d,\theta_2,\varepsilon_r',\varepsilon_r'')\right|^2 = T_{p2}, \tag{5}$$

where T_{p1} and T_{p2} are power transmission coefficients measured at the incidence angles θ_1 and θ_2, respectively, d is the thickness of a dielectric slab. Then, the proper standard deviations were determined for textile and leather samples in order to characterize angular invariances of the measured parameters. This model is used below for extracting complex permittivity from the measured data.

3. The Experimental Setup

In order to avoid an undesirable cross-polarization effect, all the measurements are suggested to perform with a parallel polarization, for which MPM is well matched as it is operating near the Brewster angle of the absorbing RF power thin-film sensor as shown in Figure 1. For absorbing thin film element of the power meter (Thomas Keating THz Power Meter, http://www.terahertz.co.uk/) this angle is equal to 55.5 degrees. The input window of the power meter must be aligned at the Brewster angle in order to reach the best matching condition.

Figure 2 shows the photo of the experimental setup assembled for the material characterization near f = 330 GHz. The THz source consists of the X-Band Synthesizer with power amplifier at its output that drives multistage multiplier (4 sections) with the total multiplication factor $32 = 4\times2\times2\times2$. Output signal about 10 mW is radiated by the horn-lens antenna (1.3 deg beam width and 42 dB gain). Other details concerning the source can be found in [1]. Energy propagating through the dielectric slab is captured by absorbing thin film of THz power meter and after processing is displayed on the PC monitor. The tested material is fixed by special frame maintained on programmable rotary joint model URS-75-BPP (Newport).

The measurement procedure includes the two steps:

(i) calibration-measurement of the transmitted signal without material under test (P_0);

(ii) measurements of the transmitted signal as a function of incident angle in a presence of material under test (P_1);

the ratio P_1/P_0 is the power transmittance needed to extract the complex permittivity.

4. Results of Transmittance Measurements and Reconstruction Complex Permittivity

Several types of the textiles have been tested. The specifications of these materials are summarized in the Table 1. Also,

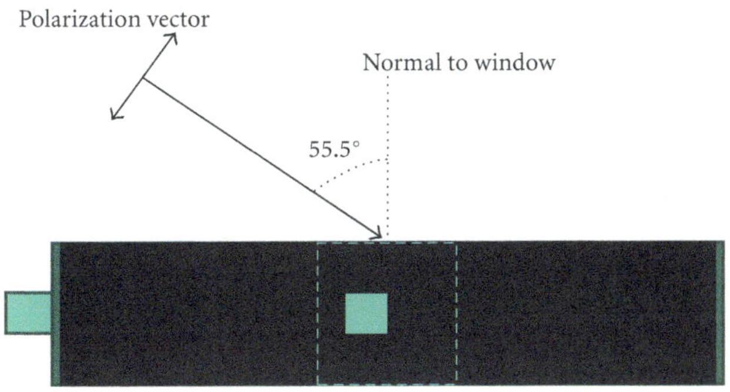

FIGURE 1: The position the MPM's head aligned at the Brewster angle 55.5 degrees to the incoming beam in order to minimize reflections from the input window.

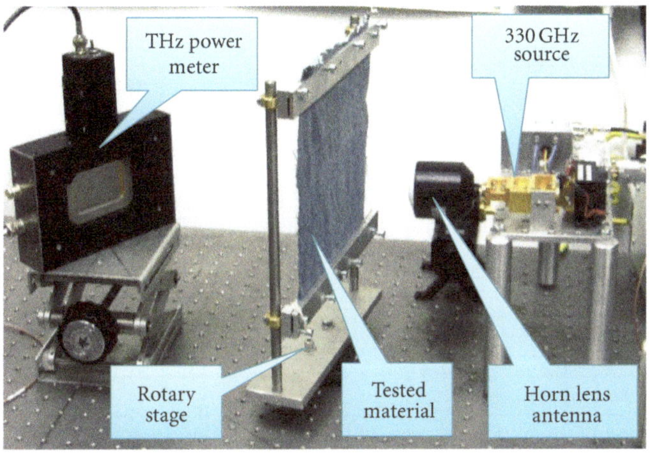

FIGURE 2: Photo of the experimental setup.

several types of leathers have been tested, see Table 2. Figures 3 and 4 show the measured transmittances as a function of incidence angle for textile and leather samples, respectively. The polarization is matched with polarization of the power meter.

The reconstructing procedure needed to restore complex permittivity from the transmittance measurements has been described in [9]. The solution of the system of nonlinear equations (5) requires knowledge of the two independent transmittances for corresponding incidence angles. The latter can be arbitrary but we have fixed one of them assuming that $\theta_1 = 0$ while the second angle was varied in range $5 \leq \theta_2 \leq 50$ degrees. Tables 3 and 4 summarize the restored real and imaginary parts of dielectric constants of the textile's samples for different incidence angles. The similar data are given in the Tables 5 and 6 for leather's samples.

Both textile and leather samples have demonstrated variations of complex permittivity for different incidence angles due to nonuniform thickness, surface roughness, specific texture and non-ideal flatness. All these factors have statistical nature and can be characterized by standard deviation presented in the last lines of the Tables 3–6.

5. Angular-Invariant Limits of the Incidence Angle

In order to estimate the angular-invariant limits of the incident angle we assume that surface the roughness of textile's and leather's samples satisfies to the Rayleigh scattering condition [11]:

$$\Delta h < \frac{\lambda}{8 \cos \theta_i}, \tag{6}$$

where Δh is roughness variation, θ_i is incidence angle, and λ is free space wavelength. Figure 5 illustrates schematically forward scattered rays "generated" by roughness surface.

Since the roughness variation Δh and incidence angle θ_i are interrelated due to expression (6), we can expect that the uncertainty in a determination of the complex permittivity should be increased with increasing θ_i, too. The question is what is the acceptable range of incidence angles θ_i invariant to the bulky complex permittivity of the material under test-textile and leather in our case? To answer the question we need to compare the measured transmittance of a real material having rough surface with the calculated transmittance of the smoothed dielectric slab made from the same bulky material. The results of such comparison are shown in Figure 6 (textile's sample number 3, $\varepsilon_r' = 2.72$ and $\varepsilon_r'' = 0.068$) and Figure 7 (leather's sample number 0, $\varepsilon_r' = 3.4$ and $\varepsilon_r'' = 0.127$), where the lines marked by symbol □□□ correspond to the measured transmittance while the solid lines depict the calculated data. The averaged values of the real and imaginary parts of the dielectric constants given in the Tables 3–6 were substituted to the theoretical model described by (4). The comparison theory and experiments reveals an existence of the critical incidence angle, θ_c. The latter is determined by the angular-invariance limit of the forward Rayleigh scattering as well as operating wavelength. The materials discussed have the critical angles about 25 degrees for leather's sample and 30 degrees for textile's one. If an incidence angle is greater than the critical one ($\theta_i > \theta_c$), the measurement of the bulk permittivity will be quite problematic due to violating the angular-invariant limit. On the other hand, there is a critical roughness variation Δh, too. Using (6), this parameter was found to be equal to about 0.15 mm for operating frequency 330 GHz and $\theta_c = 30°$. These estimates seem to be realistic for the typical kinds of textiles and leathers.

It would be important to derive an existence of the angular-invariant limit from a theoretical model. Such a model must unify both deterministic and statistic properties of real structure of materials under test. Microwave properties of the common weave architectures used in woven fabric structural composites have been investigated in [12] using deterministic approach. Other perspective way is an application of the concept of the fractional Brownian motion appropriately describing natural surfaces [13] assuming that surface profiles can be approximated by the Weierstrass-Mandelbrot (WM) fractal functions. However, some entry parameters needed for real-world calculations such as the

Table 1: The summary of textile's samples under test.

Material	Sample no. 1 (denim)	Sample no. 2	Sample no. 3	Sample no. 4	Sample no. 5	Sample no. 6 (wool scarf)	Sample no. 7 (stockinet)	Sample no. 8 (satin)
General view of samples under testing								
Thickness d, mm	0.8	0.45	0.25	0.7	0.25	1.6	0.65	0.2
Power transmittance at normal incidence	77%	87%	92%	86%	88.8%	35%	67%	91%

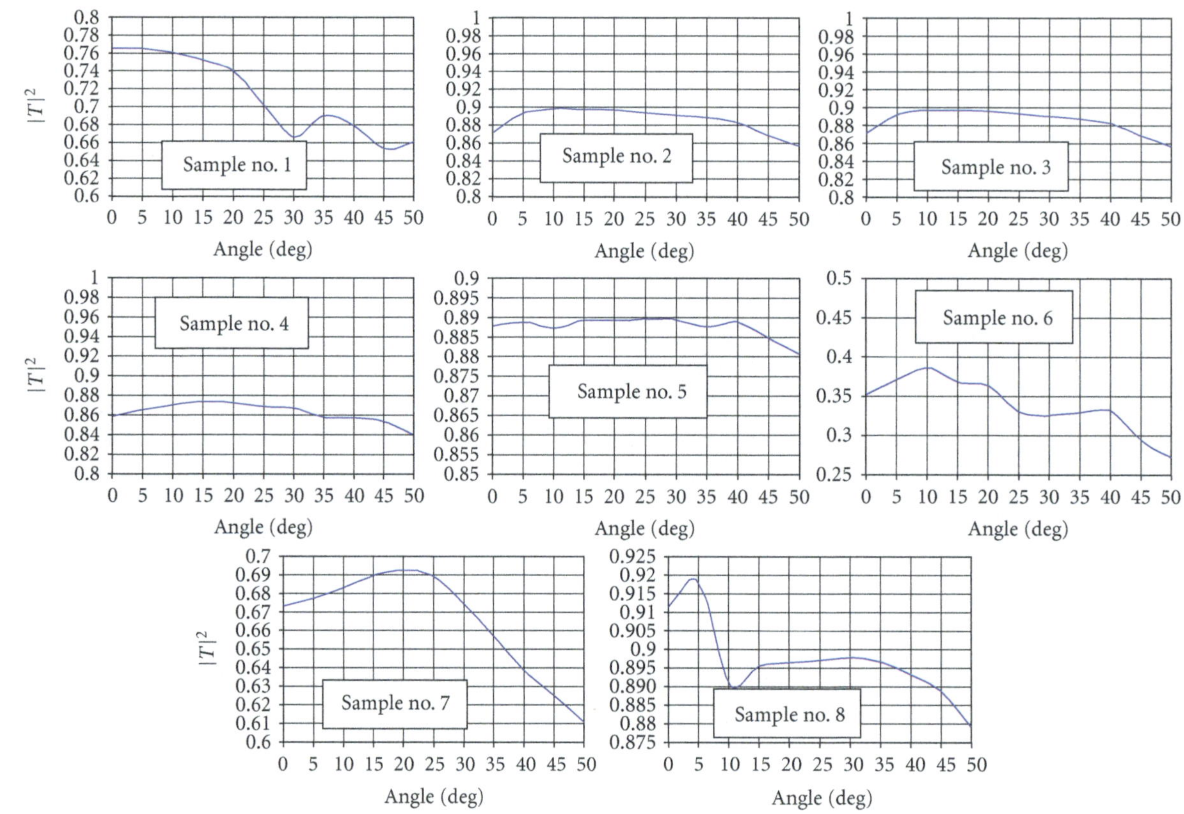

Figure 3: Measured power transmittance of the textile's samples as a function of incidence angle.

Table 2: The summary of leather's samples under test.

Material	Sample no. 0 (natural leather)	Sample no. 1 (artificial leather)	Sample no. 2 (artificial leather)	Sample no. 3 (artificial leather)	Sample no. 4 (artificial leather)	Sample no. 5 (artificial leather)
General view of samples under testing	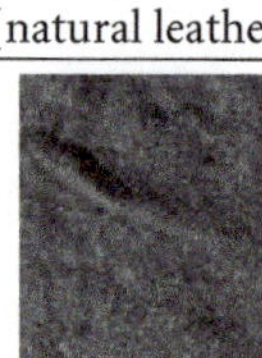		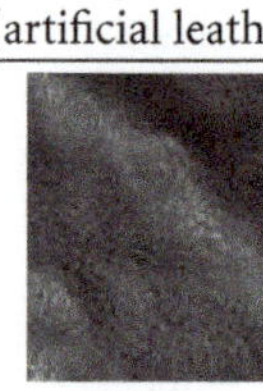	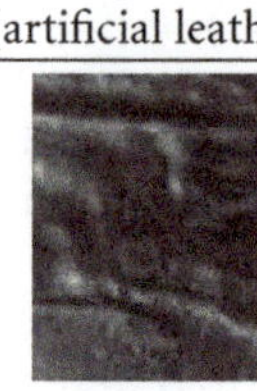		
Thickness d, mm	1.15	0.95	1	0.6	0.8	0.6
Power transmittance at normal incidence	48%	63%	68%	81.5%	71%	77%

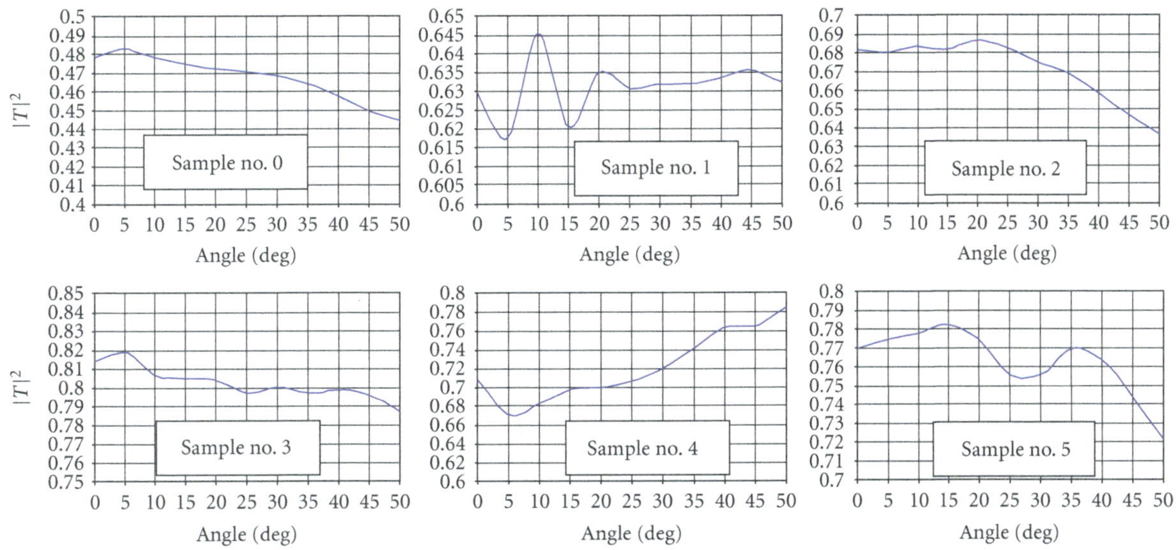

FIGURE 4: Measured power transmittance of the leather's samples as a function of incidence angle.

TABLE 3: Real part of the permittivity evaluated for the textile's samples: $\theta_1 = 0$ and different incidence angles θ_2.

Incidence angle, θ_2 degrees	Sample no. 1 (Denim)	Sample no. 2	Sample no. 3	Sample no. 4	Sample no. 5	Sample no. 6 (wool scarf)	Sample no. 7 (stock-inet)	Sample no. 8 (satin)
5	2.83	2.77	2.73	3.65	2.54	3.2	3.15	2.73
10	2.71	2.71	2.7	3.76	2.43	3.21	3.13	2.69
15	2.67	2.92	2.74	3.67	2.41	3.23	3.17	2.73
20	2.7	2.76	2.77	3.67	2.49	3.12	3.17	2.75
25	2.73	2.68	2.79	3.65	2.51	3.11	3.2	2.76
30	2.69	2.5	2.75	3.67	2.54	3.14	3.34	2.75
35	2.67	2.26	2.7	3.66	2.54	3.17	3.34	2.76
40	2.75	2.84	2.75	3.68	2.58	3.2	3.22	2.76
45	2.78	2.95	2.59	3.66	2.68	3.22	3.26	2.75
50	2.79	3.01	2.71	3.58	2.68	3.23	3.21	2.76
Averaged, ε_r'	2.73	2.74	2.72	3.66	2.54	3.18	3.22	2.74
Standard deviation	0.054	0.253	0.0623	0.0458	0.0956	0.0452	0.0737	0.0216

TABLE 4: Imaginary part of the permittivity evaluated for the textile's samples: $\theta_1 = 0$ and different incidence angles θ_2.

Incidence angle, θ_2 degrees	Sample no. 1 (cotton jeans)	Sample no. 2	Sample no. 3	Sample no. 4	Sample no. 5	Sample no. 6 (wool scarf)	Sample no. 7 (stock-inet)	Sample no. 8 (satin)
5	0.0715	0.033	0.07	0.041	0.065	0.133	0.054	0.074
10	0.0633	0.029	0.057	0.028	0.06	0.13	0.057	0.072
15	0.0603	0.0103	0.06	0.038	0.045	0.134	0.041	0.071
20	0.0634	0.033	0.062	0.043	0.048	0.145	0.042	0.072
25	0.071	0.0397	0.065	0.044	0.052	0.158	0.054	0.072
30	0.0781	0.011	0.068	0.044	0.061	0.159	0.076	0.073
35	0.074	0.011	0.071	0.044	0.072	0.159	0.1	0.074
40	0.079	0.019	0.08	0.043	0.077	0.161	0.119	0.077
45	0.0858	0.06	0.085	0.044	0.089	0.172	0.126	0.08
50	0.0866	0.065	0.069	0.047	0.092	0.175	0.13	0.086
Averaged, ε_r''	0.073	0.031	0.068	0.042	0.066	0.15	0.08	0.075
Standard deviation	0.0092	0.0222	0.0097	0.0056	0.0172	0.0162	0.0356	0.0049

Table 5: Real part of the permittivity evaluated for the leather's samples: $\theta_1 = 0$ and different incidence angles θ_2.

Incidence angle, θ_2 degrees	Sample no. 0	Sample no. 1	Sample no. 2	Sample no. 3	Sample no. 4	Sample no. 5
5	2.91	3.3	2.95	2.18	2.7	2.48
10	3.31	2.97	3.03	2.03	2.71	2.75
15	3.39	3.17	3.03	2.05	2.73	2.66
20	3.42	2.99	3.08	2.16	2.96	2.51
25	3.45	3.02	3.04	2.11	3.03	2.33
30	3.44	3.03	3.05	2.26	3.09	2.45
35	3.46	3.06	3.05	2.28	3.19	2.45
40	3.5	3.12	3.03	2.28	3.23	2.43
45	3.58	3.17	2.9	2.26	3.25	2.43
50	3.54	3.21	2.92	2.13	3.28	2.45
Averaged, ε_r'	3.4	3.104	3.008	2.17	3.02	2.49
Standard deviation	0.178	0.102	0.058	0.094	0.220	0.1221

Table 6: Imaginary part of the permittivity for the leather's samples: $\theta_1 = 0$ and different incidence angles θ_2.

Incidence angle, θ_2 degrees	Sample no. 0	Sample no. 1	Sample no. 2	Sample no. 3	Sample no. 4	Sample no. 5
5	0.118	0.093	0.078	0.066	0.088	0.084
10	0.109	0.072	0.08	0.062	0.086	0.065
15	0.115	0.079	0.081	0.063	0.083	0.075
20	0.118	0.061	0.08	0.066	0.096	0.09
25	0.12	0.067	0.079	0.065	0.096	0.088
30	0.125	0.071	0.079	0.069	0.091	0.088
35	0.131	0.077	0.078	0.069	0.084	0.087
40	0.14	0.083	0.079	0.067	0.076	0.088
45	0.148	0.089	0.079	0.0676	0.0751	0.088
50	0.15	0.098	0.083	0.067	0.069	0.095
Averaged, ε_r''	0.127	0.079	0.08	0.066	0.084	0.085
Standard deviation	0.0135	0.0111	0.0014	0.0023	0.0086	0.0084

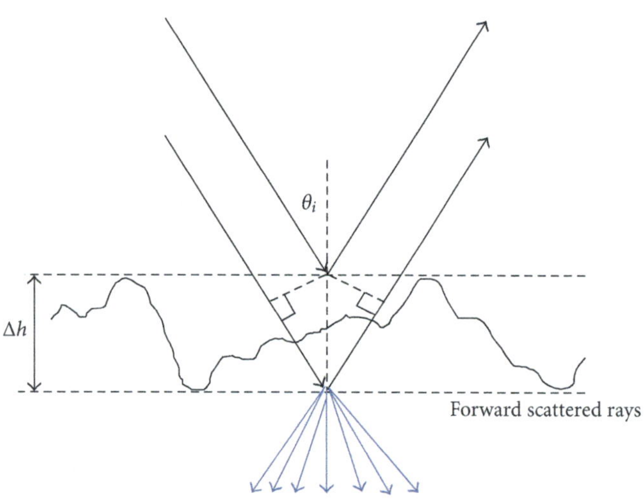

Figure 5: Illustration of the forward Rayleigh scattering effect in presence of a roughness surface.

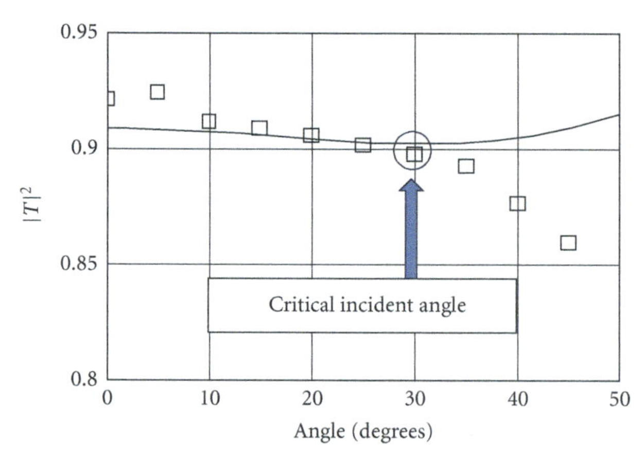

Figure 6: Comparison calculated (solid line) and measured (symbol □□□) transmittances as a function of an incidence angle (the textile's sample #3, $\varepsilon_r' = 2.72$ and $\varepsilon_r'' = 0.068$).

vertical height profile scaling factor, random variables which account for the amplitude, and the phase behavior of each tone, and others [14] must be specified for materials under test which is not trivial problem. Anyway, we hope that the above experiments will be an additional stimulus for developing proper models combining deterministic and statistic approaches, too.

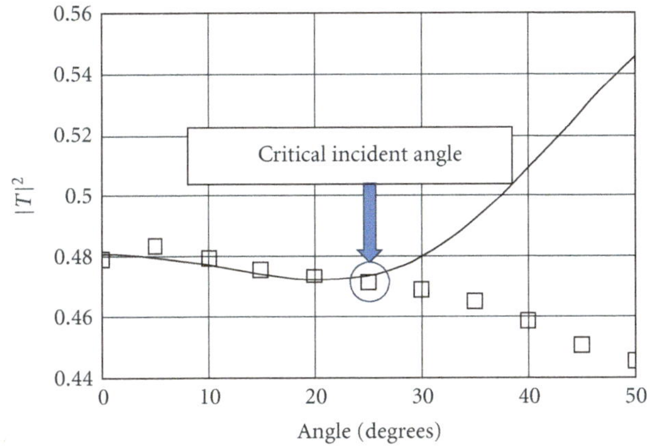

FIGURE 7: Comparison calculated (solid line) and measured (symbol □□□) transmittances as a function of an incidence angle (the leather's sample no. 0, $\varepsilon_r' = 3.4$ and $\varepsilon_r'' = 0.127$).

6. Conclusions

The complex permittivities of textiles and leathers were measured at 330 GHz using the absolute THz power meter with thin-film sensor operating near the Brewster angle in order to provide a minimum multiple-reflection effect. The incidence-angular invariance has been estimated experimentally and existence of a critical incidence angle was found out. It is not recommended to carry out free-space permittivity measurements if the incidence angle exceeds the critical one (25–30 degrees in our case) since the forward Rayleigh scattering effect may drastically degrade the measurement accuracy.

Acknowledgments

Authors would like to thank Dr. J Polivka for fruitful discussions and A. Shulzinger and S. Neporent for assistance in the design and assembling the measuring setup used in the experiments discussed.

References

[1] B. Kapilevich, Y. Pinhasi, R. Arusi et al., "330 GHz FMCW image sensor for homeland security applications," *Journal of Infrared, Millimeter, and Terahertz Waves*, vol. 31, no. 11, pp. 1370–1381, 2010.

[2] B. Kapilevich and M. Einat, "Detecting hidden objects on human body using active millimeter wave sensor," *IEEE Sensors Journal*, vol. 10, no. 11, Article ID 5482043, pp. 1746–1752, 2010.

[3] S. Van den Bulcke, A. Franchois, and D. De Zutter, "2.5 D quantitative millimeter wave imaging of a hidden object on a simplified human body model using value picking regularization," *Journal of Infrared, Millimeter, and Terahertz Waves*, vol. 31, no. 12, pp. 1478–1490, 2010.

[4] L. F. Chen, C. K. Ong, C. P. Neo, V. V. Varadan, and V. K. Varadan, *Microwave Electronics: Measurement and Material Characterization*",, John Wiley & Sons, New York, NY, USA, 2004.

[5] N. Gagnon, J. Shaker, L. Roy, A. Petosa, and P. Berini, "Low-cost free-space measurement of dielectric constant at Ka band," *IEE Proceedings: Microwaves, Antennas and Propagation*, vol. 151, no. 3, pp. 271–276, 2004.

[6] K. M. Hock, "Error correction for diffraction and multiple scattering in free-space microwave measurement of materials," *IEEE Transactions on Microwave Theory and Techniques*, vol. 54, no. 2, pp. 648–659, 2006.

[7] W. D. Burnside and K. W. Burgener, "High frequency scattering by a thin lossless dielectric slab," *IEEE Transactions on Antennas and Propagation*, vol. AP-31, no. 1, pp. 104–110, 1983.

[8] S. W. Harmer, N. Rezgui, N. Bowring, Z. Luklinska, and G. Ren, "Determination of the complex permittivity of textiles and leather in the 14–40 GHz millimetre-wave band using a free-wave transmittance only method IET Microw," *IEEE Transactions on Antennas and Propagation*, vol. 2, no. 6, pp. 606–614, 2008.

[9] B. Kapilevich, Y. Pinhasi, and B. Litvak, "Measurement of complex permittivity of lossy materials in free space using matched THz power meter?" *The International Journal on Infrared Millimeter THz Waves*, vol. 32, no. 12, pp. 1446–1456, 2011.

[10] S. V. Marshall and G. G. Skitek, *Electromagnetic Concepts and Applications*, Prentice-Hall, New York, NY, USA, 2nd edition, 1987.

[11] S. R. Saunders and A. Aragon-Zavala, *Antennas and Propagation for Wireless Communication Systems*, John Wiley, New York, NY, USA, 2007.

[12] M. S. Mirotznik, S. Yarlagadda, R. McCauley, and P. Pa, "Broadband electromagnetic modeling of woven fabric composites," *IEEE Transactions on Microwave Theory and Techniques*, vol. 60, no. 1, pp. 158–169, 2012.

[13] B. B. Mandelbrot, *The Fractal Geometry of Nature*, W. H. Freeman, New York, NY, USA, 1983.

[14] G. Franceschetti, A. Iodice, D. Riccio, and G. Ruello, "Fractal surfaces and electromagnetic extended boundary conditions," *IEEE Transactions on Geoscience and Remote Sensing*, vol. 40, no. 5, pp. 1018–1031, 2002.

14

Copper-Metalized GaAs pHEMT with Cu/Ge Ohmic Contacts

E. V. Anichenko,[1] V. S. Arykov,[1] E. V. Erofeev,[1] and V. A. Kagadei[2]

[1] *Microelectronics Department, Research and Production Company "Micran", 47 Vershinina Street, Tomsk 634045, Russia*
[2] *RF Electronics Department, Submicron Technologies, 8 Akademicheskiy Avenue, Tomsk 634021, Russia*

Correspondence should be addressed to E. V. Erofeev, erofeev@sibmail.com

Academic Editor: Rüdiger Quay

The fully Cu-metalized GaAs pHEMT using developed Cu/Ge-based ohmic contacts and T-gate Ti/Mo/Cu with length of the 150 nm has been successfully fabricated for the high-frequency applications. The fabricated Cu-metalized pHEMT has a maximum drain current of 360 mA/mm, an off-state gate-drain breakdown of 7 V, and a transconductance peak of 320 mS/mm at $V_{ds} = 3$ V. The maximum stable gain value was about 15 dB at frequency 10 GHz. The current gain cutoff frequency of the copper-metalized device is about 60 GHz at $V_{ds} = 3$ V, and maximum frequency of oscillations is beyond 100 GHz. This work investigated in detail the formation of Cu/Ge ohmic contacts to *n*-GaAs with an atomic hydrogen preannealing step. It was shown that after the first preliminary annealing is carried out in a flow of atomic hydrogen with a flow density of atoms of $10^{13}/10^{16}$ at. cm^2 s^{-1} a reduction in specific contact resistance by 2/2.5 times is observed. The reduction in specific contact resistance is apparently caused by the action of the hydrogen atoms which minimise the rate of the oxidizing reactions and activate solid phase reactions forming the ohmic contact during the thermal treatment process.

1. Introduction

Wafer production cost is an important metric that measures the profitability and success of a semiconductor fab. The current economic situation has driven the price of Au and Pt up steadily over the last couple of years. Although the precious metals prices are beyond one's control, with proper engineering, these metals can be replaced by a less expensive alternative—copper.

Copper metallization has been extensively used in the silicon integrated circuit technology since IBM announced its success in silicon very large scale integration processing [1–3]. The advantages of copper metallization for Si technology include lower resistivity and higher electromigration resistance. Even though the use of copper as metallization metal has become very popular in Si devices, there were very few reports of copper metallization of GaAs devices published in the literature [4–7]. The use of copper as the metallization metal instead of gold has several advantages such as lower resistivity, higher thermal conductivity, and lower cost. If Cu replaces Au as the metallization metal for the pHEMTs, then the resulting improvement in the electrical conductivity can increase the transmission speed of the circuits [8]. However, copper diffuses very fast into GaAs and is a deep acceptor, moreover, copper oxidized under atmosphere exposure, so copper metallization of GaAs device without a diffusion and passivation barriers degrades their electrical properties.

Other authors in [3] reported Cu-metalized compound semiconductor pHEMT with Pd/Ge/WN$_x$/Cu ohmic contacts a thickness of 500 nm, Ti/Pt/Cu gate and Cu-based interconnect air-bridge structure are presented. However, in the nanoscale-gate-length process, this thick Pd/Ge/WN$_x$/Cu ohmic contact may induce a poor edge definition after alloying which can cause the problem such as gate misalignment and the use of precious metals such as Pt, Pd increases the cost of wafer production.

The formation of nonprecious Cu-based ohmic contact with good parameters is obviously the most important problem in Cu-metalized GaAs pHEMT fabrication. In studies [9–11] it was established that on the basis of two-layer Cu/Ge it is possible to form a contact of Cu_3Ge with a low value of contact resistance, high thermal stability of the parameters, smooth surface structure, even edges of the contact pads, and also low production costs. It was shown that the ohmic behaviour was partly due to Ge occupation of Ga vacancies. Ge atoms can serve as donors and help to induce a heavily doped interfacial region which enhances tunnelling current.

The authors of study [12] proposed a two-stage annealing of the Cu/Ge contact. The first stage of annealing (preliminary low-temperature annealing) was carried out under vacuum at a temperature of 100°C *in situ* with the deposition of contact metallization, and the second high-temperature annealing was carried out *ex situ* after completing the process of forming the contact pads patterns. The use of two-stage annealing enabled the contact resistance to be reduced to 6.5×10^{-7} ohm cm^2. Thus these studies [10–12] testify to the potential of using Cu/Ge contacts in nanoscale GaAs MMIC technology.

The purposes of the present work are to develop and produce fully copper-metalized pHEMT with Cu/Ge-based ohmic contacts and Ti/Mo/Cu 150 nm T-gate; to investigate the possibility to improve the parameters of Cu/Ge ohmic contacts by two-stage annealing with the first preliminary annealing being carried out *in situ* under vacuum, or in a flow of atomic hydrogen, and the second *ex situ* in a nitrogen atmosphere; and to investigate the DC and RF parameters of fabricated Cu-metalized GaAs pHEMT.

2. Experimental

2.1. Fabrication of GaAs pHEMT. The fully Cu-metalized GaAs pHEMT was formed on the GaAs/AlGaAs/InGaAs pseudomorphic structures have been grown by using molecular beam epitaxy. A 30 nm thick undoped GaAs buffer layer and ten AlAs-GaAs superlattices are grown on 3-inch diameter semiinsulating GaAs substrates, followed by a 60 nm undoped GaAs layer. The thickness of the $In_{0.24}Ga_{0.76}As$ channel layer is 10 nm. The planar doping with a donor density of 5×10^{12} cm^{-2} is separated from the active layer by a thin undoped $Al_{0.22}Ga_{0.78}As$ spacer and an undoped $Al_{0.22}Ga_{0.78}As$ Schottky layer is grown. To reduce parasitic resistance, the GaAs cap layer is highly doped with a concentration of 5×10^{18} cm^{-3}.

After mesa isolation the source and drain ohmic metals were formed. Cu/Ge (122 nm/78 nm) ohmic contacts deposition by the electron beam evaporation under a vacuum of 1×10^{-6} torr was used. After the ohmic contact structure was formed using the liftoff method a rapid thermal annealing was carried out. Annealing was performed *ex situ* in a single high-temperature stage in a inert environment at a temperature range $T = 200$–500°C for a period of $t = 3$ min.

150 nm Ti/Mo/Cu T-shape gates were fabricated by electron beam lithography and liftoff process. There was used a trilayer resist stack of 950 PMMA/LOR 5B/495 PMMA (from bottom to top). The resists were spin-coated onto the substrate. Each resist layer was baked for a 5 min at 180°C on the hotplate. The Raith-150^{TWO} e-beam nanolithography system was used for a single exposure of the resist with 30 kV electron beam energy. The gate pattern consists of three parts. The central pattern is a footprint and was a 1 pixel line. Two head patterns have a size of 200 nm. The optimum conditions, to define T-shape gates at 150 nm level for pHEMTs were obtained when the footprint dose is 600 μC/cm^2, the head dose is 140 μC/cm^2. The top layer was developed in 1 : 1 mixture of MIBK : IPA for 60 sec and then rinsed in IPA and blown dry in nitrogen. The second layer was developed in MF-319 developer and then rinsed in water. The third layer was developed for 30 s in MIBK : IPA (1 : 3) followed by rinsing in an IPA and blown dry with nitrogen.

Then recess etching was performed using citric acid based solution as an etchant to remove GaAs cap layer. After that T-gate metallization of Ti/Mo/Cu (20 nm/20 nm/400 nm) was deposited by e-beam evaporation under a vacuum of 5×10^{-7} torr and lifted off in acetone. Finally, a layer of PECVD silicon nitride passivation was deposited; openings were etched in dielectric and the ohmic contacts were gold-plated for a good probe contact.

The device DC properties were measured with a Tektronix 370 A. The microwave S-parameter measurements have been carried out by using a ZVA-40 network analyzer.

2.2. Investigation of Cu/Ge Ohmic Contacts Improvement. To increase the specific contact resistance measurement accuracy in Cu/Ge ohmic contacts formation experiments the Si-doped samples *n-i*-GaAs ($n = 2 \times 10^{17}$ cm^{-3}, $d = 0.12\,\mu$m) were used. For liftoff lithography process a mask based on two-layer SiO_2 of thicknesses 0.3 and 0.5 μm respectively was formed on the surface. The layers were deposited by plasma-enhanced CVD at different wafer temperatures. This enabled films to be formed with different etching rates and a hard mask for liftoff process. The openings in the mask of SiO_2 were formed using etching in a buffer etching agent through the photoresistive mask. To remove the oxides of GaAs before deposition, the samples were treated in a 1 : 10 solution of H_2SO_4 for three minutes, followed by rinsing in de-ionised water and drying in a flow of nitrogen.

For deposition of the Cu/Ge contact, the method of e-beam evaporation was used, with a residual pressure of p less than 1×10^{-6} torr. The thickness was 122 nm for the Cu film and 78 nm for the Gefilm. After the ohmic contact pattern was formed using the liftoff method a rapid thermal annealing of the samples was carried out. Annealing was performed *ex situ* in a single high-temperature stage in a inert environment at a temperature $T_2 = 440$°C for a period of $t = 60$ s. In the experiments with two-stage annealing the wafer, after treatment in H_2SO_4 solution, was divided into two parts, which were loaded into the vacuum chamber of the deposition unit. Films of Cu and Ge with an overall thickness of 200 nm were deposited on both wafers by the e-beam evaporation method under a vacuum of 5×10^{-6} torr. The weight content of Ge in the two-layer composition was about 30%. Then both wafers underwent preliminary thermal treatment in a single vacuum cycle at a temperature of $T_1 = 75$°C, one under vacuum, and the other in a flow of atomic hydrogen with a flow density of 10^{15} at. cm^{-2} s^{-1}. The wafers were removed from the vacuum chamber and, after forming the ohmic contact pattern using the liftoff method, rapid annealing was carried out in an inert atmosphere at a

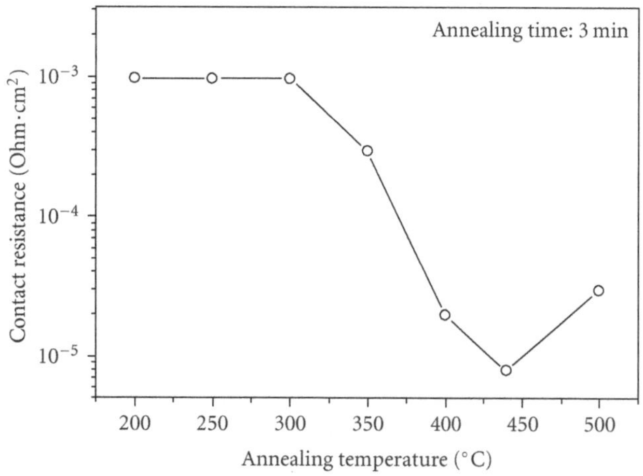

FIGURE 1: Dependence of the specific contact resistance of Cu/Ge ohmic contacts to GaAs/AlGaAs/InGaAs versus the annealing temperature (t = 3 min).

temperature of T_2 = 80–460°C for a period of t = 60 s. The process was repeated for preliminary annealing temperatures ranging from 100 to 400°C The two-layer SiO_2-based hard mask was not lifted off. The specific contact resistance ρ was measured using the transmission line method. Morphology of the contact surfaces was obtained using scanning electron microscope.

3. Results

3.1. The Cu/Ge Ohmic Contacts and T-Gate Fabrication. Figure 1 shows results of the specific contact resistance measurements of the Cu/Ge ohmic contacts produced by e-beam evaporation and RTA annealed in inert environment.

The temperature dependences have a traditionally curves shape with the ρ value minimum. With the annealing temperature raise the interdiffusion between multilayer metals film and GaAs is increased. This leads to a reduction of the specific contact resistance, down to minimum value of ρ. The subsequent growth of the temperature leads to ρ increasing. It may be caused by the copper penetration from the top layer to the GaAs surface.

The minimal value of the specific contact resistance for Cu/Ge ohmic contacts is 8×10^{-6} Ohm cm² after annealing at T = 440°C during t = 3 min in inert environment.

The Ti/Pt/Au Shottky contact on n-GaAs is the most widely used Shottky structure in the GaAs HEMT fabrication. In present study, the thick overcoat gold layer was replaced with copper. And Pt layer was replaced with the transition metal such Mo. The choice of the refractory metals was based on their resistivity and the capability as the diffusion barrier. The resistivity of Mo metal is much lower than of Pt, therefore, the fabricated T-gate Ti/Mo/Cu has a smaller gate resistance.

Figure 2 shows the SEM image of fabricated fully Cu-metalized pHEMT with Cu/Ge ohmic contacts and the T-gate Ti/Mo/Cu with a length of 150 nm.

FIGURE 2: SEM image of fabricated fully Cu-metalized pHEMT with 150 nm T-gate Ti/Mo/Cu.

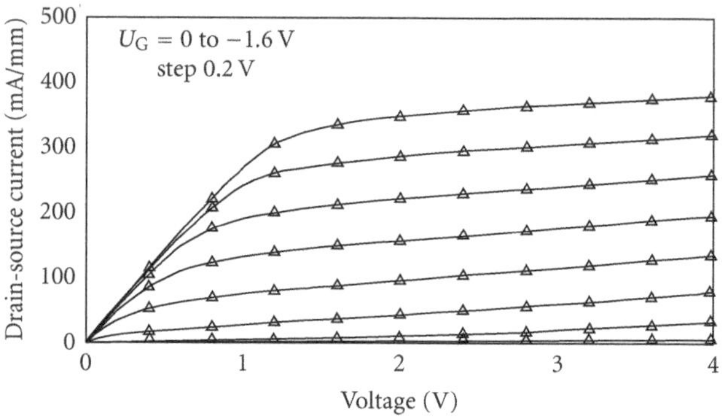

FIGURE 3: Current-voltage characteristics of the fabricated Cu-based pHEMT.

3.2. The Copper-Metalized pHEMT Performance. Figures 3 and 4 show the DC and RF characteristics of a fabricated fully Cu-metalized GaAs pHEMT.

The device performance was measured for pHEMT with 150 nm gate-length and a total gate width of 100 μm. The fabricated Cu-based pHEMT has a maximum drain current of 360 mA/mm, an off-state gate-drain breakdown of 7 V, and a transconductance peak of 320 mS/mm at V_{ds} = 3 V.

The maximum stable gain value was about 15 dB at a frequency of 10 GHz. The current gain cutoff frequency of the copper-metalized device is about 60 GHz at V_{ds} = 3 V and maximum frequency of oscillations is beyond 100 GHz.

The DC and RF performance shows little difference between devices fabricated with copper and conventional AuGeNi based metallization.

3.3. Investigation of Cu/Ge Ohmic Contacts Improvement. To study further improvement of pHEMT characteristics, Cu/Ge ohmic contacts resistance minimization was investigated and the results are presented in the following.

Figure 5 shows the dependence of specific contact resistance ρ on the temperature of preliminary thermal treatment T_1, carried out *in situ* under vacuum (1) and in a flow of atomic hydrogen (2) with a flow density of hydrogen atoms of 10^{15} at. cm^{-2} s^{-1}. The second thermal treatment was carried out *ex situ* in an inert atmosphere at T_2 = 440°C

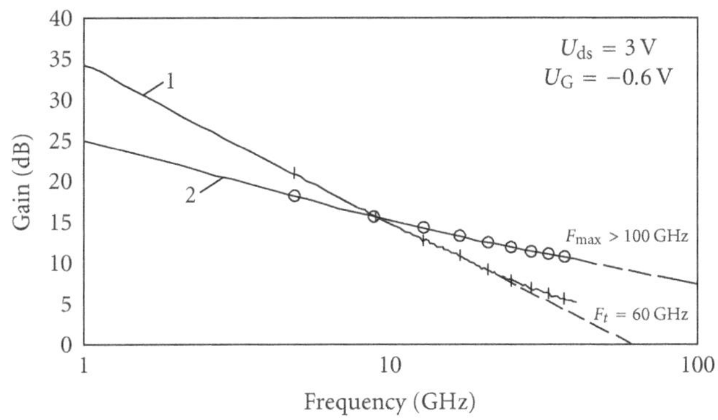

FIGURE 4: Measured current gain (1) and maximum stable gain (2) as a function of frequency of fabricated Cu-based GaAs pHEMT.

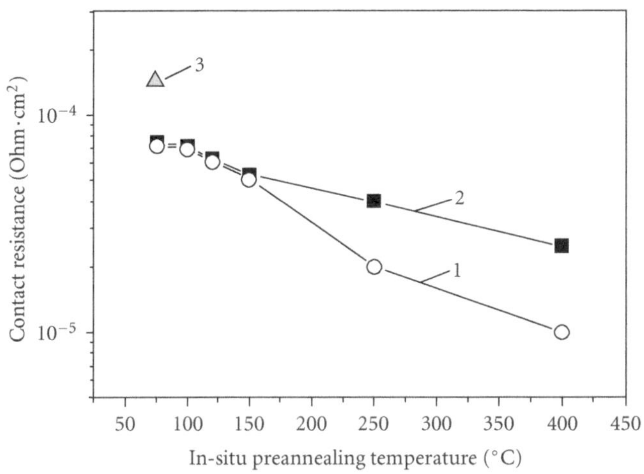

FIGURE 5: Dependence of specific contact resistance of Cu/Ge ohmic contacts to n-GaAs on the temperature of preliminary thermal treatment T_1, carried out under vacuum (1), in a flow of atomic hydrogen (2) and without preliminary thermal treatment (3).

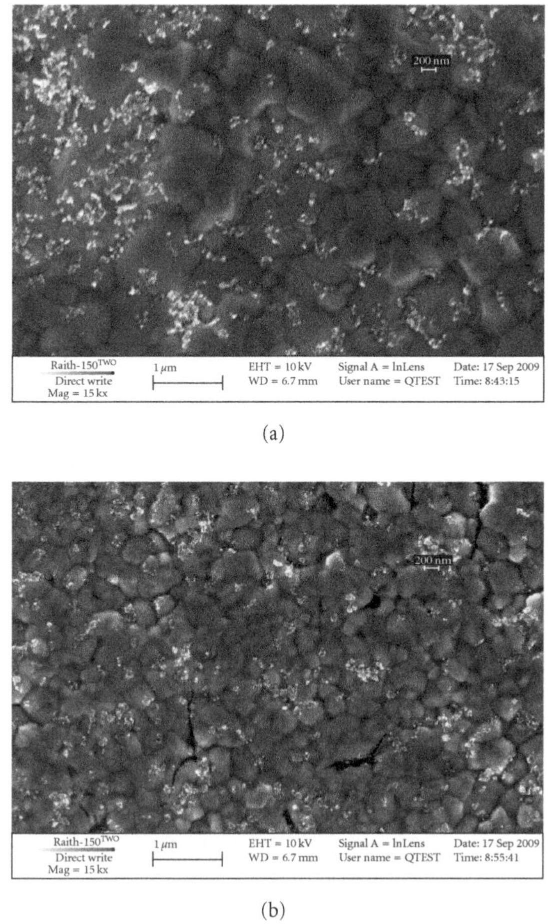

FIGURE 6: SEM images of the surface of ohmic contacts which have undergone thermal treatment at $T = 400°C$ under vacuum (a) in a flow of atomic hydrogen (b); magnification 15,000x.

for 60 s. At temperatures of the first annealing higher than $T_1 = 150°C$ for samples annealed in a flow of atomic hydrogen, smaller values of ρ are observed than for samples annealed under vacuum. The difference of the values of ρ for samples of both types increases with the temperature of the first annealing.

In Figure 6 SEM images are shown of the surface of ohmic contacts obtained using two-stage annealing under vacuum, and in a flow of atomic hydrogen at the preliminary *in situ* annealing temperature $T_1 = 400°C$. It can be seen that the samples which have undergone preliminary annealing in a flow of atomic hydrogen have smaller grain size.

As it is shown in [13] the annealing in atomic hydrogen has a significant influence on the microcrystal structure of the Cu film metallization, increasing the grain size, relative to annealing under vacuum. In the present study the reverse effect is observed in the Cu/Ge/GaAs system, which apparently is connected with the particular behaviour of solid phase reactions in a four-component system.

The minimum value of ρ for a Cu/Ge contact obtained using two-stage annealing at a temperature $T_1 = 400°C$ under vacuum and in a flow of atomic hydrogen is 2.5×10^{-5} ohm cm^2 and 1×10^{-5} ohm cm^2, respectively. These values are significantly lower than the value characteristic of single-stage annealing *ex situ* of 1.5×10^{-4} ohm cm^2.

4. Conclusions

In this work the fully Cu-based GaAs pHEMT with Cu/Ge ohmic contacts and Ti/Mo/Cu 150 nm T-gate was successfully produced. The fabricated pHEMT has a peak transconductance of 320 mS/mm at $V_{ds} = 3$ V and maximum drain current of 360 mA/mm. The maximum stable gain and current gain cutoff frequency were about 15 dB by at 10 GHz and 60 GHz at $V_{ds} = 3$ V, accordingly.

The possibility to improve the parameters of Cu/Ge ohmic contacts to n-GaAs using two-stage annealing in the case when the preliminary annealing was performed at $T_1 = 150$–$400°C$*in situ* in a flow of atomic hydrogen with flow density 10^{15} at. cm^{-2} s^{-1} was demonstrated. Using such annealing enables the minimum value of contact resistance to be reduced by a factor of 2–2.5 relative to preliminary annealing under vacuum, or by 2–15 times as compared to single-stage annealing in inert atmosphere.

Acknowledgments

This work was sponsored by the grant Carl Zeiss, Bortnik's grant U.M.N.I.K of Russia, and also by a grant of Tomsk administration. The authors wish to acknowledge the lab.7 staff of the Scientific and Research Institute of Semiconductor Devices Tomsk, Russia for support and also to the Scientific-Education Centre staff of the Tomsk University of Control Systems and Radio Electronics for microwave measurements.

References

[1] K. Holloway and P. M. Fryer, "Tantalum as a diffusion barrier between copper and silicon," *Applied Physics Letters*, vol. 57, no. 17, pp. 1736–1738, 1990.

[2] K. Holloway, P. M. Fryer, C. Cabral, J. M. E. Harper, P. J. Bailey, and K. H. Kelleher, "Tantalum as a diffusion barrier between copper and silicon: failure mechanism and effect of nitrogen additions," *Journal of Applied Physics*, vol. 71, no. 11, pp. 5433–5444, 1992.

[3] US Patent no. 0292785 A1, 2006.

[4] D. S. Yoon, H. K. Baik, and S. M. Lee, "Effect on thermal stability of a Cu/Ta/Si heterostructure of the incorporatio of cerium oxide into the Ta barrier," *Journal of Applied Physics*, vol. 83, no. 12, pp. 8074–8076, 1998.

[5] C. Y. Chen, L. Chang, E. Y. Chang, S. H. Chen, and D. F. Chang, "Thermal stability of Cu/Ta/GaAs multilayers," *Applied Physics Letters*, vol. 77, no. 21, pp. 3367–3368, 2000.

[6] C. Y. Chen, E. Y. Chang, L. Chang, and S. H. Chen, "Backside copper metallization of GaAs MESFETs," *Electronics Letters*, vol. 36, no. 15, pp. 1317–1318, 2000.

[7] C. Y. Chen, E. Y. Chang, L. Chang, and S. H. Chen, "Backside copper metallization of GaAs MESFETs using TaN as the diffusion barrier," *IEEE Transactions on Electron Devices*, vol. 48, no. 6, pp. 1033–1036, 2001.

[8] H. C. Chang, E. Y. Chang, Y. C. Lien et al., "Use of WNx as diffusion barrier for copper airbridged low noise GaAs PHEMT," *Electronics Letters*, vol. 39, no. 24, pp. 1763–1765, 2003.

[9] E. V. Erofeev and V. A. Kagadei, "A means of preparation of ohmic contacts to GaAs on the basis of thin films of Ge and Cu," Patent application no. 2009130823, 2009.

[10] M. O. Aboelfotoh, C. L. Lin, and J. M. Woodall, "Novel low-resistance ohmic contact to n-type GaAs using CuGe," *Applied Physics Letters*, vol. 65, no. 25, pp. 3245–3247, 1994.

[11] S. Oktyabrsky, M. O. Aboelfotoh, and J. Narayan, "Chemistry of Cu-Ge Ohmic Contact Layers to GaAs," *Journal of Electronic Materials*, vol. 25, no. 11, pp. 1673–1683, 1996.

[12] M. O. Aboelfotoh, S. Oktyabrsky, J. Narayan, and J. M. Woodall, "Electrical and microstructural characteristics of Ge/Cu ohmic contacts to n-type GaAs," *Journal of Materials Research*, vol. 12, no. 9, pp. 2325–2331, 1997.

[13] C. S. Zhavzharov, N. Antonchenko, and V. Matyushin, "Modification of parameters of thin copper films under the action of atomic hydrogen," *Visnyk of Lviv University*, vol. 42, pp. 39–46, 2008.

Wideband Lithium Niobate FBAR Filters

Thomas Baron,[1] Eric Lebrasseur,[1] Florent Bassignot,[1] Haixia Wang,[1] Sylvain Ballandras,[1] Ji Fan,[2] Lise Catherinot,[2] Matthieu Chatras,[2] Philippe Monfraix,[3] and Laetitia Estagerie[4]

[1] *FEM TO-ST UMR 6174 CNRS-UFC-ENSMM-UTBM, ENSMM, 26 Chemin de l'Epitaphe, 25030 Besançon Cedex, France*
[2] *Xlim, UMR 6172 CNRS, Université de Limoges, 87060 Limoges, France*
[3] *Thales Alenia Space, CCEL/LPH/EA, Toulouse, France*
[4] *CNES, 18 Avenu Edouard Belin, 31401 Toulouse Cedex 9, France*

Correspondence should be addressed to Thomas Baron; thomas.baron@femto-st.fr

Academic Editor: Priyanka Mondal

Filters based on film bulk acoustic resonators (FBARs) are widely used for mobile phone applications, but they can also address wideband aerospace requirements. These devices need high electromechanical coupling coefficients to achieve large band pass filters. The piezoelectric material $LiNbO_3$ complies with such specifications and is compatible with standard fabrication processes. In this work, simple metal—$LiNbO_3$—metal structures have been developed to fabricate single FBAR elements directly connected to each other on a single chip. A fabrication process based on $LiNbO_3$/silicon Au-Au bonding and $LiNbO_3$ lapping/polishing has been developed and is proposed in this paper. Electrical measurements of these FBAR filters are proposed and commented exhibiting filters with 8% of fractional bandwidth and 3.3 dB of insertion losses. Electrical measurements show possibilities to obtain 14% of fractional bandwidth. These devices have been packaged, allowing for power handling, thermal, and ferroelectric tests, corresponding to spatial conditions.

1. Introduction

Acoustic waves, using surface or bulk propagation, are used in numerous applications in frequency generation, control, or filtering in modern wireless communication systems [1–3]. With the growing demand for multimedia and mobile applications, new generations of telecommunication satellites require higher performances, higher functionalities, and still stronger cost and size constraints [4, 5]. In that context, bulk acoustic waves (BAWs) or film bulk acoustic resonator (FBAR) devices can offer many potentialities for smart RF components or systems. For instance, this technology is now used as alternative to surface acoustic waves (SAWs) filters in handset duplexers for UMTS and DCS standards around 2 GHz with aluminium nitride piezoelectric layers [6]. This material is mainly processed for local oscillators or narrowband filtering operations (<5%) [7–12].

To achieve large bandwidth filters with fractional bandwidth over 10%, higher coupling coefficient materials are required. These materials should be compatible with batch processes as those used for the micro-electromechanical systems (MEMS). The filter conception shows great interest for different orientation cut of $LiNbO_3$. A specific microfabrication process has been developed to achieve such resonators and filters. These devices are characterized to obtain filters behaviour. Finally, packaging of these devices allows testing power, thermal, and ferroelectric behaviours.

2. Conception

BAW filters based on electrical couplings are usually designed with series or shunt resonators. By this way, two main topologies are used: the ladder [13] and the lattice [14] ones. ladder topology (Figure 1) is the easiest method to achieve BAW filter. Such a filter exhibits high selectivity but low out-of-band rejection.

As one can see in Figure 1, separation between resonance and antiresonance frequencies has a direct impact on the bandwidth of Ladder filter. Piezoelectric materials with high electromechanical coupling coefficient allow the conception of large band pass filters. Indeed, (1) describes the dependence

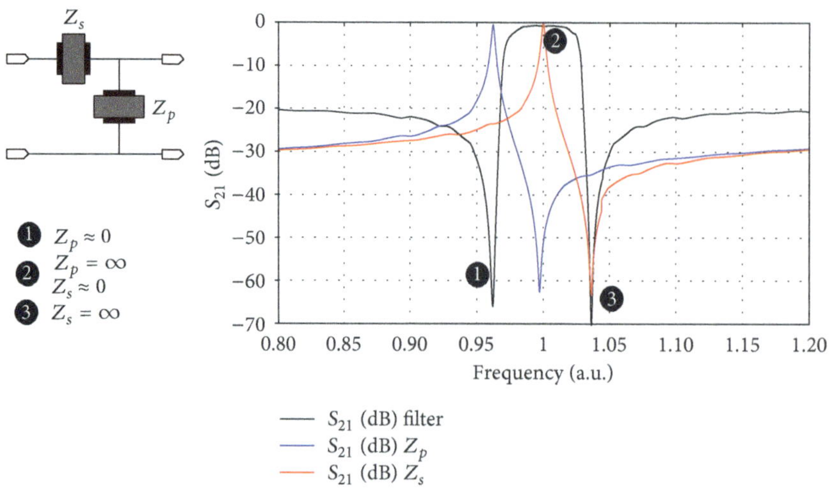

FIGURE 1: 2-pole ladder filter topology.

of k^2 according to the resonance and antiresonance frequencies. To obtain a filter, the antiresonance frequency of the parallel resonator needs to be at the same frequency as that of the resonance of the series resonator

$$k^2 = \frac{\pi^2}{4} \frac{f_p - f_s}{f_p}. \quad (1)$$

Aluminium nitride presents a coupling coefficient around 7% [9]. But these values of electromechanical coupling coefficient are too low for large band specifications. Two $LiNbO_3$ cut orientations show high coupling coefficient for only one kind of volume wave, as shown in Figure 2. $LiNbO_3$ Y⊠l/36° has a coupling coefficient for longitudinal mode of 31.7%, and $LiNbO_3$ Y⊠l/163° has a coupling coefficient for longitudinal mode of 53.3%.

Using these material properties, large band pass filter can be designed. Ladder filter using three resonators based on $LiNbO_3$ Y⊠l/36° membrane presents a fractional bandwidth of 16% (Figure 3).

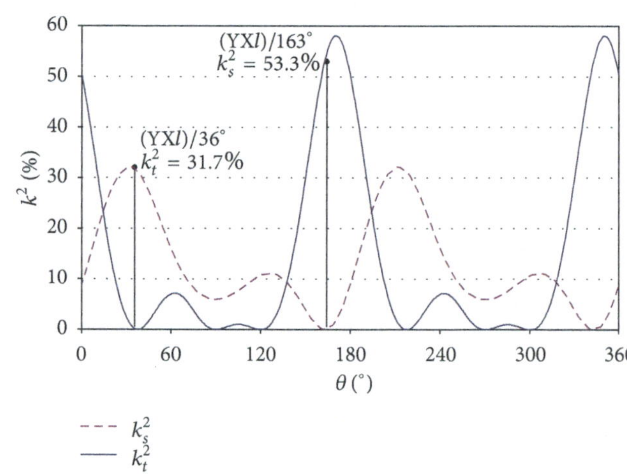

FIGURE 2: The electromechanical coupling coefficient of $LiNbO_3$ for different (Y⊠l) cut orientation and for two volume waves: the longitudinal wave (k_t^2) and the shear wave (k_s^2).

3. Microfabrication

Conception of such devices needs to have membrane of $LiNbO_3$ with bottom and top electrodes. Frequency shift between resonators is obtained with a third metallic layer, used as loading layer. To achieve this structure, the fabrication process is based on gold bonding silicon and $LiNbO_3$ wafer and lapping/polishing step of $LiNbO_3$ wafer, as described in papers mentioned in [15, 16]. This new process needs more technological developments. Technological works were focus on one hand on insertion of different technological steps before, during, and after bonding and grinding technological step, as shown in Figure 4, and on the other hand on industrialization process and stability.

Deep reaction ion etching (DRIE) of silicon is performed to start the fabrication process. On the $LiNbO_3$ wafer, titanium bottom electrodes are patterned. The mass loading layer was patterned on bottom electrode to prevent the membrane from breaking down. Chromium and gold sputtering are deposited on both silicon and $LiNbO_3$ wafers. The Gold and the chromium layers on $LiNbO_3$ wafer are patterned to avoid unwanted electrical connection between the resonators of the filter. The Gold bonding is made with a wafer bonder EVG machine. This bonding is checked by transmitted ultrasonic characterization, as shown Figure 5. In this scan, the red area corresponds to the well-bonding zone and the green areas correspond to the membranes. The lapping and polishing steps are performed to obtain thin membrane around 10 μm, and the bottom aluminium electrodes are patterned.

Development of this process allows obtaining a membrane on $LiNbO_3$, as shown Figure 6. This technology

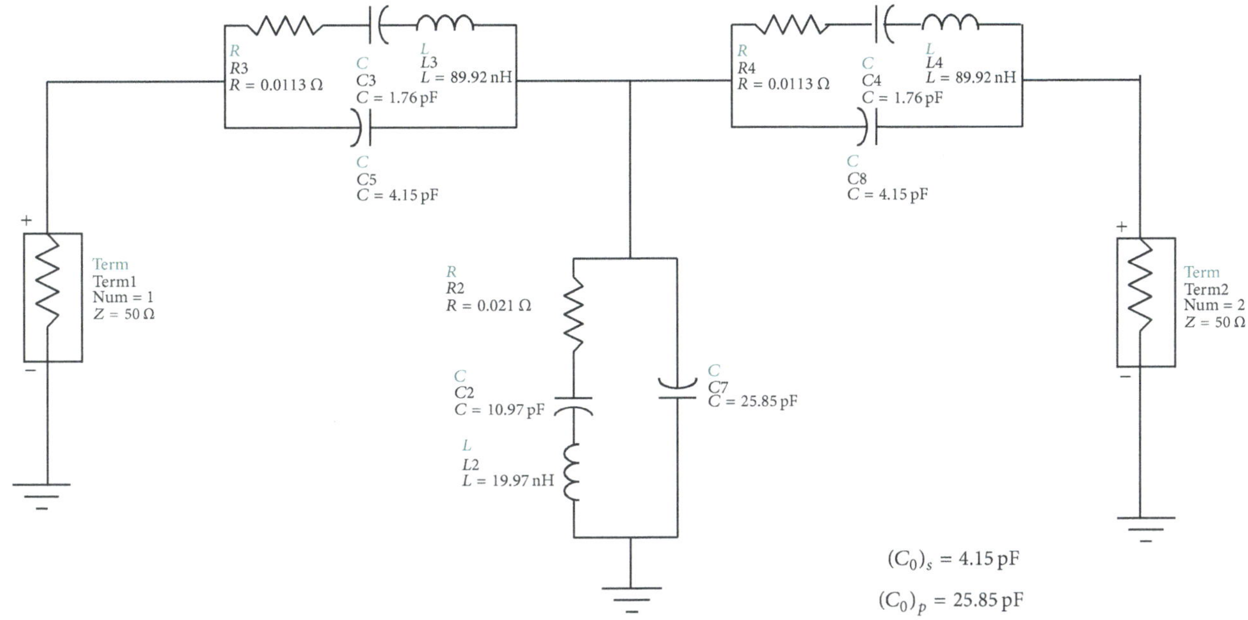

FIGURE 3: 3-pole filter simulation using MBVD model.

presents two major drawbacks: the membrane weakness and the difficulty to tune the final frequency by local etching of the top electrode.

One well-known solution is the SMR (solidly mounted resonator) solution [17]. Bragg mirror can be fabricated using dielectric sputtering layer or stack obtained by the same technology as shown in Figure 7. With Bragg mirror, filter could be easily tuned and could present better mechanical stability. However, Bragg mirror imposes the frequency range of the filter and a new conception is necessary for a new frequency.

4. Measurements and Discussion

First electrical measurements of resonators validate the conception. Figure 8 shows electrical responses of two resonators fabricated with different orientation cut of $LiNbO_3$. The first resonator is based on $LiNbO_3$ Y⊠l/163° membrane. The electromechanical coupling coefficient of the fundamental mode is around 31%, whereas the theoretical one is 37% [18] (Figure 8(a)). The second resonator is done with $LiNbO_3$ Y⊠l/36° membrane. The electromechanical coupling coefficient of the fundamental mode is 20% (Figure 8(b)). Several electrical characterizations are made after different polishing steps during the fabrication process. The membrane thickness decreases clearly during the different polishing steps.

Based on these developments, filter devices have been manufactured. Figure 9 shows the corresponding electrical behaviour for fundamental resonance and large bandwidth measurements. Although in-band ripples were observed due to electromechanical coupling overestimations, the responses of the devices were found in agreement with the expected behaviours. For the fundamental mode, insertion losses of 3.3 dB are obtained with fractional bandwidth around 14%, if we do not take into account the notch due to a wrong thickness ratio between the Pt loading layer and piezoelectric layer. Due to this mismatch of the loading layer, the parallel resonator is not at the good frequency, creating a dip in the pass band of the filter. This measured fractional bandwidth is very similar to the predicted one with our computation method (15.6%). The thickness of the $LiNbO_3$ layer is around 10 μm. Better control of the loading layer will enable the improvement of the filter behaviour.

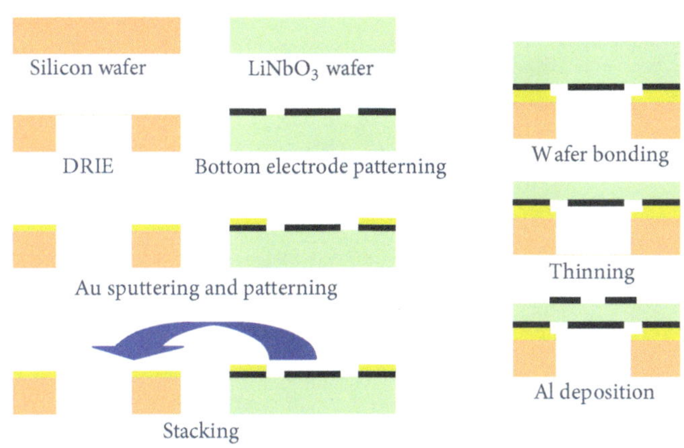

FIGURE 4: Flowchart process for the fabrication of the FBAR based on bonding and lapping technology.

Packaging of these devices allows testing behaviour under different conditions. Firstly, power tests of one resonator and one filter are done at room temperature, −20°C, and +60°C for

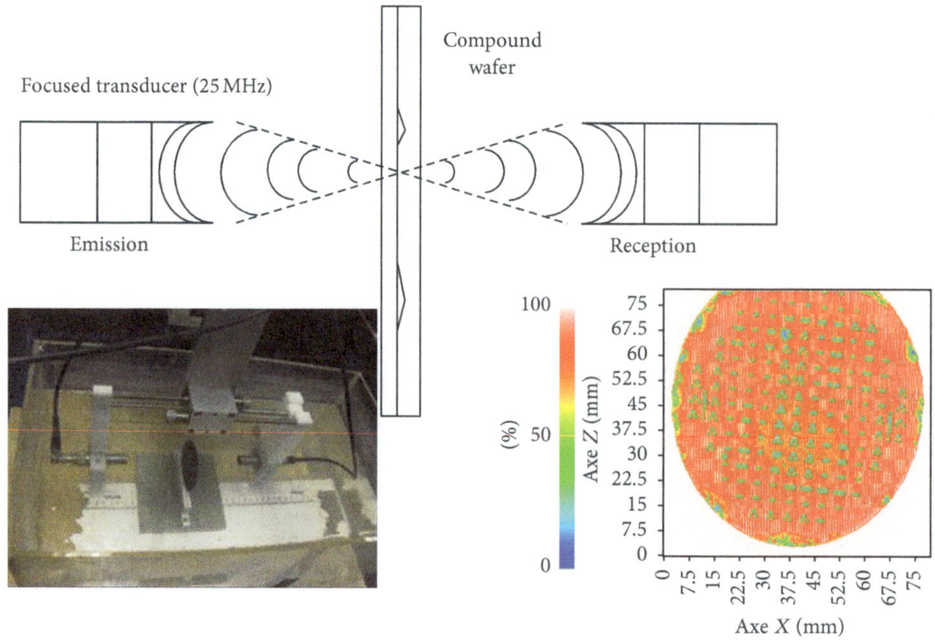

FIGURE 5: Ultrasonic bench for the characterization of silicon and $LiNbO_3$ wafer bonding and scan of the transmission characterization.

FIGURE 6: Resonator based on $LiNbO_3$ membrane.

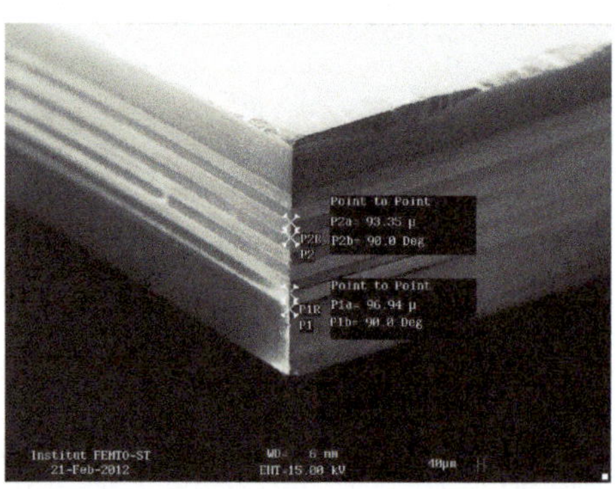

FIGURE 7: Bragg mirror is composed four pairs of quartz and lithium of tantalate.

−10 dBm, 0 dBm, +5 dBm, +10 dBm, and +15 dBm. Filter and resonator electrical responses present no change for all the tested power. Secondly, thermal shocks are performed on one resonator. One hundred thermal chocks between −50°C and 125°C are done. Electrical measurements before and after this test do not reveal the failure of the resonator. The resonator presents exactly the same electrical characteristics before and after the thermal shocks. Although, these tests do not constituted a complete and formal evaluation of reliability, these preliminary results are very promising for the proposed technology.

As shown in Figure 10, relative frequency shift with temperature between −20°C and +60°C is 61 ppm·K^{-1}. To address the radio-frequency applications, compensation of temperature effect will be required. This compensation could be done with voltage control of the frequency and temperature measurement. To check if the ferroelectric effect of $LiNbO_3$ can be used to achieve this function, DC voltage has been added on top electrodes, and frequency variation is observed. The measured relative frequency shift with DC voltage is 6 ppm·V^{-1}. To achieve higher frequency shift due to ferroelectric effect, DC voltage must be applied near Y⊠l/90° direction and in smaller distance. Top electrodes are separated by around 100 μm, whereas thickness membrane is around 10 to 30 μm. So, the temperature compensation needs to be done with DC voltage applied between bottom and top electrodes to obtain more frequency shift with the same DC value. To do that, fabrication process needs to be completed to achieve vias to the bottom electrode.

5. Conclusion

Large band filter applications can be addressed with filters based on $LiNbO_3$ membrane. Generic microfabrication of such filters shows possibilities to adapt process for other devices which can address different applications such as energy harvesting. Some test vehicles have been successfully

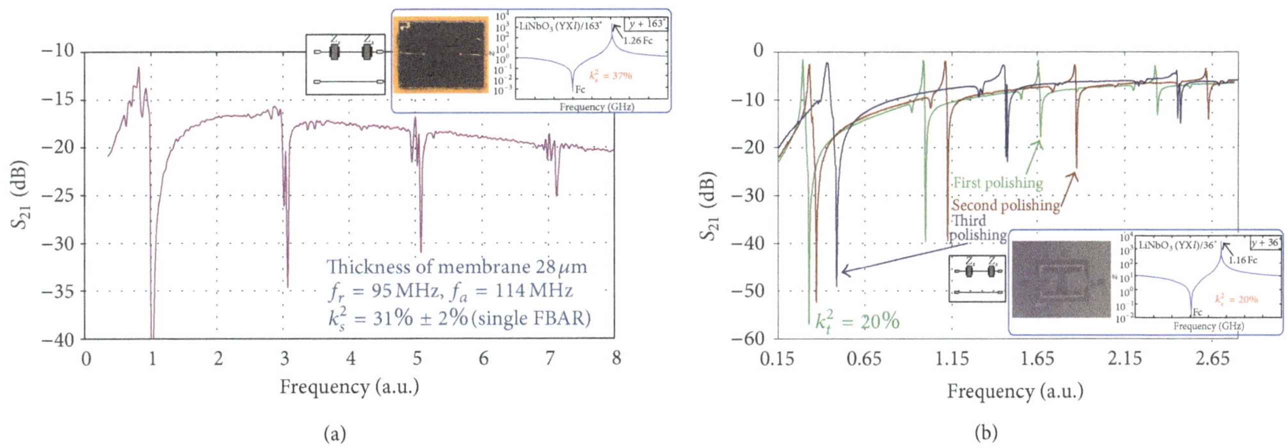

FIGURE 8: Electrical measurement of the first resonator based on $LiNbO_3$ Y⊠l/163° membrane (a) and the second resonator based on $LiNbO_3$ Y⊠l/36° membrane (b).

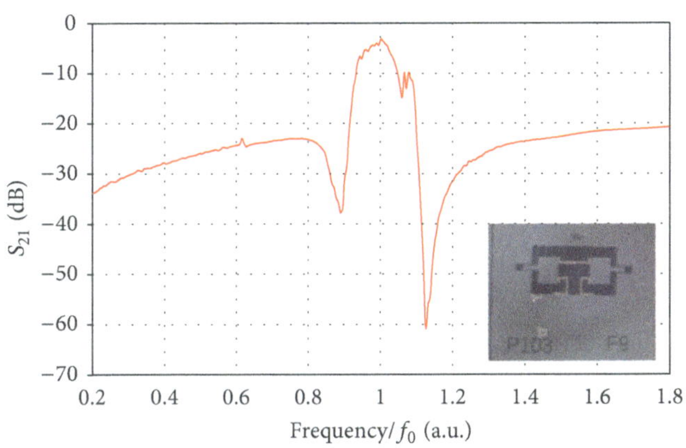

FIGURE 9: Electrical measurements of filters based on $LiNbO_3$ Y⊠l/36° membrane after several polishing steps.

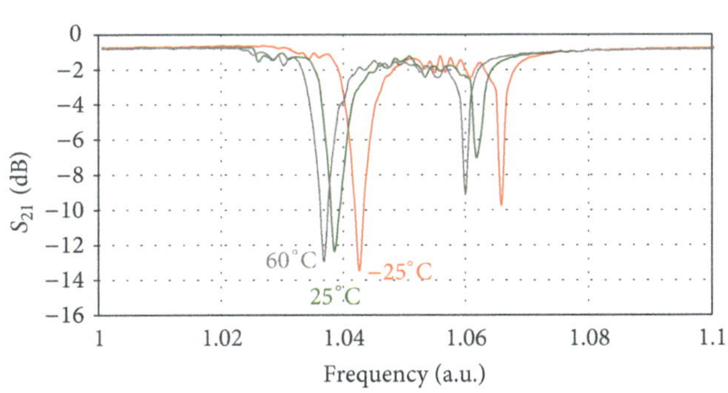

FIGURE 10: Electrical measurements of resonators at three different temperatures.

fabricated. Electrical responses of the filters fabricated confirm potentiality for space applications. Future work on technology will be done to increase frequency, and better control on thickness ratio of piezoelectric layer and mass loading layer will improve fractional bandwidth. However, the filter presents a significant thermal sensibility. This drawback can be easily improved by the use of ferroelectric properties of $LiNbO_3$.

Acknowledgments

The authors want to thank Lionel Portal for the electrical characterizations. This work was supported by the CNES under Grant no. R⊠T R-S08/TC-0001-026 through the FOVETTES Project.

References

[1] K. M. Lakin, G. R. Kline, and K. T. McCarron, "High Q microwave acoustic resonators and filters," in *Proceedings of the IEEE MTT-S International Symposium on Circuits and Systems*, pp. 1517–1520, June 1993.

[2] R. Aigner, "SAW and BAW technologies for RF filter applications: a review of the relative strengths and weaknesses," in *Proceedings of IEEE Ultrasonics Symposium*, Beijing, China, 2008.

[3] A. Flament, A. Frappe, B. Stefanelli et al., "A complete UMTS transmitter using BAW filters and duplexer, a 90 nm CMOS digital RF signal generator and a 0.25 μm BiCMOS power amplifier," *International Journal of RF and Microwave Computer-Aided Engineering*, vol. 21, no. 5, pp. 466–476, 2011.

[4] R. Aigner, J. Kaitila, J. Ellä et al., "Bulk-acoustic-wave filters: performance optimization and volume manufacturing," in *Proceedings of the IEEE MTT-S International Microwave Symposium Digest*, pp. 2001–2004, Philadelphia, Pa, USA, June 2003.

[5] P. Bar, A. Giry, P. Triolet et al., "Full-duplex receiver and PA integration with BAW devices," in *Proceedings of the IEEE Topical Meeting on Silicon Monolithic Integrated Circuits in RF Systems (SiRF '08)*, pp. 9–12, January 2008.

[6] S. Giraud, S. Bila, M. Chatras, D. Cros, and M. Aubourg, "Bulk acoustic wave filters synthesis and optimization for multi-standard communication terminals," *IEEE Transactions on Ultrasonics, Ferroelectrics, and Frequency Control*, vol. 57, no. 1, pp. 52–58, 2010.

[7] A. Reinhardt, G. Parat, E. Defaÿ, M. A⊠l, and F. Domingue, "Acoustic technologies for advanced RF architectures," in *Proceedings of the 8th IEEE International NEWCAS Conference (NEWCAS '10)*, pp. 161–164, June 2010.

[8] R. Lanz and P. Muralt, "Solidly mounted BAW filters for 8 GHz based on AlN thin films," in *Proceedings of the IEEE Ultrasonics Symposium*, pp. 178–181, October 2003.

[9] Akyiama, *Journal of Advanced Materials*, 2009.

[10] ⊠. Zhu, V. Lee, J. Phillips, and A. Mortazawi, "An intrinsically switchable FBAR filter based on barium titanate thin films," *IEEE Microwave and Wireless Components Letters*, vol. 19, no. 6, pp. 359–361, 2009.

[11] H. Yu, W. Pang, H. Zhang, and E. S. Kim, "Ultra temperature-stable BAW resonators with SiO2 compensation layer," *IEEE Transactions on Ultrasonics Ferroelectrics and Frequency Control*, vol. 54, no. 10, pp. 2102–2109, 2007.

[12] K. Y. Hashimoto, "Demands of highly piezoelectric materials for radio frequency acoustic wave devices," *Physica Status Solidi A*, vol. 208, no. 5, pp. 1084–1092, 2011.

[13] L. Catherinot, S. Giraud, M. Chatras et al., "A general procedure for the desing of BAW filters," *International Journal of RF and Microwave Computer-Aided Engineering*, vol. 21, no. 5, pp. 458–465, 2011.

[14] A. A. Shirakawa, J. B. David, P. Vincent, M. El Hassan, E. Kerherv⊠, and A. Cathelin, "A mixed ladder-lattice bulk acoustic wave duplexer for W-CDMA handsets," in *Proceedings of the 14th IEEE International Conference on Electronics, Circuits and Systems (ICECS '07)*, pp. 554–557, December 2007.

[15] T. Baron, J. Masson, J.-P. Romand et al., "BAW pressure sensor on $LiNbO_3$ membrane lapping," in *Proceedings of the 24th European Frequency and Time Forum (EFTF '10)*.

[16] T. Baron, E. Lebrasseur, J. P. Romand et al., "Temperature compensated radio-frequency harmonic bulk acoustic resonators pressure sensors," in *Proceedings of the IEEE Ultrasonics Symposium (IUS '10)*, San Diego, Calif, USA, 2010.

[17] M. Gilles and M. Thomas, "New trends in piezoelectric devices for RF application in mobile phones," in *Proceedings of the 2nd International Workshop on Piezoelectric MEMS*, 2011.

[18] M. Chatras, L. Catherinot, S. Bila et al., "Large band-pass BAW filter for space applications," in *Proceedings of the IEEE Frequency Control and the European Frequency and Time Forum (FCS '11)*, pp. 472–474, 2011.

A Simple Microstrip Bandstop Filter Using Cross-Coupling Stubs

Leung Chiu and Quan Xue

Department of Electronic Engineering, City University of Hong Kong, Hong Kong

Correspondence should be addressed to Leung Chiu, eechiuleung@yahoo.com.hk

Academic Editor: Samir Trabelsi

This paper presents an alterative implementation of a shunt open-circuited stub for multilayer microwave circuit. With making use of the proposed implementation, a simple and compact bandstop filter with a district bandstop characteristics and a first spurious at the third harmonic is proposed. The proposed filter exhibits about 12.5% fractional bandwidth of −10 dB signal rejection at the center frequency of 2.04 GHz and flat group delay at the pass bands. Besides, by cascading a number of the proposed bandstop filters designed at the different frequencies, multiband bandstop filters can be easily realized. A dual- (tri-) band design at the center frequencies of 2 GHz and 3 GHz (and 4 GHz) is designed, realized, and measured. The proposed bandstop filter well suits the nowadays multilayer and compact radio frequency integrated circuit design.

1. Introduction

Bandstop filter is one of the indispensable components for radio frequency and microwave circuits' applications. This filter rejects unwanted signals such as harmonic and spurious signals and allows desired signal to pass through. Recently, a kind of bandstop filter has been integrated into nonlinear circuits such as mixer, oscillator, and amplifier to enhance their performances [1–3].

Electromagnetic band gap (EBG) and resonant structures are found interesting in designing bandstop filter. EBG structure exhibits bandstop effect because of the band gap effect [4, 5]. EBG structure is a periodic structure; hence, the strength of the bandstop effectiveness is determined by the number of stage of the unit cells. It occupies large circuit area if a strong or a significant bandstop effect is required.

Patterned ground structures and resonance cells belong to the family of the resonant structures. They provide bandstop effect by its nature resonance characteristics. For the patterned ground structure, the slotted pattern on the ground plane can be modeled as an *LC* resonator to have strong bandstop effect; however, strong radiation is found on the pattern ground to introduce electromagnetic compatibility and interference issues. Besides, the reported works on the resonance cells cannot achieve low insertion losses at the both lower and higher pass bands at the same time leading to the limited applications [6].

A shunt open-circuited stub is the simplest bandstop structure for microwave circuit design. Figure 1 shows the schematic diagram of the proposed bandstop structure modified from a simple shunt open-circuited stub and its frequency response. It exhibits distinct bandstop characteristics with sharp roll-off and the first spurious response at the third harmonic. It consists of a pair of quarter wavelength open-circuited stubs separated by quarter wavelength, where Z_0 is the port impedance, Z_S is the characteristics impedance of the shunt stub, and C is capacitance between the two stubs. Cross-coupling between the two stubs helps to enhance the bandstop effect by introducing two transmission zeros within the stop band. A bandstop filter with similar schematic diagram was reported in [7]. The idea is realized by planar microstrip lines; however, it is still found that it occupies large circuit area.

In this paper, a multilayer structure to realize the modified bandstop structure as shown in Figure 1(a) is proposed. Recently, low-temperature cofire ceramic and CMOS techniques are found particularly suitable to realize multilayer structure to further reduce the circuit size. The proposed concept provides an alternative solution for the integrated circuit to realize shunt open-circuited stubs and other related circuit elements.

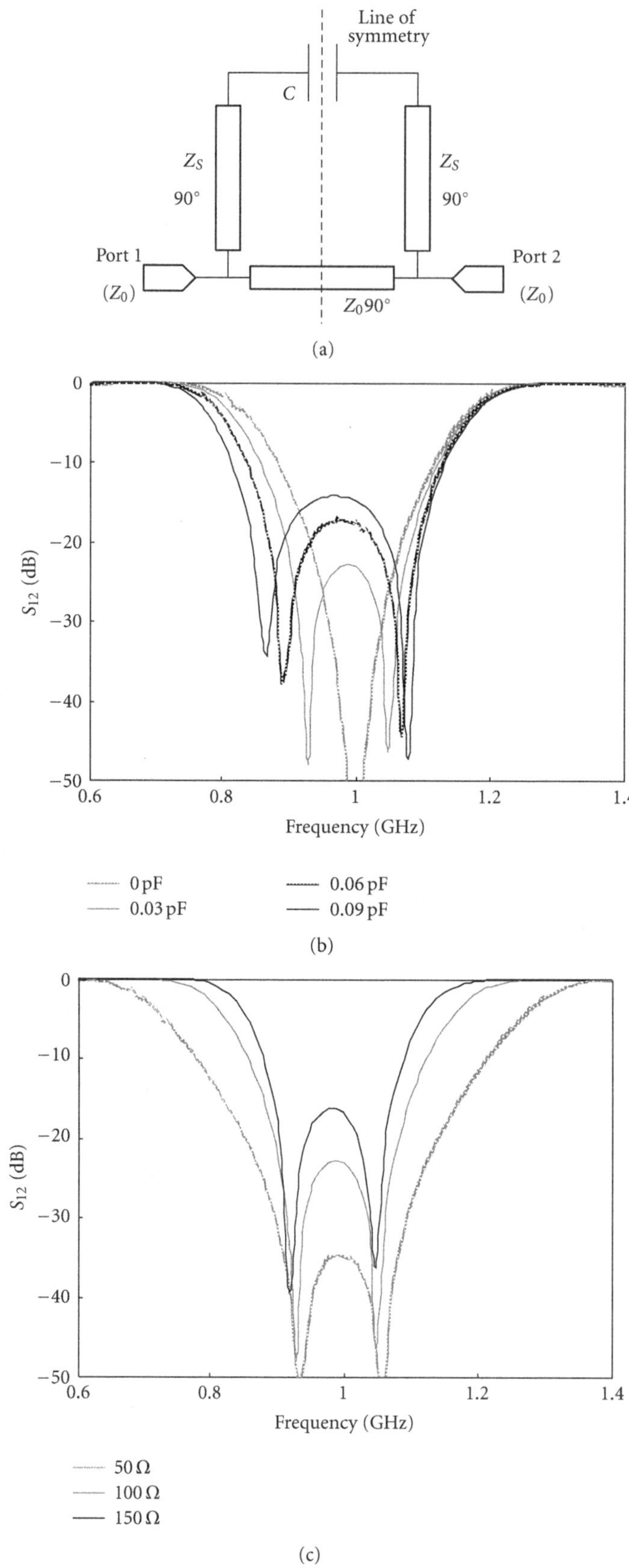

FIGURE 1: (a) Schematic diagram of the proposed bandstop filter. (b) Calculated frequency response of the $|S_{21}|$ with different C and fixed $Z_S = 100\,\Omega$. (c) Calculated frequency response of the $|S_{21}|$ with different Z_S and fixed $C = 0.03$ pF.

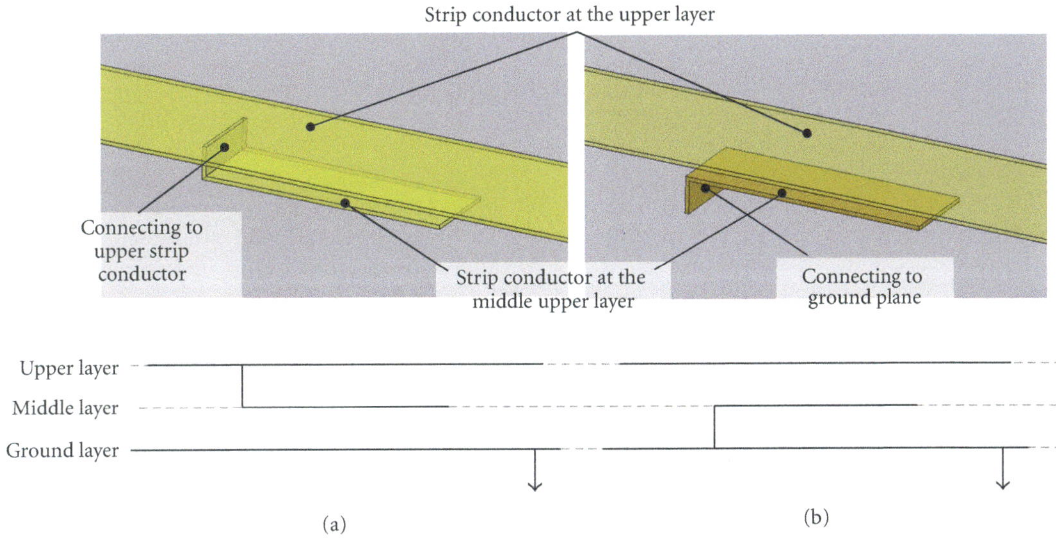

FIGURE 2: 3D and side views of the proposed open-circuit stubs (conductor only). (a) Version 1: the middle conductor connected to the upper-layer conductor. (b) Version 2: the middle conductor connected to the ground layer.

2. Analysis

A simple open circuit shunt stub is the simplest way to realize bandstop filter that stops the signal at the frequency with stub's electrical length of 90°. The roll-off of the filter can be improved by increasing the number of the shunt stub; however, the circuit size will be increased too. Cross-coupling is introduced into the design with two shunt stubs and maintained roll-off performance. The characteristic impedances of the two shunt stubs and the section of transmission line are Z_S and Z_0, respectively. The electrical lengths of all lines are 90°. The cross-coupling is modeled by the capacitor C connecting between the ends of the two stubs.

With the symmetric structure of the proposed filter, even- and odd-mode analysis can be applied. The S-parameters of the structure shown in Figure 1(a) are given by

$$\begin{aligned} S_{11} = S_{22} &= \frac{1}{2}\left\{\frac{1 - j((Z_0/Z_S)\tan 2\theta + \tan\theta)}{1 + j((Z_0/Z_S)\tan 2\theta + \tan\theta)}\right. \\ &\left. + \frac{1 + j[(Z_0/Z_S)((2\omega CZ_S + tan2\theta)/(2\omega CZ_S tan2\theta - 1)) + (1/tan\theta)]}{1 - j[(Z_0/Z_S)((2\omega CZ_S + tan2\theta)/(2\omega CZ_S tan2\theta - 1)) + (1/tan\theta)]}\right\} \\ S_{12} = S_{21} &= \frac{1}{2}\left\{\frac{1 - j((Z_0/Z_S)\tan 2\theta + \tan\theta)}{1 + j((Z_0/Z_S)\tan 2\theta + \tan\theta)}\right. \\ &\left. - \frac{1 + j[(Z_0/Z_S)((2\omega CZ_S + tan2\theta)/(2\omega CZ_S tan2\theta - 1)) + (1/tan\theta)]}{1 - j[(Z_0/Z_S)((2\omega CZ_S + tan2\theta)/(2\omega CZ_S tan2\theta - 1)) + (1/tan\theta)]}\right\}, \end{aligned} \quad (1)$$

where $\theta = (\pi f/4f_0)$. f and f_0 are the center frequency and operating frequency, respectively. $Z_0 = 50\,\Omega$ is chosen and is the same as port impedance. Parametric studies with different values of C and Z_S are shown in Figures 1(b) and 1(c), respectively. The cross-coupling introduces two transmission zeros at the stop band. The larger C results in the stronger cross-coupling between the two stubs. The stop band bandwidth and roll-off can be improved by introducing suitable cross-coupling. With keeping the same C, the stop band bandwidth is adjusted by the characteristic impedance Z_S of the stubs. The bandwidth is reduced with increasing Z_S.

3. Multilayer Shunt Open-Circuited Stub Implementation

Figure 2 shows the geometry of the proposed implementation of microstrip shunt open-circuited stub. It consists of three conductor layers, namely, ground plane, middle strip layer, and upper strip layer. Dielectric substrate is completely filled between the layers. The dielectric substrate used has dielectric constant of 2.65, substrate thickness of each dielectric layer of 0.8 mm, and metal thickness of 0.02 mm. The thicknesses of the two dielectric substrates are

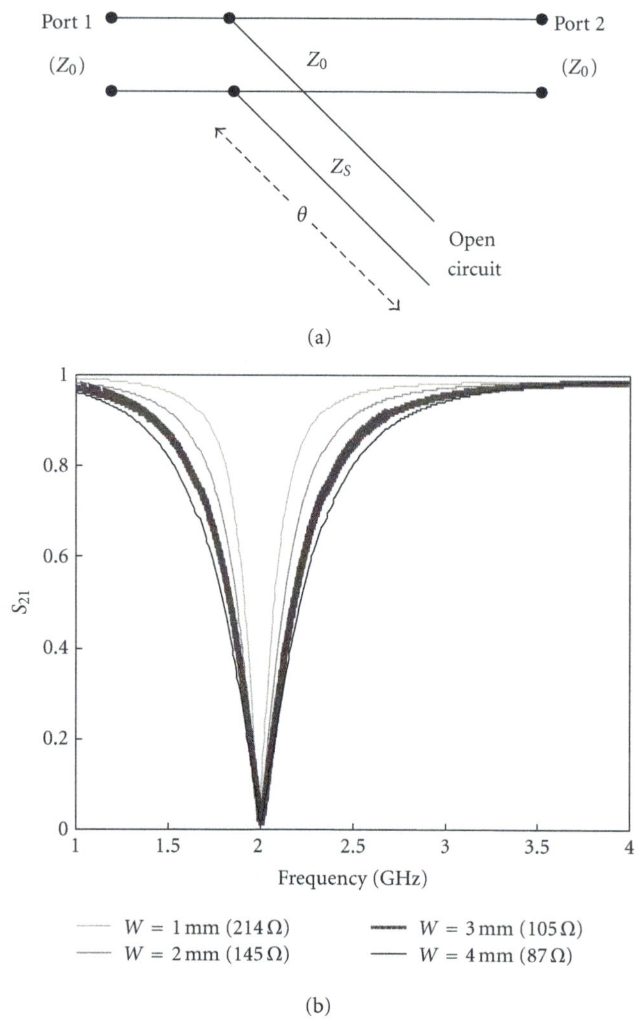

FIGURE 3: (a) The equivalent circuit model of the structure as shown in Figure 2(b). (b) Simulated frequency responses of $|S_{21}|$ of the structure as shown in Figure 2(b) with different width at middle layer and the corresponding equivalent Z_S, where θ is set to be 90° at 2 GHz.

not necessarily equal, while these thicknesses are assigned to be equal in this paper.

The proposed implementation has no difference from the conventional microstrip line, except that there are a middle conductor layer and a metallic via between middle and either upper or ground layer. The two different versions of the implementations as shown in Figures 2(a) and 2(b) are exactly the same theoretically. It is preferable to connect the middle and ground layers by the metallic via as shown in Figure 2(b), since it can have smaller insertion loss and is easier to integrate lumped element or active device on the upper layer than that as shown in Figure 2(a) in the practical implementation.

Figure 3(a) shows the equivalent circuit model of the structure as shown in Figure 2(b). The length of the conductor at the middle is set to be quarter wavelength. By tuning its width, the characteristics impedance of shunt stub in the equivalent circuit can be changed. Figure 3

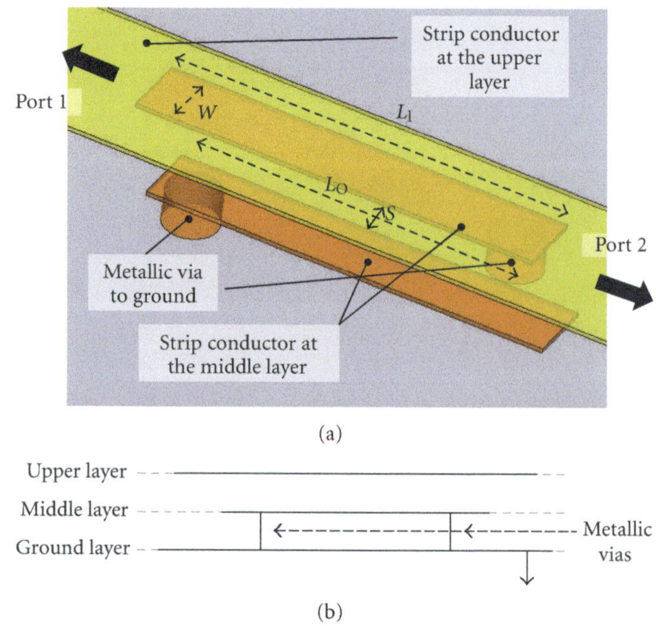

FIGURE 4: 3D and side views of the proposed bandstop filter using the proposed implementation with dimensions of the line width (W = 1.5 mm), the line length (L_l = 23 mm), the line separation (S = 1.44 mm), and the overlapping line (L_O = 20.2 mm).

shows the bandstop characteristics with different conductor width at the middle layer. The parametric study showing the relationship between the conductor width and the corresponding characteristics impedance (Z_S) is shown in Figure 3(b).

4. Application: Bandstop Filter

4.1. Single-Band Design. A working prototype was realized by a standard printed circuit board fabrication technique. The dielectric substrate used is the same as reported in previous section. By arranging two shunt open-circuited stubs in the antiparallel configuration as shown in Figure 4, a bandstop filter is realized. With keeping uniform width of a 50 Ω line, we just tune the line width (W = 1.5 mm), the line length (L_l = 23 mm), the line separation (S = 1.44 mm), and the overlapping line (L_O = 20.2 mm). The entire printed circuit board size is 10 cm by 5 cm.

Figure 5 shows the both simulated and measured S-parameters and group delays of the proposed bandstop filter. The corresponding values of the circuit model as shown in Figure 1(a) are Z_0 = 50 Ω, Z_S = 209 Ω, and C = 3.55 fF, and the center frequency is set to be 2 GHz. It exhibits distinct stop band at the center frequency of 2.04 GHz with the first spurious response at about 6 GHz. The measured −10 dB signal rejection bandwidth is 12.5%. Acceptable agreement between the model, the simulation, and the measurement is obtained. However, the soldering between the via and the conductors are not modeled resulting in the inaccurate prediction especially of the return loss performance as shown in Figure 5(b).

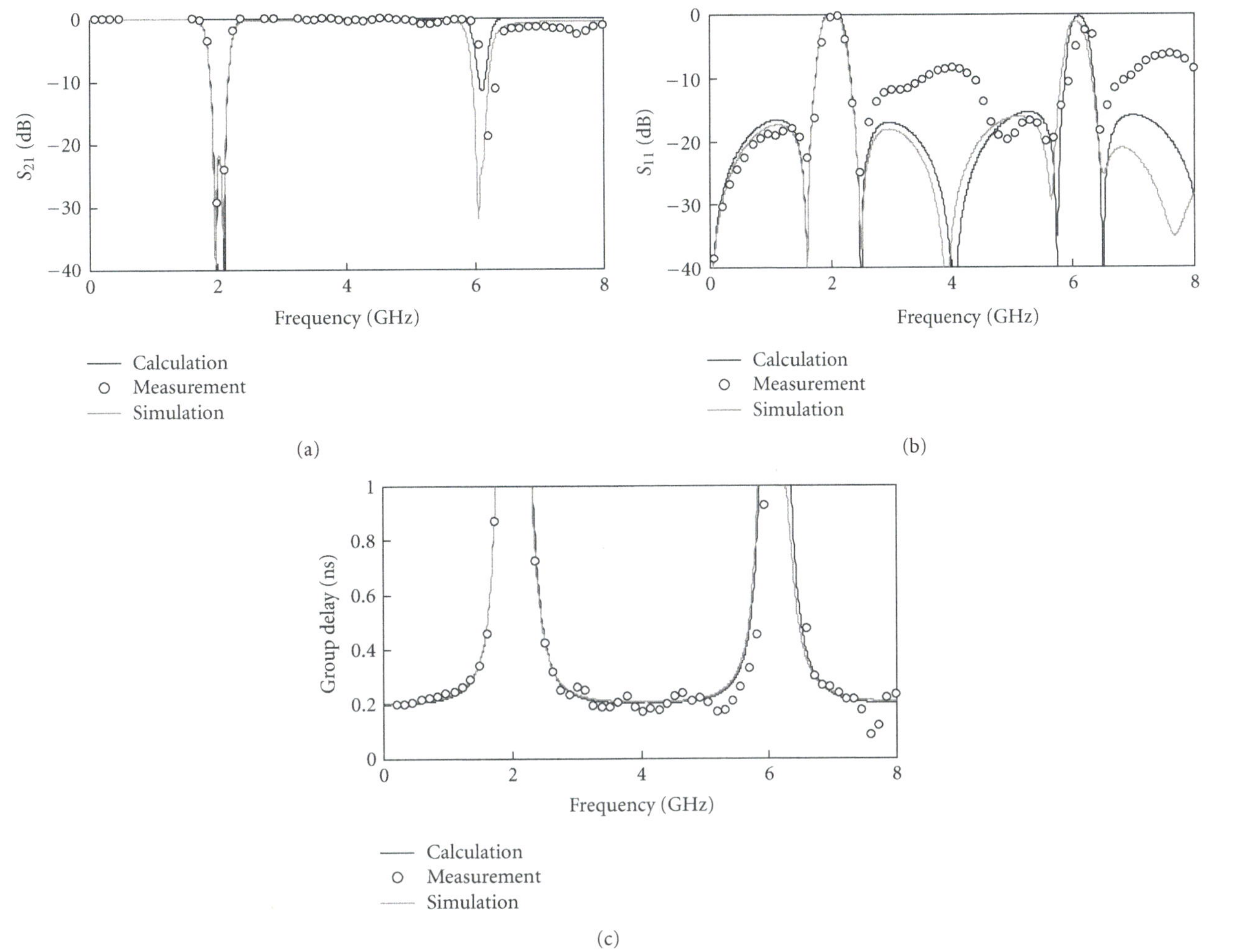

FIGURE 5: Calculated, simulated, and measured performances of the proposed single-bandstop filter. (a) S_{21}, (b) S_{11}, and (c) group delay.

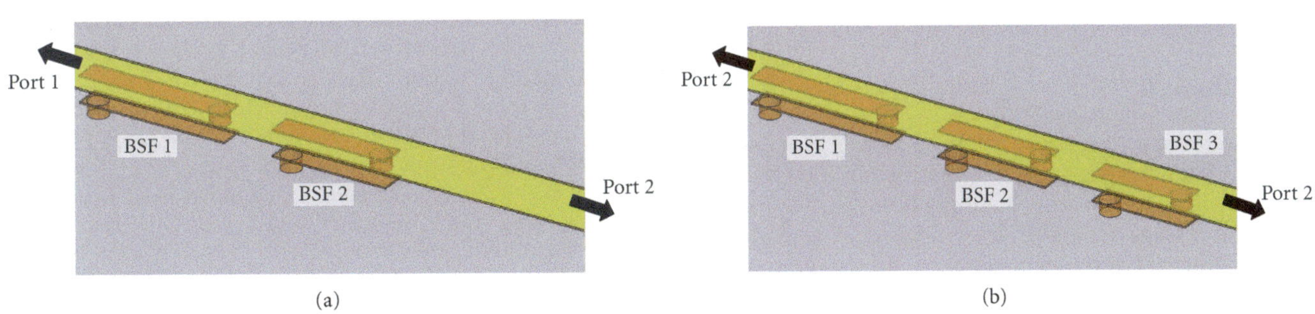

FIGURE 6: Geometries of the proposed multiband bandstop filters consisting a number of subbandstop filters (BSFs). (a) Dual-band bandstop filter. (b) Triband bandstop filter.

4.2. Multiband Design. In order to meet different standards simultaneously, dual-band or multiband designs should be employed to meet the various specializations. By connecting a number of the proposed bandstop filters designed at different frequencies in series connection as shown in Figure 6, a multiband design can easily be realized. A dual-band design at the center frequencies of about 2 GHz and 3 GHz and a triband design at the center frequencies of about 2 GHz, 3 GHz, and 4 GHz are realized to demonstrate the multiband concept. Their simulated and measured performances are shown in Figure 7. The dual-band design offers two distinct stop bands at the center frequencies of 2.04 GHz and 2.87 GHz with fractional bandwidths of 12.5% and 11.8%, respectively, while the triband design offers

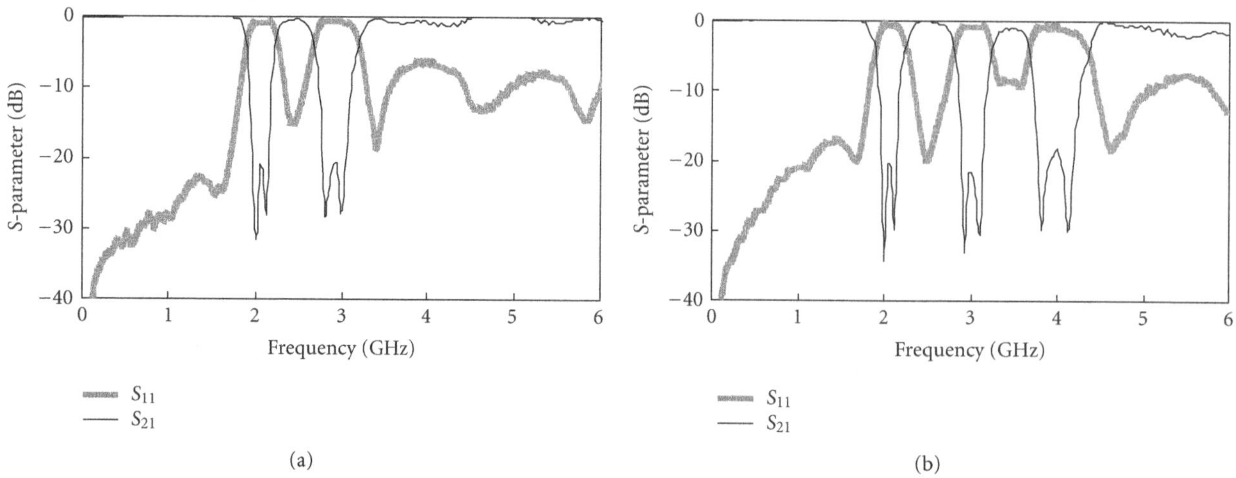

Figure 7: Measured S-parameters of the proposed multiband bandstop filters. (a) Dual-band bandstop filter. (b) Triband bandstop filter.

an extra stop band at the center frequencies of 3.97 GHz with fractional bandwidths of 13.9%. The dimension of each section of the multiband bandstop filter can be designed separately and each section is almost independent to others. The 3 GHz bandstop filter (BSF2) has physical dimensions of $W = 1.50$ mm, $L_l = 16.0$ mm, $S = 1.44$ mm, and $L_O = 13.2$ mm and electrical parameters of $Z_0 = 50\,\Omega$, $Z_S = 209\,\Omega$, and $C = 2.37$ fF. The 4 GHz bandstop filter (BSF3) has physical dimensions of $W = 1.50$ mm, $L_l = 11.5$ mm, $S = 1.44$ mm, and $L_O = 8.70$ mm and electrical parameters of $Z_0 = 50\,\Omega$, $Z_S = 209\,\Omega$, and $C = 1.78$ fF. The two filters were built on the two separated printed circuit boards with physical size of 10 cm by 5 cm. The proposed bandstop filters achieve very compact design without changing the signal line, while some of the previously proposed filters such as [5–7] achieve bandstop performance by etching 2D patterns on the single line or dielectric substrate.

5. Conclusion

A new shunt stub implementation scheme is proposed for bandstop filter design. Distinct bandstop ability with sharp roll-off is achieved by cross-coupling between two quarter-wave shunt stubs separated by a section of transmission line with quarter wavelength long. A flat response of the group delay at the pass bands is achieved. By connecting several bandstop filters designed at the different frequencies, a multiband design can be easily realized. It is a simple bandstop structure, which well suits multilayer integrated circuit design. The proposed filter achieves 1D design, while the previous bandstop filter requires 2D patterns on the signal lines.

Acknowledgments

This work was supported in part by the Science and Technology Development Fund of Macau SAR under Grant 020/2009/A1 and by Shenzhen Science and Technology Planning Project for the Establishment of Key Laboratory in 2009 (project CXB200903090021A).

References

[1] L. Chiu, T. Y. Yum, Q. Xue, and C. H. Chan, "A low-conversion-loss ring-hybrid mixer using TS-CMRC circuitry," *Microwave and Optical Technology Letters*, vol. 41, no. 3, pp. 206–209, 2004.

[2] Q. Xue, K. M. Shum, and C. H. Chan, "Novel oscillator incorporating a compact microstrip resonant cell," *IEEE Microwave and Wireless Components Letters*, vol. 11, no. 5, pp. 202–204, 2001.

[3] T. Y. Yum, Q. Xue, and C. H. Chan, "Amplifier linearization using compact microstrip resonant cell—theory and experiment," *IEEE Transactions on Microwave Theory and Techniques*, vol. 52, no. 3, pp. 927–934, 2004.

[4] V. Radisic, Y. Qian, and T. Itoh, "Broad-band power amplifier using dielectric photonic bandgap structure," *IEEE Microwave and Guided Wave Letters*, vol. 8, no. 1, pp. 13–14, 1998.

[5] V. Radisic, Y. Qian, R. Coccioli, and T. Itoh, "Novel 2-D photonic bandgap structure for microstrip lines," *IEEE Microwave and Guided Wave Letters*, vol. 8, no. 2, pp. 69–71, 1998.

[6] Q. Xue, K. M. Shum, and C. H. Chan, "Novel 1-D microstrip PBG cells," *IEEE Microwave and Wireless Components Letters*, vol. 10, no. 10, pp. 403–405, 2000.

[7] H. Shaman and J. S. Hong, "Wideband bandstop filter with cross-coupling," *IEEE Transactions on Microwave Theory and Techniques*, vol. 55, no. 8, pp. 1780–1785, 2007.

17

Swarm Optimization Methods in Microwave Imaging

Andrea Randazzo

Department of Naval, Electrical, Electronic, and Telecommunication Engineering, University of Genoa, Via Opera Pia 11A, 16145 Genova, Italy

Correspondence should be addressed to Andrea Randazzo, randazzo@dibe.unige.it

Academic Editor: Kristen M. Donnell

Swarm intelligence denotes a class of new stochastic algorithms inspired by the collective social behavior of natural entities (e.g., birds, ants, etc.). Such approaches have been proven to be quite effective in several applicative fields, ranging from intelligent routing to image processing. In the last years, they have also been successfully applied in electromagnetics, especially for antenna synthesis, component design, and microwave imaging. In this paper, the application of swarm optimization methods to microwave imaging is discussed, and some recent imaging approaches based on such methods are critically reviewed.

1. Introduction

Microwave imaging denotes a class of noninvasive techniques for the retrieval of information about unknown conducting/dielectric objects starting from samples of the electromagnetic field they scatter when illuminated by one or more external microwave sources [1]. Such techniques have been acquiring an ever growing interest thanks to their ability of directly retrieving the distributions of the dielectric properties of targets in a safe way (i.e., with nonionizing radiation) and with quite inexpensive apparatuses. In recent years, several works concerned those systems. In particular, their ability to provide excellent diagnostic capabilities has been assessed in several areas, including civil and industrial engineering [2], nondestructive testing and evaluation (NDT&E) [3], geophysical prospecting [4], and biomedical engineering [5].

The development of effective reconstruction procedures is, however, still a quite difficult task. The main difficulties are related to the underlying mathematical problem. In fact, the information about the target are contained in a complex way inside the scattered electric field. In particular, the governing equations turn out to be highly nonlinear and strongly ill posed. Consequently, inversion procedures are usually quite complex and time consuming, especially when high resolution images are needed.

In the literature several approaches have been proposed for solving this problem. In particular, two main classes of algorithms can be identified. Deterministic [6–21] and stochastic strategies [22–33]. Deterministic methods are usually fast and, when converge, they produce high quality reconstructions. However, their main drawback is that they are local approaches, that is, they usually require to be started with an initial guess "near" enough to the correct solution. Otherwise, such approaches can be trapped in local minima corresponding to false solutions. Moreover, in most cases, it is difficult to introduce a priori information in the reconstruction process. On the contrary, stochastic approaches are global optimization methods, that is, they are able to find the global solution of the problem. Furthermore, thanks to their flexibility, they usually easily allow the introduction of a priori information on the unknowns. The main drawback of this class of approaches is their computational burden. However, it should be noted that with the recent growth of computational resources, it can be envisioned that future generation computers will allow faster reconstructions.

Stochastic approaches are usually based on a population of trial solutions that is iteratively updated. Depending on how the population is modified at each iteration, different class of methods can be identified. The "classical" approaches have been developed in order to simulate the evolutionary processes of biological entities. Such kind of methods are now very common in several areas of electromagnetic engineering and, in particular, in nondestructive testing and imaging. Among them, the most successful ones are the genetic algorithm (GA) [34] and the differential evolution (DE) method [35].

A new class of stochastic approaches has been recently introduced by mimicking the collective behavior of real entities such as particles, birds, and ants. Such approaches, usually referred as swarm methods [36, 37], have been proven to be quite effective in several applications, where they outperform standard evolutionary methods. Recently, they are becoming very popular in electromagnetics, too. Several applications, ranging from antenna synthesis to microwave component design, have been proposed in the literature. Moreover, several different approaches, such as the particle swarm optimization (PSO) [38, 39] and the ant colony algorithm (ACO) [40, 41], have been successfully applied to microwave imaging.

In the present paper, the application of swarm algorithms to microwave imaging is discussed, and some of the recent literature results are critically reviewed. The paper is organized as follows. In Section 2, the mathematical framework of optimization problems for microwave imaging is briefly recalled. Section 3 describes the considered swarm methods. Section 4 reviews the applications of such algorithms in the framework of microwave imaging. Finally, conclusions are drawn in Section 5.

2. Microwave Imaging as an Optimization Problem

Microwave imaging approaches aim at retrieving information about unknown objects (e.g., the full distribution of dielectric properties, the shape in the case of conducting targets, the position and size of an inclusion, etc.) starting from measures of the electromagnetic field, they scatter when illuminated by a known incident electric field.

Despite different equations are needed for modeling different problems (e.g., two-dimensional or three-dimensional problems, dielectric or perfectly conducting objects, and so on), it is usually possible to write a relationship between the desired unknown quantities and the measured field (at least in implicit form), that is, an equation of the form

$$F(x) = e, \tag{1}$$

where x is the unknown function describing the searched features of the object (e.g., $x = \epsilon_r$ when dealing with the reconstruction of the distribution of the dielectric permittivity), and e is the measured (vector or scalar) electric field, that is, the known data of the equation. Consequently, the inverse problem can be recast as an optimization problem by defining a cost function of the form

$$f(x) = w\|e - F(x)\|_2^2, \tag{2}$$

where w is a constant normalization parameter. In (2), the standard Euclidean 2-norm has been considered. This is often a common choice in microwave imaging and allows the use of widely studied mathematical tools for the analysis of the convergence and regularization behaviors. However, different norms have been recently proposed, too (e.g., the norm of L^p Banach spaces [42]).

For illustrative purposes, the case of cylindrical dielectric targets embedded in free space is explicitly described in the following. The cylinder axis is assumed to be parallel to the z-axis. A time-harmonic (with angular frequency ω) transverse magnetic (TM) incident field is assumed. Similar expressions can be derived for other configurations (e.g., half-space and multilayer media, three-dimensional vector problems, etc.).

When dealing with inhomogeneous dielectric objects embedded in an infinite and homogeneous medium, the electromagnetic inverse scattering problem is governed by the following two operator equations [1]:

$$\begin{aligned} e_{\text{scatt}}(\mathbf{r}) &= G_{\text{ext}}(ce_{\text{tot}})(\mathbf{r}), \quad \mathbf{r} \in D_{\text{obs}}, \\ e_{\text{inc}}(\mathbf{r}) &= e_{\text{tot}}(\mathbf{r}) - G_{\text{int}}(ce_{\text{tot}})(\mathbf{r}), \quad \mathbf{r} \in D_{\text{inv}}, \end{aligned} \tag{3}$$

where $G_{\text{ext}}(\cdot)(\mathbf{r}) = -k^2 \int_{D_{\text{inv}}}(\cdot)g_0(\mathbf{r},\mathbf{r}')d\mathbf{r}'$, $\mathbf{r} \in D_{\text{obs}}$ (being D_{obs} the observation domain where the scattered electric field is collected), and $G_{\text{int}}(\cdot)(\mathbf{r}) = -k^2 \int_{D_{\text{inv}}}(\cdot)g_0(\mathbf{r},\mathbf{r}')d\mathbf{r}'$, $\mathbf{r} \in D_{\text{inv}}$ (being D_{inv} the investigation area where the target is located), are data and state operators whose kernel is the free-space Green's function g_0 (being $k = \omega\sqrt{\epsilon_0\mu_0}$ the free-space wavenumber), $c(\mathbf{r}) = \epsilon_r(\mathbf{r}) - 1$ is the contrast function (being ϵ_r the space dependent relative complex dielectric permittivity of the investigation area D_{inv}), e_{inc} and e_{tot} are the z-components of the incident and total electric fields inside the investigation area, and e_{scatt} is the z-component of the scattered electric field in the points of the observation domain D_{obs} [1].

In discrete setting, the two equations in (3) can be replaced by the following matrix equations:

$$\begin{aligned} \mathbf{e}_{\text{scatt}} &= \mathbf{G}_{\text{ext}} \operatorname{diag}(\mathbf{c})\mathbf{e}_{\text{tot}}, \\ \mathbf{e}_{\text{inc}} &= \mathbf{e}_{\text{tot}} - \mathbf{G}_{\text{int}} \operatorname{diag}(\mathbf{c})\mathbf{e}_{\text{tot}}, \end{aligned} \tag{4}$$

where $\mathbf{c}$ is an array containing the N unknown coefficients of the expansion of c in a given set of basis functions, $\mathbf{e}_{\text{inc}}$ and $\mathbf{e}_{\text{tot}}$ are arrays of dimensions N containing the coefficients of the incident and total electric fields, and $\mathbf{e}_{\text{scatt}}$ is an array of M elements containing the coefficients used to represent the known scattered field in the measurement domain.

As in the continuous case, the equations in (4) represent an ill-conditioned nonlinear problem. Directly solving this problem is very difficult. However, as previously introduced, it is possible to recast its solution as the minimization of a proper cost function. Usually, the following functional is considered

$$\begin{aligned} f(\mathbf{x}) &= w_D\|\mathbf{e}_{\text{scatt}} - \mathbf{G}_{\text{ext}} \operatorname{diag}(\mathbf{c})\mathbf{e}_{\text{tot}}\|_2^2 \\ &\quad + w_S\|\mathbf{e}_{\text{inc}} - \mathbf{e}_{\text{tot}} + \mathbf{G}_{\text{int}} \operatorname{diag}(\mathbf{c})\mathbf{e}_{\text{tot}}\|_2^2, \end{aligned} \tag{5}$$

where w_D and w_S are weighting parameters, often chosen equal to $w_D = \|\mathbf{e}_{\text{scatt}}\|_2^{-2}$ and $w_S = \|\mathbf{e}_{\text{inc}}\|_2^{-2}$. In this case, the unknown array $\mathbf{x}$ is composed by the elements of $\mathbf{c}$ and $\mathbf{e}_{\text{tot}}$.

Clearly, other forms can be used, too. For example, the two equations can be combined together in order to obtain the following cost function:

$$f(\mathbf{x}) = w_D \left\| \mathbf{e}_{\text{scatt}} - \mathbf{G}_{\text{ext}} \operatorname{diag}(\mathbf{c}) \left(\mathbf{I} - \mathbf{G}_{\text{int}} \operatorname{diag}(\mathbf{c})\right)^{-1} \mathbf{e}_{\text{inc}} \right\|_2^2, \tag{6}$$

which has the advantage of considering only the contrast function coefficients as unknowns. As a drawback, it needs a matrix inversion, leading to a high computational cost.

The cost function can be modified by introducing multiview and multifrequency information. In this case, the target is illuminated by several incident fields (e.g., generated by S sources located all around the objects and operating at F different frequencies), and the cost function becomes

$$\begin{aligned} f(\mathbf{x}) &= w_D \sum_{f=1}^{F} \sum_{s=1}^{S} \left\| \mathbf{e}_{\text{scatt}}^{f,s} - \mathbf{G}_{\text{ext}}^{f,s} \operatorname{diag}(\mathbf{c}) \mathbf{e}_{\text{tot}}^{f,s} \right\|_2^2 \\ &\quad + w_S \sum_{f=1}^{F} \sum_{s=1}^{S} \left\| \mathbf{e}_{\text{inc}}^{f,s} - \mathbf{e}_{\text{tot}}^{f,s} + \mathbf{G}_{\text{int}}^{f,s} \operatorname{diag}(\mathbf{c}) \mathbf{e}_{\text{tot}}^{f,s} \right\|_2^2, \end{aligned} \tag{7}$$

where the indexes f and s denote that the corresponding quantities are related to the fth operating frequency and the sth illumination.

3. Swarm Optimization Algorithms

Swarm algorithms belong to the class of optimization methods, that is, they find the minimum of a given cost function $f(x)$. Similarly to other evolutionary approaches, they usually allow reaching the global optimum, thus avoiding to find a suboptimal solution corresponding to a local minima. Moreover, they are able to easily incorporate constraints on the search space. However, while evolutionary algorithms are inspired by the genetic adaptation of organisms, swarm methods exploit their collective social behavior.

In order to define a general framework for swarm methods, let us consider a cost function $f : S \subseteq \mathbb{R}^G \rightarrow [0, +\infty)$ to be minimized (or maximized). For sake of simplicity, the case of bound constraints (i.e., $l_g \leq x_g \leq u_g$, being l_g and u_g the lower and upper bounds on the gth component of $\mathbf{x}$) is considered in the following. However, the unknown array $\mathbf{x} = [x_1, x_2, \ldots, x_G]^t \in S \subseteq \mathbb{R}^G$ can be subjected to arbitrary constraints.

Swarm algorithms are usually iterative methods based on a population of P trial solutions $\wp_k = \{\mathbf{x}_p^{(k)}, p = 1, \ldots, P\}$ (being k the iteration number) representing P agents inspired from real world (e.g., particles, birds, ants, etc.). The population is iteratively modified according to rules aimed at mimicking the natural behavior of those agents.

In the framework of electromagnetic imaging, the following swarm algorithms have been mainly considered.

(i) Particle swarm optimization.

(ii) Ant colony optimization.

(iii) Artificial bee colony optimization.

Apart from the standard approaches, hybridization with other methods (e.g., local optimization methods, other evolutionary approaches, machine learning algorithms, etc.) has also been proposed in the literature.

In the following sections, some information about the basic versions of those methods are briefly recalled.

3.1. Particle Swarm Optimization (PSO). PSO is inspired by the behavior of flocks of birds and shoals of fish [38, 39]. Each entity moves through the space of solutions with a velocity that is related to the locations and cost function values of the members of the swarm. In particular, the basic PSO algorithm [38] considers a set of $p = 1, \ldots, P$ "particles" characterized, at each iteration k, by their positions $\mathbf{x}_p^{(k)}$ and velocity $\mathbf{v}_p^{(k)}$. If no a priori information is available, usually the algorithm is initialized by using random values, that is,

$$\begin{aligned} x_{p,g}^{(0)} &= l_g + \left(u_g - l_g\right) U(0,1), \\ v_{p,g}^{(0)} &= l_g^v + \left(u_g^v - l_g^v\right) U(0,1), \\ g &= 1, \ldots, G, \quad p = 1, \ldots, P, \end{aligned} \tag{8}$$

where l_g and u_g are the lower and upper bounds for the gth component of particles' positions, l_g^v and u_g^v are the lower and upper bounds for the gth component of the particles' velocities, and $U(0,1)$ is a function returning a random variable uniformly distributed between 0 and 1. Clearly, if some a priori information is available, the initialization scheme can be modified for taking it into account (e.g., in microwave imaging, when the aim is the identification of one or more localized objects, it is possible to generate random targets and use them as starting random solutions).

The trial solutions are iteratively updated by using the following two-step scheme (as also shown in the flow chart in Figure 1):

$$\begin{aligned} \mathbf{v}_p^{(k+1)} &= \omega^{(k)} \mathbf{v}_p^{(k)} + \eta_1 U(0,1) \left(\mathbf{p}_p - \mathbf{x}_p^{(k)}\right) \\ &\quad + \eta_2 U(0,1) \left(\mathbf{g} - \mathbf{x}_p^{(k)}\right), \\ \mathbf{x}_p^{(k+1)} &= \mathbf{x}_p^{(k)} + \mathbf{v}_p^{(k)}, \end{aligned} \tag{9}$$

where $\omega^{(k)}$ is the inertia parameter, η_1 and η_2 are acceleration coefficients, and $\mathbf{p}_p$ and $\mathbf{g}$ are the best solution achieved by the pth particle and by the whole swarm so far, respectively. The two acceleration terms in the velocity update can be thought as two elastic forces with random magnitude attracting the particles to the best solutions achieved so far by each entity and by the whole swarm, respectively. After the new solutions are generated the values of $\mathbf{p}_p$ and $\mathbf{g}$ are updated.

The procedure is iterated until some predefined stopping criteria is fulfilled. In particular, the stopping criteria can be composed by several conditions. Some of the most commonly used are the following.

(i) Maximum number of iterations: the method is stopped when a given number of iteration $k_{\max}$ is reached.

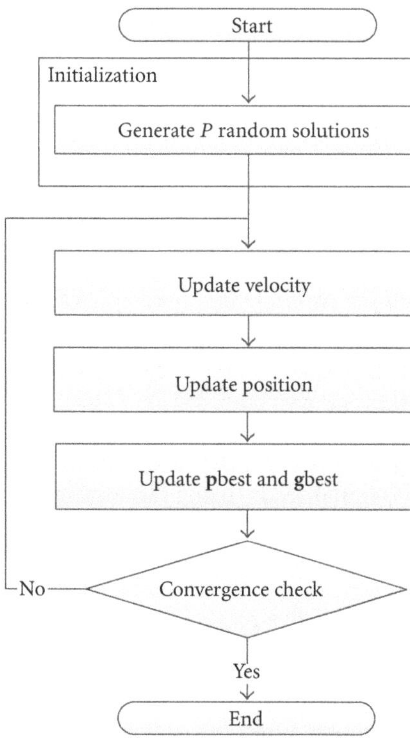

FIGURE 1: Flow chart of the PSO algorithm.

(ii) Cost function threshold: the method is stopped when the value of cost function of the best trial solution falls below a given threshold f_{th}.

(iii) Cost function improvement threshold: the method is stopped if the improvement of the cost function of the best individual after k_{th} iterations is below a fixed threshold Δf_{th}.

3.2. Ant Colony Optimization. Ant colony optimization (ACO) is a recently developed swarm optimization method based on the behavior of ants, and, in particular, on how they find the optimal path for reaching the food starting from their nest [40]. Initially, ants explore the area around their nest in a random manner searching for food. When a food source is found, ants evaluate it and bring back some food to the nest depositing a pheromone trail on the ground during the trip. The amount of pheromone depends on the quantity and quality of the food, and it is used to guide other ants to the food source. It has been found that the pheromone trails allow ants to find the shortest path between nest and food sources. On the basis of such behavior, ACO was initially designed for solving hard combinatorial problems, such as the traveling salesman problem [41]. Successively, the original algorithm has been extended for considering different applications. In particular, several efforts have been devoted to the extension to continuous domains, and some different versions of the algorithm have been proposed [43–45]. In the following, the $\text{ACO}_{\mathbb{R}}$ version [46] is described. Such algorithm can be considered as composed by three functional blocks, as shown in Figure 2: initialization, solution construction, and pheromone update. The algorithm iterates until a predefined stopping criteria is satisfied.

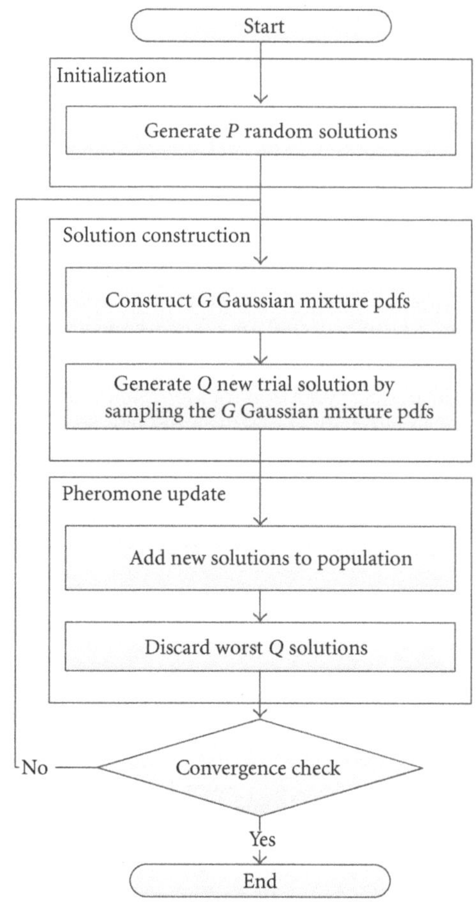

FIGURE 2: Flow chart of the $\text{ACO}_{\mathbb{R}}$ algorithm.

In the initialization block, the initial population $\wp_0 = \{\mathbf{x}_p^{(0)}, p = 1, \ldots, P\}$ is created by generating P random trial solutions $\mathbf{x}_p^{(0)} = [x_{p,1}^{(0)}, \ldots, x_{p,G}^{(0)}]^t$. Similarly to the PSO algorithm, when no a priori information is available and assuming boundary constraints, the gth components of the pth trial solution are generated by sampling a uniform distribution as defined in (8). In this case, too, if additional a priori information is available, it can be included in the initialization procedure.

The solution construction block is used to generate new trial solutions. In particular, at the kth iteration, Q new solutions $\widetilde{\mathbf{x}}_q^{(k)}$, $q = 1, \ldots, Q$ are generated by sampling a set of Gaussian mixture probability density functions. In particular, the probability density function of the gth component is built as

$$G_g^{(k)}(x) = \sum_{p=1}^{P} w_p \frac{1}{s_{p,g}^{(k)}\sqrt{2\pi}} e^{-(x-m_{p,g}^{(k)})^2/2(s_{k,p,g}^{(k)})^2}, \quad (10)$$

where the weighting parameters w_p, $p = 1, \ldots, P$, are given by

$$w_p = \frac{1}{\rho P\sqrt{2\pi}} e^{-(p-1)^2/2\rho^2P^2}, \quad p = 1, \ldots, P, \quad (11)$$

and the mean, $m_{p,g}^{(k)}$, and standard deviation, $s_{p,g}^{(k)}$, of the Gaussian kernels are given by

$$m_{p,g}^{(k)} = x_{p,g}^{(k)}, \quad p = 1, \ldots, P, \; g = 1, \ldots, G,$$
$$s_{p,g}^{(k)} = \xi \sum_{i=1}^{P} \frac{\left| x_{p,g}^{(k)} - x_{i,g}^{(k)} \right|}{P-1}, \quad p = 1, \ldots, P, \; g = 1, \ldots, G. \tag{12}$$

In the previous equations, ρ is the pheromone evaporation rate, and ξ is a scaling parameter. Such quantities are key parameters of the ACO algorithm [46], and the best choice for their values depends on the specific application.

The pheromone update block is responsible of the population update. In particular, two mechanism are used to build the new population $\wp_{k+1}$ of the $(k+1)$th iteration:

(i) positive update: a temporary population is created by adding newly created solutions to the solution archive. A pool of $P + Q$ trial solutions $\widetilde{\wp}_{k+1} = \{x_{k,p}, p = 1, \ldots, P\} \bigcup \{\widetilde{x}_{k,q}, q = 1, \ldots, Q\}$ is then obtained,

(ii) negative update: the worst Q elements of $\widetilde{\wp}_{k+1}$ (i.e., those characterized by the higher values of the cost function) are discarded. The remaining P solutions constitute the new population $\wp_{k+1}$.

In order to speed up this stage, usually the solution archive is ordered on the basis of the cost function, that is, $f(x_{k,1}) \leq f(x_{k,2}) \leq \cdots \leq f(x_{k,P})$. The two previous blocks are iteratively applied since some predefined stopping criteria is fulfilled. In this case, too, the stopping criteria can be a combination of several different conditions.

3.3. Artificial Bee Colony Optimization. The artificial bee colony algorithm is a swarm optimization method introduced by Karaboga [47] and Karaboga and Basturk [48] and inspired by the foraging behavior of honey bees. In particular, ABC is based on the model proposed in [49], which defines two main self-organizing and collective intelligence behaviors: recruitment of foragers for working on rich food sources and abandonment of poor sources. In ABC, a colony of artificial bees search for rich artificial food sources (representing solutions of the considered optimization problem) by iteratively employing the following two strategies: movement towards better solutions by means of a neighbor search mechanism and abandonment of poor solutions that cannot be further improved.

In the following, an artificial colony, composed by C entities, is considered. Bees can be classified in three groups: employed (i.e., already working on a known food source), onlooker (i.e., waiting for a food source), and scout (i.e., randomly searching for new sources) bees. The trial solutions are the food sources associated with the employed bees. Let us denote by P the number of employed bees (corresponding to the number of trial solutions). Often, such number is set equal to $P = C/2$, and the number of onlooker bees is chosen equal to P. A block diagram of the method is shown in Figure 3.

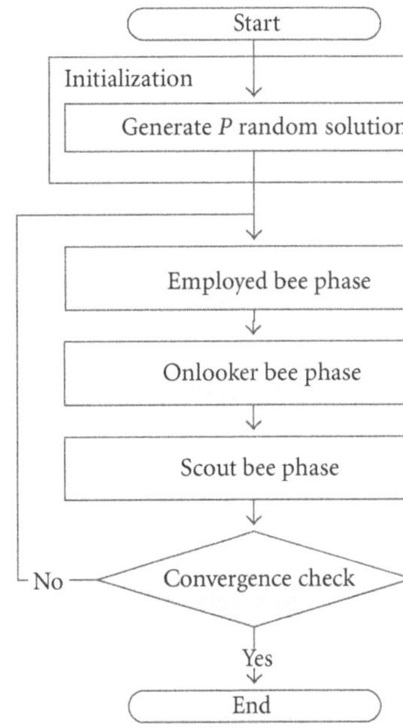

FIGURE 3: Flow chart of the *ABC* algorithm.

In the initialization phase, the P trial solution are randomly generated, for example, by using a relationship similar to that in (8). The cost function of all trial solutions is evaluated and stored. In the employed bee phase, the employed bees modify their trial solutions according to

$$x_{p,g}^{(k+1)} = x_{p,g}^{(k)} + U(0,1)\left(x_{p,g}^{(k)} - x_{h,g}^{(k)}\right), \tag{13}$$

where g and h are randomly chosen.

In the onlooker bee phase, each onlooker bee select a food source in a probabilistic way. In particular, the probability of choosing the pth food source is given by

$$p_p = \frac{f_{\text{fit}}\left(\mathbf{x}_p^{(k)}\right)}{\sum_{p=1}^{P} f_{\text{fit}}\left(\mathbf{x}_p^{(k)}\right)}, \tag{14}$$

where f_{fit} is a fitness function defined as $f_{\text{fit}}(\mathbf{x}) = 1/(f(\mathbf{x})+1)$. The food sources selected by the onlooker bees are further improved by using (13).

If a food source is not improved after a predefined number of iterations K_{lim} (food source limit), that is, $f(\mathbf{x}_p^{(k)}) \geq f(\mathbf{x}_p^{(k-q)})$ for $q = 1, \ldots, K_{\text{lim}}$, it is abandoned by its employed bee. Such bee becomes a scout bee and starts searching for a new food source randomly. Consequently, in the scout bee phase, employed bees working on abandoned sources generate new trial solution in a random way (i.e., by using the same relationship used in the initialization phase). After a new solution is found, the scout bees are reverted to employed bees, and they start working on the new food sources. Clearly, at the beginning of the optimization process, no scout bees are present and consequently, for the first iterations, only the solutions initially discovered in the initialization phase are processed.

TABLE 1: Overview of swarm algorithms applications in microwave imaging.

Object type	Material	Illumination type	Method	Reference
1D profile	Dielectric	Time domain	PSO	[50–52]
2D cylinders	Dielectric	Time domain	PSO	[53, 54]
2D cylinders	Dielectric	Single frequency	PSO	[55–58]
2D cylinders	PEC	Single frequency	PSO	[59, 60]
3D objects	Dielectric	Single frequency	PSO	[58, 61]
3D objects	Dielectric	Single frequency	μPSO	[62]
2D cylinders	Dielectric	Time domain	APSO	[63]
2D cylinders	PEC	Time domain	APSO	[64, 65]
2D cylinders	PEC	Singe frequency	PSO-SA	[66]
2D cylinders	Dielectric	Single frequency	HSPO	[67]
2D cylinders	Dielectric	Single frequency	PSO-RBF	[68]
2D cylinders	Dielectric	Single frequency	IMSA-PSO	[69–73]
3D objects	Dielectric	Single frequency	IMSA-PSO	[74, 75]
2D cylinders	Dielectric	Single frequency	ACO	[28, 76]
2D cylinders	Dielectric	Single frequency	ACO-LSM	[77]
3D objects	Dielectric	Single frequency	ACO-LSM	[78, 79]
3D objects	Dielectric	Time domain	ABC	[80]
2D cylinders	Dielectric	Single frequency	ABC	This paper

4. Application of Swarm Optimization to Microwave Imaging

Swarm algorithms have been used for solving different types of microwave imaging problems. In particular, two-dimensional and three-dimensional dielectric and PEC targets have been considered in the literature. Moreover, both single-frequency, multifrequency, and time-domain incident radiation has been used. An overview of papers proposing swarm optimization methods in microwave imaging is given in Table 1. Specific details are provided in the following sections.

4.1. PSO. PSO has been extensively used in electromagnetic problems [39]. Several works proposed PSO, also with enhancement to the standard algorithm, for microwave imaging problems. In [50, 51], the reconstruction of one-dimensional dielectric profiles, illuminated by a Gaussian pulse (plane waves are assumed), is considered. Both noiseless and noisy data have been used. Two size of the populations are considered equal and twice the number of unknowns. The acceleration coefficients of the PSO are $\eta_1 = \eta_2 = 0.5$, and the inertia parameters is initially set to 1 and decreased linearly to 0.7 in 500 iterations (maximum number of iterations). A comparison with the DE is also provided. In the reported numerical results, the PSO shows slightly better convergence rate, but the DE allows obtaining a more precise reconstruction. A similar approach is proposed in [52], too.

In [53], the authors propose the use of the PSO for the localization of dielectric circular cylinders under a TM time-domain formulation. The unknowns are the position, size, and dielectric properties of the target. Good reconstructions are obtained by using five particles.

In [54], the reconstruction of homogeneous dielectric cylinders is considered. The external shape of the cylinder is described by using a spline representation. The unknowns are the parameters of the spline and the dielectric permittivity. The proposed approach is able to reconstruct such quantities with errors less than about 7% (shape) and 3% (permittivity) for signal-to-noise ratio above 10 dB.

In [55], the reconstruction of the distribution of the relative dielectric permittivity of 2D dielectric structures is concerned. A single-frequency multiview TM illumination is considered. The cost function defined in (7) is used. A population size of twice the unknowns number is used; the acceleration coefficients are randomly generated in the range [0, 2], and the inertia parameter is set equal to 0.4. The reported results show that the PSO is able to provide good reconstruction of the considered objects. Moreover, comparisons with a genetic algorithm and with the conjugate gradient (CG) method are provided. For the considered case, PSO outperforms both GA and CG (the mean relative reconstruction errors are 1.7%, 1.8%, and 5.7%, resp.). A similar approach is proposed in [56], too, providing similar conclusions. In [57], the PSO is used to find the parameters of a crack in the outer layer of a two-layer dielectric cylinder. The direct solver is based on a finite difference frequency Domain (FDFD) scheme. Good agreements are obtained for several positions of the cracks. The same approach is also applied in [58] to the detection of a tumor inside a model of breast. Both 2D and 3D simplified configurations are assumed. The unknowns are the position and shape of the malignant inclusion.

The reconstruction of the shape of PEC cylinders is considered in [59, 60]. A cubic-spline-based representation is used for defining the shape of the cylinder. The target is illuminated by TM waves impinging from 7 directions uniformly distributed around it and, for each illumination, the scattered field is collected in 32 points. Single-frequency operation is assumed. The maximum radius is approximately 1.5λ (being λ the wavelength in the background medium). Under such assumptions, good reconstructions have been obtained by considering 10 control points for the splines, $\eta_1 = \eta_2 = 0.5$, and population sizes ranging from 10 to 60. The inertia is initialized to 1.0 and decreased to 0.7 after 200 iterations (maximum number of iterations). A comparison with DE is also performed. Both algorithms provide good results, although DE, for the considered cases, usually produces lower values of the cost function and of the reconstruction error.

In [61], 3D objects are reconstructed. The measurement configuration simulates those employed for breast cancer detection, that is, four circular arrays of 36 antennas are located around a cubic investigation domain. In this reference, the PSO-based approach is able to retrieve a small centered inclusions.

In [62], a μPSO algorithm is proposed for tackling the high dimensionality of the microwave imaging problems. Satisfactory reconstructions are obtained in the reconstruction of a 3D model of breast with a malignant inclusion by using only 5 particles. Comparisons with the standard PSO (with swarm size of 25 particles) show that the new approach is able to obtain comparable results, but with a smaller population size.

In [63], an asynchronous PSO (APSO) is proposed for the reconstruction of the location, shape, and permittivity of a dielectric cylinder illuminated by TM pulses. The direct solver is based on a finite difference time-domain (FDTD) scheme. The main difference with respect to the standard PSO is the population updating mechanism. In the APSO, the new best position is computed after every particle update, and it is used in the following updates immediately. Consequently, the swarm reacts more quickly. The same approach is used in [64, 65] for the identification of the shape of PEC cylinders in free space and inside a dielectric slab. In such cases, the APSO is able to correctly identify the shape of the targets with an error of about 5% (in presence of noise on the data) and it shows better performance with respect to standard PSO.

Some hybrid versions of the PSO have also been proposed in the literature. In [66], PSO is combined with simulated annealing (SA) for exploiting the exploration properties of PSO and the exploitation ability of SA. The reported results concern the reconstruction of the external shape of a cylinder under multiview TM illumination and show that the hybrid approach allows reaching better results than the standard one (reconstructions errors were 0.0014 and 0.072, resp.). In [67], a hybrid PSO (HPSO) is used for reconstructing dielectric cylinders under TM illuminations. The difference from the standard PSO is mainly related to the use of a particle swarm crossover for enhancing trial solutions. In [68], the PSO is combined with radial basis function (RBF) networks. In particular, the RBF is used to obtain an estimate of the dielectric properties of two-dimensional cylinders under single-frequency TM illumination. The PSO is employed to efficiently training the RBF starting from a set of simulated configurations.

In [69–71], an integrated multiscaling approach (IMSA) relying upon PSO is presented. In such technique, the investigation area is iteratively reconstructed at different scales. At every scale, the PSO is used to obtain a quantitative reconstruction of the distribution of the dielectric properties. A clustering techniques is used to identify the scatterers, and then the investigation area is refined in order to focus only on the objects. The proposed approach is tested by using several different two-dimensional targets illuminated by monochromatic TM incident waves. The reported results confirm that the integrated strategy is able to outperform its standard counterpart. Moreover, the new approach is also able to provide better results than those obtained by using the CG and GA (both in their standard form and inserted in a IMSA framework). As an example, a square hollow cylinder ($\epsilon_r = 1.5$, sides $L_{\text{in}} = 0.8\lambda$ and $L_{\text{out}} = 1.6\lambda$) contained in an investigation area of side 2.4λ and illuminated by plane waves impinging from 4 different directions (with the electric field measured in 21 points located on a circumference of radius 1.8λ for every view) is efficiently reconstructed in 4 steps. At each step, the PSO is executed with 20 particles, $\eta_1 = \eta_2 = 2.0$, $\omega^{(k)} = 0.4$, and $k_{\max} = 2000$ (the mesh used to discretized the investigation area has size 36). The obtained reconstruction error is 3.8%. For the same configuration, the multiscaling version of CG provides an error of 4.6%. An experimental validation with the Fresnel data [81] is also provided, confirming the capabilities of the approach when working in real environments. In [72, 73], the IMSA-PSO is tested on phaseless data (i.e., amplitude-only scattered field measures are available). In this case, too, good agreement with the actual profile are obtained. In [74, 75], the IMSA-PSO is also extended to the reconstruction of 3D objects.

4.2. Ant Colony Optimization-Based Imaging Algorithms. ACO has been used in several electromagnetic applications, in particular for design of antennas and microwave components [82, 83], for the allocation of base stations [84], and for microwave imaging. Concerning electromagnetic imaging, ACO has been applied both to the reconstruction of 2D cylindrical structures and 3D objects.

In [76], the ACO algorithm is applied to the reconstruction of multiple dielectric lossless cylinders under TM illumination. A two-dimensional formulation is assumed. Two representations of the dielectric properties are considered, pixelbased and splinebased. The number of unknowns, in the two cases, are 256 and 169, respectively. The population size is set equal to the number of unknowns and, at every iteration, $Q = P/10$ new solutions are created. The parameters of the ACO are set equal to $\rho = 0.1$ and $\xi = 0.85$ (according to the suggestions available in the literature). The iterations are stopped when a maximum number of iterations, $k_{\max} = 2000$, is reached. The provided results show that in all cases the ACO-based approach is able to correctly reconstruct the targets with a mean relative error lower than 5%. Moreover, as expected, the spline representation allows for a faster convergence (thanks to the lower number of unknowns).

In [28], ACO is applied to a similar configuration (two-dimensional imaging of a two-layer dielectric cylinders). A comparison with GA and DE is provided. The reconstruction results show that in the considered case ACO "reaches a more accurate reconstruction than those obtained by the other two methods" and "requires a lower number of function evaluations."

In [77], a hybrid method is proposed for two-dimensional imaging of multiple homogeneous dielectric targets. First, the shape of the targets is estimated by using the linear sampling method (LSM), which is a fast and efficient qualitative approach able to retrieve the support of the scatterers starting from the scattered field data. After this step, the ACO is applied to retrieve the values of the relative dielectric permittivity and the electric conductivity. In this way, the ACO only needs to find a few parameters (two for every objects identified by the LSM). The approach has been extended to three-dimensional targets in [78, 79], where the full vector problem with arbitrary homogeneous dielectric targets is taken into account. Single and multiple dielectric objects are reconstructed with good accuracy and with low computational efforts. As an example, for a rectangular parallelepiped of dimensions $0.66\lambda \times 0.33\lambda \times 0.5\lambda$, after the support estimation, mean relative reconstruction errors equal to 2.5% (dielectric permittivity) and 7% (electric conductivity) are obtained with 45.6 cost function evaluations (mean value).

4.3. Artificial Bee Colony-Based Imaging Algorithms. The ABC has been applied for breast cancer detection in [80]. A full three-dimensional configuration is considered. The breast is modeled both by using simplified structures and a realistic MRI-based phantom. ABC is employed for the reconstruction of the position, size, and dielectric properties of the malignant inclusion (supposed of spherical size and homogeneous). The problem size is thus $G = 6$. A population of $P = 10$ bees is employed. In the reported test cases, ABC is able to reach the convergence in less than 30 iteration. Moreover, the algorithm is able to estimate the position and size of the tumor with satisfactory accuracy (e.g., in the simplified case, the localization error is less than 10% and the size estimation error is less than 1 mm). A comparison with PSO, DE, and GA is also provided. In the considered case, ABC outperforms the other approaches (GA provided the worst results PSO and DE gave similar results).

An example of use of the ABC for the reconstruction of the full distribution of the dielectric properties of unknown objects is reported in the following. Cylindrical scatterers under TM illumination are considered. A multiview configuration is assumed, that is, $S = 8$ line-current sources uniformly spaced on a circumference of radius 1.5λ are sequentially used for illuminating the objects. The scattered electric field is collected in 51 points uniformly spaced on an angular sector of 270 degrees on the same circumference (positioned such that the source lies in the sector without measurement probes). The investigation domain is a square area of side 2λ. The input data (scattered electric field) are computed by using a numerical code based on the method of moments [85] with pulse basis and Dirac's delta weighting functions. A finer mesh is used for solving the forward problem in order to avoid inverse crimes. Moreover, the computed electric field is corrupted with a Gaussian noise with zero mean value and variance corresponding to a signal-to-noise ratio of 25 dB. In the inversion procedure, the investigation area is discretized into $N = 256$ square subdomains, and the unknowns are the values of the relative dielectric permittivity in such cells. Two separate targets are located in the investigation area: a circular cylinder (radius 0.25λ, center $(-0.25\lambda, 0.25\lambda)$, relative dielectric permittivity 2.0) and a square cylinder (side 0.5λ, center $(0.25\lambda, -0.5\lambda)$, relative dielectric permittivity 1.5). The cost function (6), in its multiview version, is employed. The parameters of the ABC have been set equal to as follows: $C = 50$, $P = 25$, $K_{\text{lim}} = 100$, and $k_{\text{max}} = 2000$.

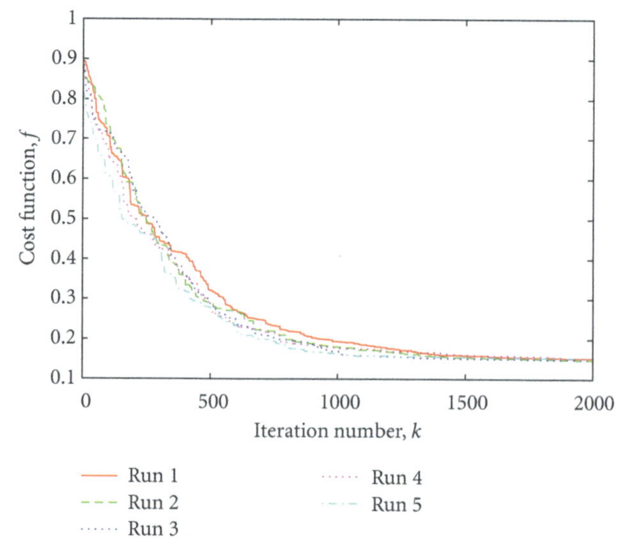

FIGURE 4: Cost function versus the iteration number. ABC-based inversion algorithm.

Some examples of the behavior of the cost function versus the iteration number are shown in Figure 4, which reports five different runs of the algorithms. As can be seen, in all cases, the method converges to a value of about 0.15. The corresponding mean relative reconstruction error is 3%. The same configuration has been considered in [76] and solved by using an ACO-based inversion approach. In the results provided in that paper, a mean relative error of about 4.5% is achieved with the ACO-based approach. Finally, an example of the reconstructed distribution of the dielectric properties obtained by the ABC approach is shown in Figure 5. As can be seen, the two objects are correctly shaped, and their permittivity is identified with quite good accuracy.

5. Conclusions

In this paper, the application of swarm intelligence algorithms, that is, stochastic algorithms inspired by the collective social behavior of agents (e.g., birds, ants, etc.), to the solution of microwave imaging problem has been reviewed.

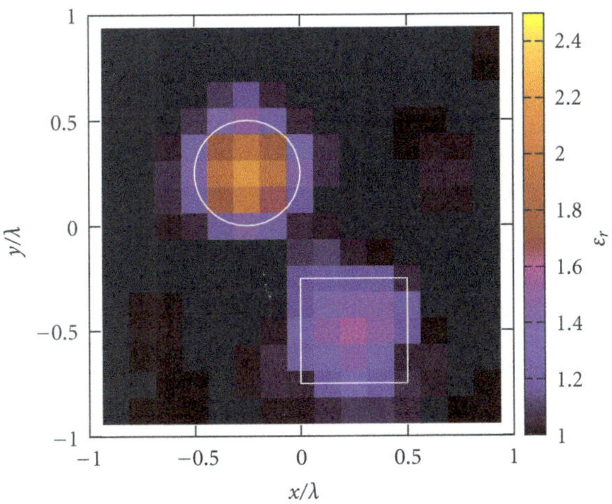

FIGURE 5: Example of reconstructed distribution of the dielectric permittivity. ABC-based inversion algorithm.

Such approaches have been proven to be very effective in several applications. The results available in the literature confirm the suitability of this class of optimization methods for microwave imaging, too.

References

[1] M. Pastorino, *Microwave Imaging*, John Wiley, Hoboken, NJ, USA, 2010.

[2] Y. J. Kim, L. Jofre, F. De Flaviis, and M. Q. Feng, "Microwave reflection tomographic array for damage detection of civil structures," *IEEE Transactions on Antennas and Propagation*, vol. 51, no. 11, pp. 3022–3032, 2003.

[3] M. Benedetti, M. Donelli, G. Franceschini, M. Pastorino, and A. Massa, "Effective exploitation of the a priori information through a microwave imaging procedure based on the SMW for NDE/NDT applications," *IEEE Transactions on Geoscience and Remote Sensing*, vol. 43, no. 11, pp. 2584–2591, 2005.

[4] L. Chommeloux, C. Pichot, and J. C. Bolomey, "Electromagnetic modeling for microwave imaging of cylindrical buried inhomogeneities," *IEEE Transactions on Microwave Theory and Techniques*, vol. 34, no. 10, pp. 1064–1076, 1986.

[5] M. Pastorino, S. Caorsi, and A. Massa, "Numerical assessment concerning a focused microwave diagnostic method for medical applications," *IEEE Transactions on Microwave Theory and Techniques*, vol. 48, no. 1, pp. 1815–1830, 2000.

[6] G. Bozza, C. Estatico, M. Pastorino, and A. Randazzo, "An inexact Newton method for microwave reconstruction of strong scatterers," *IEEE Antennas and Wireless Propagation Letters*, vol. 5, no. 1, pp. 61–64, 2006.

[7] A. G. Tijhuis, K. Belkebir, A. C. S. Litman, and B. P. De Hon, "Theoretical and computational aspects of 2-D inverse profiling," *IEEE Transactions on Geoscience and Remote Sensing*, vol. 39, no. 6, pp. 1316–1330, 2001.

[8] A. G. Tijhuis, K. Belkebir, A. C. S. Litman, and B. P. De Hon, "Multiple-frequency distorted-wave Born approach to 2D inverse profiling," *Inverse Problems*, vol. 17, no. 6, pp. 1635–1644, 2001.

[9] G. Bozza, C. Estatico, A. Massa, M. Pastorino, and A. Randazzo, "Short-range image-based method for the inspection of strong scatterers using microwaves," *IEEE Transactions on Instrumentation and Measurement*, vol. 56, no. 4, pp. 1181–1188, 2007.

[10] G. Bozza, C. Estatico, M. Pastorino, and A. Randazzo, "Application of an inexact-Newton method within the second-order Born approximation to buried objects," *IEEE Geoscience and Remote Sensing Letters*, vol. 4, no. 1, pp. 51–55, 2007.

[11] C. Estatico, G. Bozza, A. Massa, M. Pastorino, and A. Randazzo, "A two-step iterative inexact-Newton method for electromagnetic imaging of dielectric structures from real data," *Inverse Problems*, vol. 21, no. 6, pp. S81–S94, 2005.

[12] Z. Q. Zhang and Q. H. Liu, "Three-dimensional nonlinear image reconstruction for microwave biomedical imaging," *IEEE Transactions on Biomedical Engineering*, vol. 51, no. 3, pp. 544–548, 2004.

[13] P. Mojabi and J. LoVetri, "Overview and classification of some regularization techniques for the Gauss-Newton inversion method applied to inverse scattering problems," *IEEE Transactions on Antennas and Propagation*, vol. 57, no. 9, pp. 2658–2665, 2009.

[14] A. Litman, D. Lesselier, and F. Santosa, "Reconstruction of a two-dimensional binary obstacle by controlled evolution of a level-set," *Inverse Problems*, vol. 14, no. 3, pp. 685–706, 1998.

[15] R. Autieri, G. Ferraiuolo, and V. Pascazio, "Bayesian regularization in nonlinear imaging: reconstructions from experimental data in nonlinearized microwave tomography," *IEEE Transactions on Geoscience and Remote Sensing*, vol. 49, no. 2, pp. 801–813, 2011.

[16] C. Gilmore, P. Mojabi, and J. LoVetri, "Comparison of an enhanced distorted born iterative method and the multiplicative-regularized contrast source inversion method," *IEEE Transactions on Antennas and Propagation*, vol. 57, no. 8, pp. 2341–2351, 2009.

[17] M. El-Shenawee, O. Dorn, and M. Moscoso, "An adjoint-field technique for shape reconstruction of 3-D penetrable object immersed in lossy medium," *IEEE Transactions on Antennas and Propagation*, vol. 57, no. 2, pp. 520–534, 2009.

[18] P. Lobel, L. Blanc-Féraud, C. Pichet, and M. Barlaud, "A new regularization scheme for inverse scattering," *Inverse Problems*, vol. 13, no. 2, pp. 403–410, 1997.

[19] T. M. Habashy and A. Abubakar, "A general framework for constraint minimization for the inversion of electromagnetic measurements," *Progress In Electromagnetics Research*, vol. 46, pp. 265–312, 2004.

[20] P. M. Van Den Berg and A. Abubakar, "Contrast source inversion method: state of art," *Journal of Electromagnetic Waves and Applications*, vol. 15, no. 11, pp. 1503–1505, 2001.

[21] A. Randazzo, G. Oliveri, A. Massa, and M. Pastorino, "Electromagnetic inversion with the multiscaling inexact Newton method-experimental validation," *Microwave and Optical Technology Letters*, vol. 53, no. 12, pp. 2834–2838, 2011.

[22] S. Caorsi, A. Costa, and M. Pastorino, "Microwave imaging within the second-order born approximation: stochastic optimization by a genetic algorithm," *IEEE Transactions on Antennas and Propagation*, vol. 49, no. 1, pp. 22–31, 2001.

[23] C.-C. Chiu and P. T. Liu, "Image reconstruction of a perfectly conducting cylinder by the genetic algorithm," *IEE Proceedings-Microwaves, Antennas and Propagation*, vol. 143, no. 3, p. 249, 1996.

[24] A. Qing, C. K. Lee, and L. Jen, "Electromagnetic inverse scattering of two-dimensional perfectly conducting objects by real-coded genetic algorithm," *IEEE Transactions on Geoscience and Remote Sensing*, vol. 39, no. 3, pp. 665–676, 2001.

[25] S. Caorsi, A. Massa, M. Pastorino, M. Raffetto, and A. Randazzo, "Detection of buried inhomogeneous elliptic cylinders

by a memetic algorithm," *IEEE Transactions on Antennas and Propagation*, vol. 51, no. 10, pp. 2878–2884, 2003.
[26] M. Pastorino, S. Caorsi, A. Massa, and A. Randazzo, "Reconstruction algorithms for electromagnetic imaging," *IEEE Transactions on Instrumentation and Measurement*, vol. 53, no. 3, pp. 692–699, 2004.
[27] L. Garnero, A. Franchois, J. P. Hugonin, C. Pichot, and N. Joachimowicz, "Microwave imaging-complex permittivity reconstruction by simulated annealing," *IEEE Transactions on Microwave Theory and Techniques*, vol. 39, no. 11, pp. 1801–1807, 1991.
[28] M. Pastorino, "Stochastic optimization methods applied to microwave imaging: a review," *IEEE Transactions on Antennas and Propagation*, vol. 55, no. 3, pp. 538–548, 2007.
[29] Y. Rahmat-Samii and E. Michielssen, *Electromagnetic Optimization by Genetic Algorithms*, John Wiley & Sons, New York, NY, USA, 1999.
[30] J. M. Johnson and Y. Rahmat-Samii, "Genetic algorithms in engineering electromagnetics," *IEEE Antennas and Propagation Magazine*, vol. 39, no. 4, pp. 7–21, 1997.
[31] R. L. Haupt, "Introduction to genetic algorithms for electromagnetics," *IEEE Antennas and Propagation Magazine*, vol. 37, no. 2, pp. 7–15, 1995.
[32] A. Massa, M. Pastorino, and A. Randazzo, "Reconstruction of two-dimensional buried objects by a differential evolution method," *Inverse Problems*, vol. 20, no. 6, pp. S135–S150, 2004.
[33] D. S. Weile and E. Michielssen, "genetic algorithm optimization applied to electromagnetics: a review," *IEEE Transactions on Antennas and Propagation*, vol. 45, no. 3, pp. 343–353, 1997.
[34] D. E. Goldberg, *Genetic Algorithms in Search, Optimization, and Machine Learning*, Addison-Wesley Professional, Reading, Mass, USA, 1st edition, 1989.
[35] K. V. Price, "An introduction to differential evolution," in *New Ideas in Optimization*, pp. 79–108, McGraw-Hill, Maidenhead, UK, 1999.
[36] E. Bonabeau, M. Dorigo, and G. Theraulaz, *Swarm Intelligence?: From Natural to Artificial Intelligence*, Oxford University Press, New York, NY, USA, 1999.
[37] C. Blum and D. Merkle, *Swarm Intelligence Introduction and Applications*, Springer, London, UK, 2008.
[38] J. F. Kennedy, R. C. Eberhart, and Y. Shi, *Swarm Intelligence*, Morgan Kaufmann Publishers, San Francisco, Calif, USA, 2001.
[39] J. Robinson and Y. Rahmat-Samii, "Particle swarm optimization in electromagnetics," *IEEE Transactions on Antennas and Propagation*, vol. 52, no. 2, pp. 397–407, 2004.
[40] M. Dorigo, V. Maniezzo, and A. Colorni, "Ant system: optimization by a colony of cooperating agents," *IEEE Transactions on Systems, Man, and Cybernetics B*, vol. 26, no. 1, pp. 29–41, 1996.
[41] M. Dorigo and L. M. Gambardella, "Ant colony system: a cooperative learning approach to the traveling salesman problem," *IEEE Transactions on Evolutionary Computation*, vol. 1, no. 1, pp. 53–66, 1997.
[42] C. Estatico, M. Pastorino, and A. Randazzo, "A novel microwave imaging approach based on regularization in Lp Banach spaces," *IEEE Transactions on Antennas and Propagation*, vol. 60, no. 7, pp. 3373–3381, 2012.
[43] G. Bilchev and I. C. Parmee, "The ant colony metaphor for searching continuous design spaces," in *Proceedings of the Selected Papers from AISB Workshop on Evolutionary Computing*, pp. 25–39, Sheffield, UK, 1995.
[44] J. Dréo and P. Siarry, "A new ant colony algorithm using the heterarchical concept aimed at optimization of multiminima continuous functions," in *Proceedings of the 3rd International Workshop on Ant Algorithms (ANTS '02)*, pp. 216–221, London, UK, 2002.
[45] N. Monmarché, G. Venturini, and M. Slimane, "On how Pachycondyla apicalis ants suggest a new search algorithm," *Future Generation Computer Systems*, vol. 16, no. 8, pp. 937–946, 2000.
[46] K. Socha and M. Dorigo, "Ant colony optimization for continuous domains," *European Journal of Operational Research*, vol. 185, no. 3, pp. 1155–1173, 2008.
[47] D. Karaboga, "An idea based on honey bee swarm for numerical optimization," Tech. Rep. TR06, Erciyes University, Kayseri, Turkey, 2005.
[48] D. Karaboga and B. Basturk, "Artificial Bee Colony (ABC) optimization algorithm for solving constrained optimization problems," in *Foundations of Fuzzy Logic and Soft Computing*, P. Melin, O. Castillo, L. T. Aguilar, J. Kacprzyk, and W. Pedrycz, Eds., vol. 4529, pp. 789–798, Springer, Berlin, Germany.
[49] V. Tereshko and A. Loengarov, "Collective decision making in honey-bee foraging dynamics," *Computing and Information Systems*, vol. 9, no. 3, pp. 1–7, 2005.
[50] A. Semnani, M. Kamyab, and I. T. Rekanos, "Reconstruction of one-dimensional dielectric scatterers using differential evolution and particle swarm optimization," *IEEE Geoscience and Remote Sensing Letters*, vol. 6, no. 4, pp. 671–675, 2009.
[51] A. Semnani and M. Kamyab, "Comparison of differential evolution and particle swarm optimization in one-dimensional reconstruction problems," in *Proceedings of the Asia Pacific Microwave Conference (APMC '08)*, pp. 1–4, Hong Kong, December 2008.
[52] A. M. Emad Eldin, E. A. Hashish, and M. I. Hassan, "Inversion of lossy dielectric profiles using particle swarm optimization," *Progress In Electromagnetics Research M*, vol. 9, pp. 93–105, 2009.
[53] A. Fhager, A. Voronov, C. Chen, and M. Persson, "Methods for dielectric reconstruction in microwave tomography," in *Proceedings of the 2nd European Conference on Antennas and Propagation (EuCAP '07)*, pp. 1–6, Edinburgh, UK, November 2007.
[54] C.-H. Huang, C. C. Chiu, C. L. Li, and K. C. Chen, "Time domain inverse scattering of a two-dimensional homogenous dielectric object with arbitrary shape by particle swarm optimization," *Progress in Electromagnetics Research*, vol. 82, pp. 381–400, 2008.
[55] S. Caorsi, M. Donelli, A. Lommi, and A. Massa, "Location and imaging of two-dimensional scatterers by using a particle swarm algorithm," *Journal of Electromagnetic Waves and Applications*, vol. 18, no. 4, pp. 481–494, 2004.
[56] H. Zhang, X. D. Zhang, J. Ji, and Y. J. Yao, "Electromagnetic imaging of the 2-D media based on Particle Swarm algorithm," in *Proceedings of the 6th International Conference on Natural Computation (ICNC '10)*, pp. 262–265, Yantai, China, August 2010.
[57] S. H. Zainud-Deen, W. M. Hassen, and K. H. Awadalla, "Techniques, crack detection using a hybrid finite difference frequency domain and particle swarm optimization," in *Proceedings of the National Radio Science Conference (NRSC '09)*, pp. 1–8, Cairo, Egypt, March 2009.
[58] S. H. Zainud-Deen, W. M. Hassen, E. M. Ali, K. H. Awadalla, and H. A. Sharshar, "Breast cancer detection using a hybrid finite difference frequency domain and particle swarm optimization techniques," in *Proceedings of the 25th National Radio Science Conference (NRSC '08)*, pp. 1–8, Tanta, Egypt, March 2008.

[59] I. T. Rekanos and M. Kanaki, "Microwave imaging of two-dimensional conducting scatterers using particle swarm optimization," in *Electromagnetic Fields in Mechatronics, Electrical and Electronic Engineering*, A. Krawczyk, S. Wiak, and L. M. Fernandez, Eds., pp. 84–89, IOS Press, Amsterdam, The Netherlands, 2006.

[60] I. T. Rekanos, "Shape reconstruction of a perfectly conducting scatterer using differential evolution and particle swarm optimization," *IEEE Transactions on Geoscience and Remote Sensing*, vol. 46, no. 7, pp. 1967–1974, 2008.

[61] T. Huang and A. S. Mohan, "Application of particle swarm optimization for microwave imaging of lossy dielectric objects," in *Proceedings of the IEEE Antennas and Propagation Society International Symposium and USNC/URSI Meeting*, vol. 1B, pp. 852–855, Washington, DC, USA, July 2005.

[62] T. Huang and A. Sanagavarapu Mohan, "A microparticle swarm optimizer for the reconstruction of microwave images," *IEEE Transactions on Antennas and Propagation*, vol. 55, no. 3, pp. 568–576, 2007.

[63] C.-L. Li, C. C. Chiu, and C. H. Huang, "Time domain inverse scattering for a homogenous dielectric cylinder by asynchronous particle swarm optimization," *Journal of Testing and Evaluation*, vol. 39, no. 3, 2011.

[64] C. Wei, C. H. Sun, J. O. Wu, C. C. Chiub, and M. K. Wu, "Inverse scattering for the perfectly conducting cylinder by asynchronous particle swarm optimization," in *Proceedings of the 3rd IEEE International Conference on Communication Software and Networks*, pp. 410–4412, Xi'an, China, 2011.

[65] C.-H. Sun, C. C. Chiu, and C. L. Li, "Time-domain inverse scattering of a two-dimensional metallic cylinder in slab medium using asynchronous particle swarm optimization," *Progress In Electromagnetics Research M*, vol. 14, pp. 85–100, 2010.

[66] B. Mhamdi, K. Grayaa, and T. Aguili, "Microwave imaging for conducting scatterers by hybrid particle swarm optimization with simulated annealing," in *Proceedings of the 8th International Multi-Conference on Systems, Signals and Devices (SSD '11)*, pp. 1–6, Sousse, Tunisia, March 2011.

[67] G. R. Huang, W. J. Zhong, and H. W. Liu, "Microwave imaging based on the AWE and HPSO incorporated with the information obtained from born approximation," in *Proceedings of the International Conference on Computational Intelligence and Software Engineering (CiSE '10)*, pp. 1–4, Wuhan, Chin, December 2010.

[68] B. Mhamdi, K. Grayaa, and T. Aguili, "An inverse scattering approach using hybrid PSO-RBF network for microwave imaging purposes," in *Proceedings of the 16th IEEE International Conference on Electronics, Circuits and Systems (ICECS '09)*, pp. 231–234, Hammamet, Tunisia, December 2009.

[69] M. Donelli, G. Franceschini, A. Martini, and A. Massa, "An integrated multiscaling strategy based on a particle swarm algorithm for inverse scattering problems," *IEEE Transactions on Geoscience and Remote Sensing*, vol. 44, no. 2, pp. 298–312, 2006.

[70] M. Donelli and A. Massa, "Computational approach based on a particle swarm optimizer for microwave imaging of two-dimensional dielectric scatterers," *IEEE Transactions on Microwave Theory and Techniques*, vol. 53, no. 5, pp. 1761–1776, 2005.

[71] D. Franceschini and A. Massa, "An integrated stochastic multi-scaling strategy for microwave imaging applications," in *Proceedings of the IEEE Antennas and Propagation Society International Symposium and USNC/URSI Meeting*, pp. 209–212, Washington, DC, USA, July 2005.

[72] G. Franceschini, M. Donelli, R. Azaro, and A. Massa, "Inversion of phaseless total field data using a two-step strategy based on the iterative multiscaling approach," *IEEE Transactions on Geoscience and Remote Sensing*, vol. 44, no. 12, pp. 3527–3539, 2006.

[73] R. Azaro, D. Franceschini, G. Franceschini, L. Manica, and A. Massa, "A two-step strategy for a multi-scaling inversion of phaseless measurements of the total field," in *Proceedings of the IEEE Antennas and Propagation Society International Symposium, APS 2006*, pp. 1073–1076, Albuquerque, NM, USA, July 2006.

[74] M. Donelli, D. Franceschini, P. Rocca, and A. Massa, "Three-dimensional microwave imaging problems solved through an efficient multiscaling particle swarm optimization," *IEEE Transactions on Geoscience and Remote Sensing*, vol. 47, no. 5, pp. 1467–1481, 2009.

[75] D. Franceschini, G. Franceschini, M. Donelli, P. Rocca, and A. Massa, "Imaging three-dimensional bodies by processing multi-frequency data through a multiscale swarm intelligence based method," in *Proceedings of the IEEE Antennas and Propagation Society International Symposium (AP-S '07)*, pp. 409–412, Honolulu, Hawaii, USA, June 2007.

[76] M. Pastorino and A. Randazzo, "Nondestructive analysis of dielectric bodies by means of an Ant Colony Optimization method," in *Swarm Intelligence for Electric and Electronic Engineering*, G. Fornarelli and L. Mescia, Eds., IGI Global, Hershey, Pa, USA, 2012.

[77] M. Brignone, G. Bozza, A. Randazzo, R. Aramini, M. Piana, and M. Pastorino, "Hybrid approach to the inverse scattering problem by using ant colony optimization and no-sampling linear sampling," in *Proceedings of the IEEE International Symposium on Antennas and Propagation and USNC/URSI National Radio Science Meeting (APSURSI '08)*, pp. 1–4, San Diego, Calif, USA, July 2008.

[78] G. Bozza, M. Brignone, M. Pastorino, M. Piana, and A. Randazzov, "An inverse scattering based hybrid method for the measurement of the complex dielectric permittivities of arbitrarily shaped homogenous targets," in *Proceedings of the IEEE Intrumentation and Measurement Technology Conference (I2MTC '09)*, pp. 719–723, Singapore, May 2009.

[79] M. Brignone, G. Bozza, A. Randazzo, M. Piana, and M. Pastorino, "A hybrid approach to 3D microwave imaging by using linear sampling and ACO," *IEEE Transactions on Antennas and Propagation*, vol. 56, no. 10, pp. 3224–3232, 2008.

[80] M. Donelli, I. Craddock, D. Gibbins, and M. Sarafianou, "A three-dimensional time domain microwave imaging method for breast cancer detection based on an evolutionary algorithm," *Progress In Electromagnetics Research M*, vol. 18, pp. 179–195, 2011.

[81] K. Belkebir and M. Saillard, "Special section: testing inversion algorithms against experimental data," *Inverse Problems*, vol. 17, no. 6, p. 1565, 2001.

[82] K. Tenglong, Z. Xiaoying, W. Jian, and D. Yihan, "A modified ACO algorithm for the optimization of antenna layout," in *Proceedings of the 2011 International Conference on Electrical and Control Engineering*, pp. 4269–4272, Yichang, China, 2011.

[83] G. Weis, A. Lewis, M. Randall, and D. Thiel, "Pheromone pre-seeding for the construction of RFID antenna structures using ACO," in *Proceedings of the 6th IEEE International Conference on e-Science, eScience 2010*, pp. 161–167, Brisbane, Australia, December 2010.

[84] I. Vilovic, N. Burum, Z. Sipus, and R. Nad, "PSO and ACO algorithms applied to location optimization of the WLAN base station," in *Proceedings of the 19th International Conference on Applied Electromagnetics and Communications (ICECom '07)*, pp. 1–5, Dubrovnik, Croatia, September 2007.

[85] R. Harrington, *Field Computation by Moment Methods*, IEEE Press, Piscataway, NJ, USA, 1993.

18

An Inductorless Cascaded Phase-Locked Loop with Pulse Injection Locking Technique in 90 nm CMOS

Sang-yeop Lee, Hiroyuki Ito, Shuhei Amakawa, Noboru Ishihara, and Kazuya Masu

Solutions Research Laboratory, Tokyo Institute of Technology, 4259-S2-14 Nagatsuta, Midori-ku, Yokohama 226-8503, Japan

Correspondence should be addressed to Sang-yeop Lee; lee.s.af@m.titech.ac.jp

Academic Editor: Leonid Belostotski

An inductorless phase-locked loop with subharmonic pulse injection locking was realized (PLL area: $0.11\,mm^2$) by adopting 90 nm Si CMOS technology. The proposed circuit is configured with two cascaded PLLs; one of them is a reference PLL that generates reference signals to the other one from low-frequency external reference signals. The other is a main PLL that generates high-frequency output signals. A high-frequency half-integral subharmonic locking technique was used to decrease the phase noise characteristics. For a 50 MHz input reference signal, without injection locking, the 1 MHz offset phase noise was −88 dBc/Hz at a PLL output frequency of 7.2 GHz (⊠ 144 × 50 MHz); with injection locking, the noise was −101 dBc/Hz (spur level: −31 dBc; power consumption from a 1.0 V power supply: 25 mW).

1. Introduction

Conventional multistandard wireless mobile terminals contain multiple RFICs. To reduce production costs, one-chip wideband RF LSI systems are desired. A great effort is being made to develop wideband and/or multiband RF solutions using highly scaled advanced CMOS processes. The use of such processes is beneficial to A/D and D/A converters and digital baseband circuits. However, it is very difficult to reduce the scale of RF/analog circuit blocks, especially power amplifiers and oscillator circuits, including voltage-controlled oscillators (VCOs) and phase-locked loops (PLLs), because of the presence of inductors that do not scale with advancements in technology.

In designing VCOs which generate signals in RF systems, ring-type VCOs (ring VCOs) are more attractive than LC-resonant-type VCOs (LC VCOs) in terms of their small area and wide frequency tuning range since they do not use large passive devices. However, they have poor phase noise with relatively high power consumption. Nevertheless, low-phase-noise ring VCO is still a possibility if some noise-suppression mechanism is applied. One of available options would be injection locking.

In the early days, Adler [1] and many other authors studied the behavior of VCOs with injection locking. Also, there are numerous papers published in reference to VCOs with injection locking in order to achieve phase locking and high performances. Moreover, recently, PLLs with an injection-locked frequency divider and frequency multiplier, and a clock and data recovery circuit (CDR) were presented.

This paper describes a study on a ring-VCO-based PLL with pulse injection locking as a potential solution to realize a scalable inductorless PLL, which can generate wideband frequency signal with low supply voltage. Usually, the frequency range utilized consumer RF applications, such as wireless LAN a/b/g/n, Bluetooth, and digital TV (DTV), is very wide and spreading from 400 MHz to 6 GHz. Table 1 shows target performance of the proposed PLL. Generally, in RF systems using high transmitting power, a frequency synthesizer should generate higher-frequency signals up to 12 GHz to avoid injection pulling from a power amplifier. Then, some methods, such as using frequency dividers and mixers, are applied to widen frequency range [2, 3].

In addition, the proposed PLL is augmented with high-frequency half-integral subharmonic locking in order to improve its phase-noise performance. In Section 2, brief features of the proposed PLL are explained. In Section 2.1, high-frequency half-integral subharmonic locking is shown as a method of reducing phase noise. Also, the proposed

Table 1: Target performance.

VCO	Ring
Frequency range	6–12 GHz
Phase noise at 1 MHz offset	−100 dBc/Hz
CMOS process	90 nm
Supply voltage	1.0 V

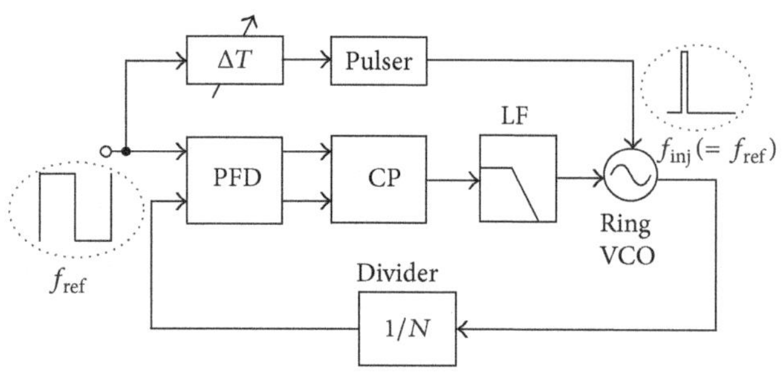

Figure 1: Injection-locked PLL with pulse injection.

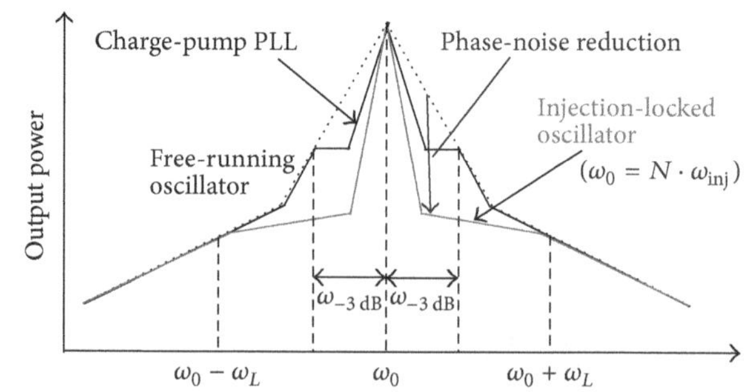

Figure 2: Phase-noise reduction with a charge-pump PLL and injection locking.

cascaded PLL (CPLL) that can achieve injection locking at high frequencies from low-frequency reference is presented. Detailed circuit designs, such as a VCO and a charge pump which are able to realize wide-band operation, and the measurement results obtained from an implementation in 90 nm CMOS process are presented in Sections 3 and 4, respectively. Finally, we conclude this work in Section 5.

2. Injection Locking in Frequency Synthesizers

Figure 1 shows an injection-locked PLL (ILPLL). The PLL is based on a ring VCO that is able to generate high-frequency outputs across wide frequency range, as well as I/Q outputs. The PLL also consists of a phase frequency detector (PFD), a charge pump (CP), a loop filter (LF), a variable delay unit (ΔT), and a pulser.

PLLs that use ring type VCOs are required to have a wide loop bandwidth of the phase-locked loop for lowering their poor phase noise characteristics. However, there is a trade-off between the loop bandwidth and the stability of PLLs. In general, the loop bandwidth ($\omega_{-3\,dB}$) must be narrower than $\omega_{ref}/20$~$\omega_{ref}/10$, where ω_{ref} is the reference-signal frequency [4]. Consequently, there is a limitation on lowering the phase noise in ring-VCO-based charge-pump PLLs (CP PLLs). Figure 2 shows phase noise characteristics of the PLL. In this case, the charge-pump noise of the PLL is assumed to be small enough and can be neglected. In Figure 2, phase noise is suppressed up to the loop bandwidth ($\omega_{-3\,dB}$) by the noise filtering of the loop. On the other hand, pulse injection locking is effective to reduce phase noise of ring VCOs since ring VCOs have a wider lock range with injection locking than that of LC VCOs because of their low quality factors. In designing subharmonically injection-locked oscillators (ILOs), N times frequency-multiplied signals as to the reference frequency can be achieved. The lock range is decided by the power of Nth superharmonics of the reference signal as follows [1, 5]:

$$\omega_L = \frac{\omega_{out}}{2Q} \cdot \sqrt{\frac{P_{injN}}{P_0}}, \tag{1}$$

where Q represents the open-loop quality factor of an oscillator (calculated by using the open-loop transfer function of the oscillator [6]), ω_{out} is the output frequency of the oscillator under injection locked condition, P_{injN} is the Nth harmonic power of the reference signal, and P_0 is the free-running output power of the oscillator. P_{injN} is approximately given by

$$P_{injN} \propto \left\{A \cdot D \frac{\sin(\pi D \cdot N)}{\pi D \cdot N}\right\}^2, \tag{2}$$

where A is the pulse amplitude, D is the duty cycle of pulses ($D = \Delta T/T$, ΔT: pulse width, T: period of pulses). From (1) and (2), the lock range ω_L can be rewritten as follows:

$$\omega_L \propto \frac{\omega_{inj}}{2Q} \cdot \frac{|\sin(\pi D \cdot N)|}{\sqrt{P_0}}, \tag{3}$$

where ω_{inj} ($= \omega_0/N$) is the injection-signal frequency [7].

The overall ILO output phase noise is obtained by adding the noise contributions in an ILO. Assuming that $S_{ILO}(\omega)$, $S_{REF}(\omega)$, and $S_{VCO}(\omega)$ are phase noise power functions of an injection-locked VCO, a reference signal, and a free-running VCO, respectively, the phase noise of an ILO, $S_{ILO}(\omega)$, simply can be expressed as

$$\begin{aligned} S_{ILO}(\omega) &= S_{REF,M}(\omega) \cdot |H_{LPF}(j\omega)|^2 + S_{VCO}(\omega) \cdot |H_{HPF}(j\omega)|^2 \\ &= M^2 \cdot S_{REF}(\omega) \cdot \frac{1}{1+(\omega/\omega_L)^2} + S_{VCO}(\omega) \cdot \frac{(\omega/\omega_L)^2}{1+(\omega/\omega_L)^2}, \end{aligned} \tag{4}$$

where $S_{REF,M}$ is the normalized phase noise power function with respect to the output frequency of f_{out}, and M means the ratio between the output frequency (f_{out}) and the input frequency (f_{inj}). $H_{LPF}(j\omega)$ and $H_{HPF}(j\omega)$ are low-pass and

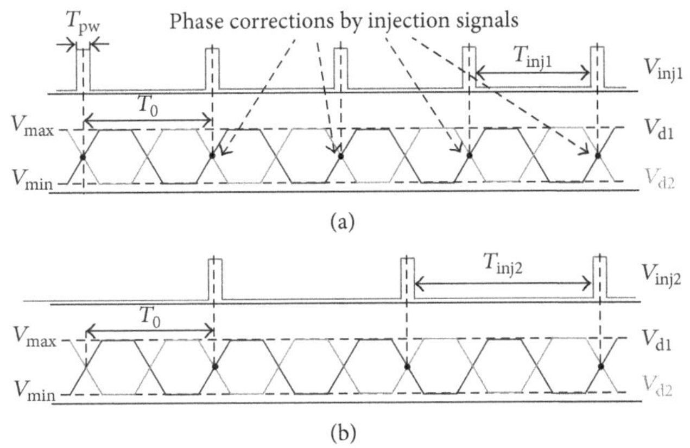

FIGURE 3: Voltage and current waveforms at differential output nodes of a VCO; (a) integral subharmonic locking ($f_0 = f_{inj}$), (b) half-integral subharmonic locking ($f_0 = 1.5f_{inj}$).

high-pass transfer functions, respectively [8]. Supposed that $H_{HPF}(j\omega)$ and $H_{LPF}(j\omega)$ have the first-order transfer functions and they have the same cutoff frequency of ω_L, the simple equation of (4) would be achieved [5, 9, 10].

In the proposed PLL, there are two kinds of phase locking mechanism: one is a phase-locked loop, and the other is pulse injection locking. In general, either of them is enough for phase locking. However, those two mechanisms are combined to get a wide frequency range operation with a low-phase-noise performance. The phase-locked loop, which uses a charge pump for controlling the oscillation frequency, is implemented to ensure correct frequency locking over the entire VCO tuning range. The final phase locking is done by injection locking to reference signal [11–13].

2.1. High-Frequency Half-Integral Subharmonic Locking Topology for Noise Reduction. A paper on half-integral subharmonic injection locking based on the use of a ring VCO has been presented [8]. A differential VCO can be easily designed to lock to half-integral subharmonics by giving its necessary symmetry properties. Suppose that a VCO consists of differential circuits and has a certain symmetry. As a method to achieve injection locking, a direct injection technique is applied, which uses nMOS switches that short the differential outputs for phase corrections. Figure 3 shows differential waveforms (V_{d1}, V_{d2}) of the VCO in the case of both integral ($f_0 = f_{inj}$) and half-integral subharmonic locking ($f_0 = 1.5f_{inj}$). The two output nodes are shorted when the injection signal (V_{inj1}, V_{inj2}) is input into the nMOS switches. Phase corrections may occur at the time and the jitter is reduced. Generally, there are two points of time during the period of the output signal when two output nodes can be shorted because of topological symmetry as shown in Figure 3. Consequently, the differential VCO is capable of both integral and also half-integral subharmonic locking.

One advantage of using half-integral subharmonic locking is to be able to use high-frequency reference signal and can make the locking range of injection locking, ω_L wide as shown in (1). Figure 2 and Equation (4) also show that

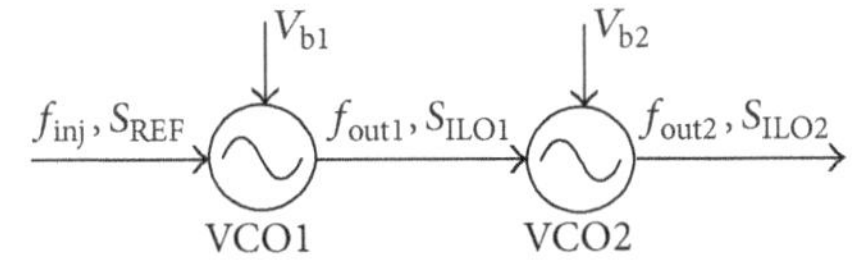

FIGURE 4: Concept of the cascaded ILOs.

the phase noise of the reference signal mainly affects the output phase noise at low offset frequencies and that the phase noise of the PLL becomes dominant as the offset frequency approaches the edge of the locking range [11]. Therefore, it will improve phase noise characteristics to the edge of the locking range to use high-frequency reference signals.

2.2. High-Frequency Signal Generation with Cascaded ILOs. As shown in (3), the lock range is proportional to the input frequency of ω_{inj}. However, narrower pulses are required to achieve smaller D with increasing the multiplication ratio (N). Unfortunately, it is difficult to achieve sufficiently narrow pulses even with the use of nm-scale CMOS processes since the reference inputs also have certain jitter and parasitic components of the pulser limit the pulse width. In other words, there is limitation to generate high-frequency (over 5 GHz) injection-locked signals with low-frequency reference such as *XTALs*.

One solution is to employ cascaded oscillators [11], which make each multiplication ratio (N) smaller by using two multiplication processes. Figure 4 shows the concept of the cascaded ILOs. Firstly, the input signal, which has sufficiently high-power superharmonics, is injected into VCO1. Then, M_1 multiplied frequency signal ($f_{out1} = M_1 \cdot f_{inj}$) of the reference frequency can be achieved by tuning the VCO1 oscillation frequency properly. In this case, the output phase noise of VCO1 with injection locking can be expressed as follows:

$$S_{ILO1}(\omega) = M_1^2 \cdot S_{REF}(\omega) \cdot \frac{1}{1+(\omega/\omega_{L1})^2} + S_{VCO1}(\omega) \cdot \frac{(\omega/\omega_{L1})^2}{1+(\omega/\omega_{L1})^2}, \tag{5}$$

where ω_{L1} is the lock range that is proportional to the input frequency (f_{inj}) and can be calculated from (3).

The output signal of VCO1 is injected into VCO2 and locked to the output of VCO2 with the same process occurred in VCO1. Also, the output phase noise of VCO2 with injection locking can be expressed as follows:

$$S_{ILO2}(\omega) = M_2^2 \cdot S_{ILO1}(\omega) \cdot \frac{1}{1+(\omega/\omega_{L2})^2} + S_{VCO2}(\omega) \cdot \frac{(\omega/\omega_{L2})^2}{1+(\omega/\omega_{L2})^2} \tag{6}$$

$$\approx M_2^2 \cdot \frac{1}{1+(\omega/\omega_{L2})^2} \cdot S_{ILO1}(\omega), \tag{7}$$

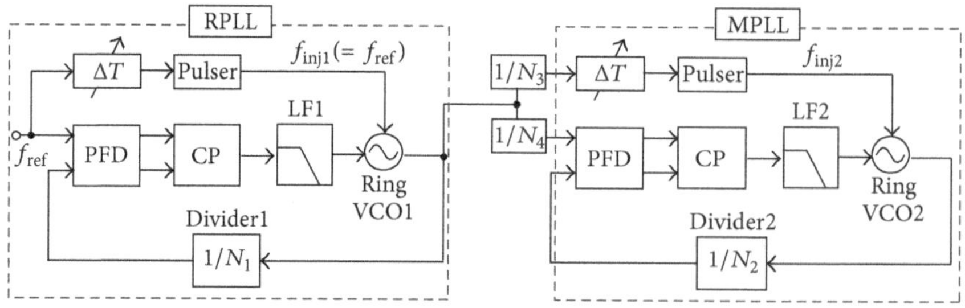

FIGURE 5: Proposed cascaded injection-locked PLL (CPLL) with pulse injection.

TABLE 2: Dividers ratio in Figure 5.

N_1	N_2	N_3	N_4
32	24, 36	1, 8	4, 8

where M_2 is the ratio between the output frequency of VCO2 (f_{out2}) and input frequency (f_{out1}) and ω_{L2} is the lock range of VCO2. When the offset frequency of ω is sufficiently lower than ω_{L2} (i.e., $\omega \ll \omega_{L2}$), (7) is held. In other words, sufficiently wide lock range makes it possible to neglect the secondary VCO phase noise up to the lock range in cascaded ILOs.

3. Proposed Injection-Locked PLL Topology

Figure 5 shows the configuration of the proposed PLL that enables the use of half-integral subharmonic locking, which was proposed in our previous work [13]. The proposed PLL consists of two injection-locked PLLs. A reference PLL, namely, RPLL generates reference signals to a main PLL, namely, MPLL from low-frequency external reference signals. In this topology, when we choose divider ratios (Table 2), respectively as, $N_2 = 36$, $N_3 = 1$, and $N_4 = 8$, the ratio between the reference signal to MPLL and the output frequency of MPLL may be 4.5 and high-frequency half-integral subharmonic locking can be applied. Variable time delay cells ΔTs are implemented to control the time when injection signals are input because phase corrections can occur easily when differential output nodes are shorted in the direct injection locking scheme (Figure 3).

3.1. Main PLL. Figure 6(a) shows the topology of the proposed delay cell that composes a ring VCO [13]. The delay cell contains an inverter latch as a negative conductance circuit that generates delay by positive feedback in order to satisfy the oscillation condition [14]. To tune the VCO output frequency widely, variable pMOS resistive loads are used. However, in the commonly used delay cells with pMOS resistive loads, the range of control voltage is limited from 0 V to the pMOS threshold voltage. In the proposed delay cell, a pMOS transistor is added into which the subcontrol voltage (V_{bn}) is input in order to make the range of sensitive voltages identical to the rail-to-rail voltage range (0 V to V_{DD}). For this purpose, the bias level shifted by about $V_{DD}/2$, V_{bn}, is input to the added pMOS transistor. As a result,

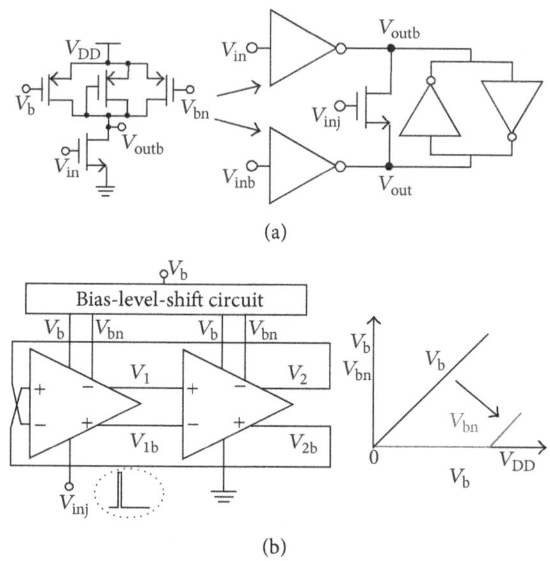

FIGURE 6: (a) Proposed differential delay cell, (b) two-stage differential VCO of MPLL with a bias-level-shift circuit.

the total equivalent resistance of the two pMOS transistors in parallel changes almost linearly versus the main control voltage, V_b. Consequently, the VCO output frequency can be tuned linearly across the wide tuning range [12, 13]. An nMOS switches are connected at the nodes between the differential nodes to achieve injection locking [15].

The proposed ring VCO is shown in Figure 6(b). It is based on a two-stage pseudo differential ring oscillator. Pulses which are generated by the on-chip pulser are injected into the left delay cell in the form of rail-to-rail pulses for injection locking. To maintain topological symmetry, an nMOS switch biased to 0 V is also applied in the right-side delay cell. We achieved the VCO tuning range of 6.02 GHz to 11.1 GHz across the rail-to-rail control voltage from the postlayout simulation of the VCO core with output buffers (90 nm CMOS process, $V_{DD} = 1.0$ V).

A tristate phase/frequency detector (PFD) is implemented, which consists of two D-flip flops, delay-path inverters, and an AND logic. The PFD detects phase and frequency difference between the reference signal and the divided VCO output and generates output pulses of V_{UP} and V_{DN} which are input into the charge pump to reduce the difference.

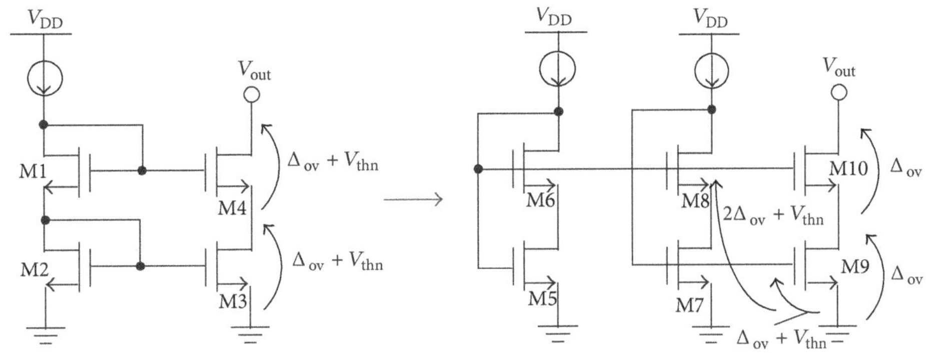

FIGURE 7: Sooch cascode current mirror circuit.

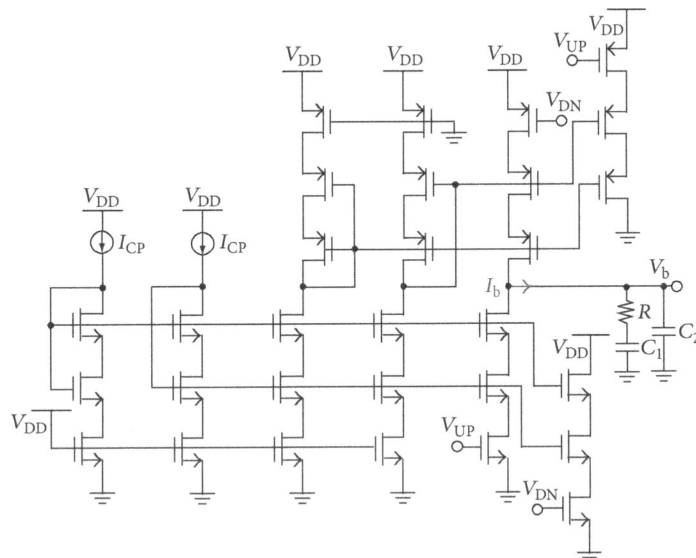

FIGURE 8: Proposed charge-pump (CP) circuit and loop filter (LF).

Figure 7 shows an implemented current mirror circuit to generate stable constant current from the charge pump. Usually, stacked current mirrors design can obtain better DC headroom and linearity with longer channel lengths as shown in the left side of Figure 7. In this case, DC headroom of the output voltage (V_{out}) is expressed as $2(\Delta_{ov} + V_{thn})$, where Δ_{ov} is the overdrive voltage of MOS transistors (M3, M4), and V_{thn} is an nMOS threshold voltage. In the case of Sooch cascode current mirror as shown in the right side of Figure 7, the MOS transistor, M5, is forced to operate in the triode region. The DC headroom can be reduced as $2\Delta_{ov}$ since MOS transistors operate in the saturation region except for M5 [16]. Consequently, low voltage operation can be achieved.

Proposed current switching charge pump (CP) that employs Sooch cascode current mirror is shown in Figure 8. Dummy switches are also implemented to maintain the balance between PFD outputs. Two external current sources ranged from $10\,\mu$A to $150\,\mu$A are used.

Figure 9 shows postlayout simulation results of the proposed charge pump, when $I_{CP} = 20\,\mu$A (90 nm CMOS process, 1.0 V supply). It shows that the charge pump can generate quite constant output current across the wide range

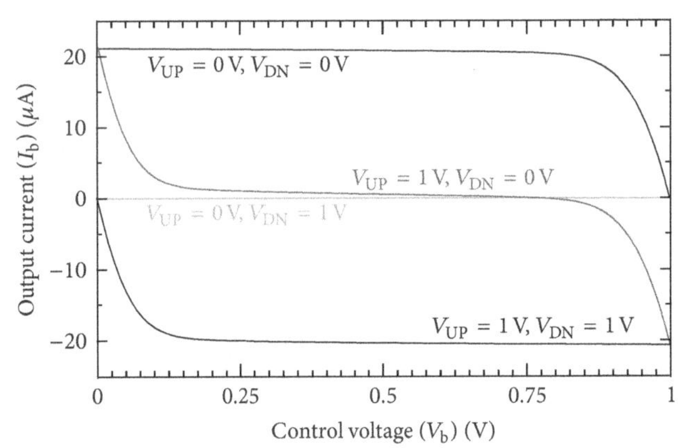

FIGURE 9: Charge-pump output current versus the control voltage (V_b), when $I_{CP} = 20\,\mu$A.

of the output voltage (V_b). When $V_{UP} = 1$ V and $V_{DN} = 0$ V, the result shows current mismatch between the up and down output currents as a function of the output voltage (V_b). The percentage mismatch error for $0.48\,V \le V_b \le 0.81$ V is less

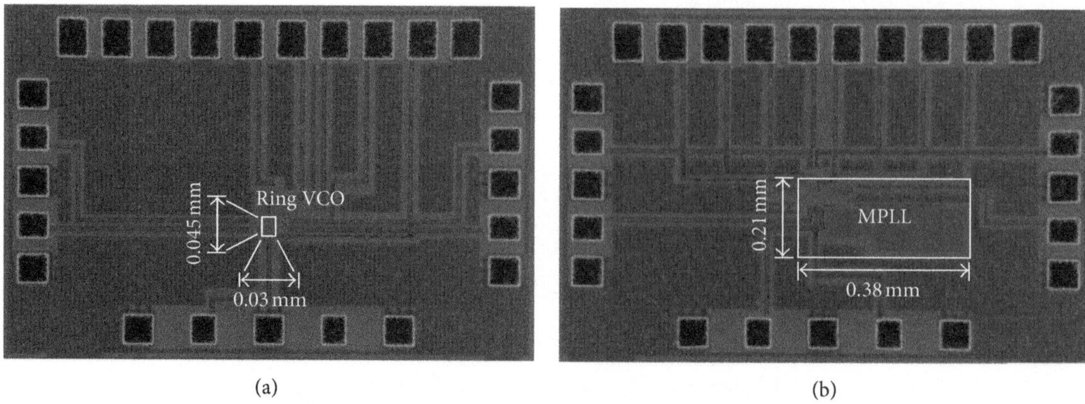

FIGURE 10: Chip micrographs; (a) proposed VCO and (b) proposed PLL.

than 2% and increases to less than 5% for $0.21\,V \leq V_b \leq 0.85\,V$.

A second-order lag-lead filter that consists of a register and two capacitors is implemented as a loop filter (LF) of the loop to suppress the charge-pump ripple. ($R = 16\,k\Omega$, $C_1 = 41\,pF$, $C_2 = 12\,pF$). In this case, on-chip MIM capacitors were used.

The frequency divider consists of differential pseudo-nMOS latches to minimize chip area and achieve low power consumption [17]. The frequency divider chain consists of three divide-by-2 circuits and one divide-by-2/3 circuit. As a result, it can divide by 24 and 36 in the loop (i.e., divider ratio $N_1 = 24, 36$).

The loop dynamic characteristics are designed to have the unity-gain bandwidth of 2.8 MHz and phase margin of 16° (VCO gain: 5 GHz/V, $I_{CP} = 20\,\mu A$, divider ratio $N_1 = 24$). When the divider ratio N_1 equals to 36, the unity-gain bandwidth of 2.2 MHz and phase margin of 19° are achieved. The PLL has poor phase margin that is related to the low damping factor and the slow settling time, because final phase locking is done not only by the phase-locked loop but also by injection locking. Injection locking that is applied into a phase-locked loop helps the phase margin to be improved [21]. In this case, large capacitance of C_2 is required to suppress the reference spur level due to the control voltage ripple. A loop bandwidth of the PLL is designed to be small enough compared to the lock range of injection locking to avoid the interference between two phase locking but can still achieve frequency locking.

To achieve subharmonic locking, an AND-based pulser is used, which is able to tune the pulse width below 40 ps by the analog control. Also, a variable time-delay unit (ΔT) which consists of inverters and tristate inverters was applied to match the zero-crossing points of differential VCO outputs to the pulses for effective injection locking.

3.2. Reference PLL. The proposed ring VCO used in RPLL is based on a four-stage pseudo differential ring oscillator. The same delay cell shown in MPLL (Figure 6(a)) is applied to widen frequency tuning range linearly. Also, long-gate channel MOS transistors are equipped in the delay cell to decrease VCO oscillation frequencies and reduce flicker noise characteristics as a reference signal into MPLL. Pulses which are generated by the on-chip pulser are injected into the left delay cell in the form of rail-to-rail pulses for injection locking. To maintain topological symmetry, an nMOS switch biased to 0 V is also applied in the other delay cells. We achieved the VCO tuning range of 0.805 GHz to 2.85 GHz across the rail-to-rail control voltage from the postlayout simulation of the VCO core with output buffers (90 nm CMOS process, $V_{DD} = 1.0\,V$).

The tristate PFD and CP presented in Figure 8 are implemented in RPLL. With postlayout simulation results of the charge pump (90 nm CMOS process, 1.0 V supply), the percentage mismatch error ($I_{CP} = 100\,\mu A$) for $0.32\,V \leq V_b \leq 0.70\,V$ is less than 2% and increases to less than 5% for $0.24\,V \leq V_b \leq 0.76\,V$ ($V_{UP} = 1\,V$, $V_{DN} = 0\,V$).

As a loop filter (LF), a second-order lag-lead filter is implemented. The filter consists of a register ($R = 16\,k\Omega$), and two on-chip capacitors ($C_1 = 41\,pF$, $C_2 = 1\,pF$). The frequency divider chain in RPLL consists of five divide-by-2 circuits. As a result, it can divide by 32 (i.e., divider ratio $N_2 = 32$). Finally, the AND-based pulser and the variable time-delay unit (ΔT) were implemented for effective injection locking. In RPLL, an injection frequency of f_{inj1} is same to a reference frequency of f_{ref}.

4. Measurement Results

4.1. Main PLL (MPLL). Figures 10(a) and 10(b) show chip micrograph of the differential ring VCO and a PLL, respectively. To clear the effectiveness of the proposed PLL, the VCO cell used in the PLL was also fabricated. They were fabricated by a 90 nm Si CMOS process. The area of the ring VCO core is $0.030 \times 0.045\,mm^2$ including the bias-level-shift circuit and the pulser. The PLL circuit occupies an area of $0.38 \times 0.21\,mm^2$. They were measured in 1.0 V supply condition. Also, the PLL circuit was measured using $20\,\mu A$-current-sources (I_{CP}) into the charge pump.

During free-running operation, the frequency tuning range of the VCO was from 6.35 GHz to 11.5 GHz as shown in Figure 11. It was measured by using an Agilent Technologies E5052B signal source analyzer. It also shows that the VCO

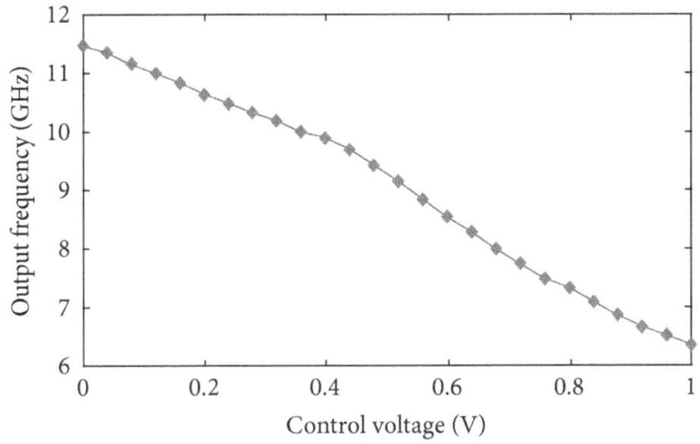

FIGURE 11: Measured frequency tuning range of the VCO.

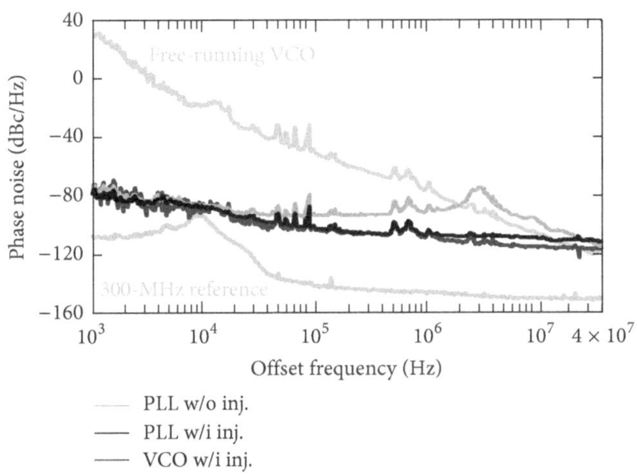

FIGURE 12: Measured phase noise characteristics at 7.2 GHz from the VCO and PLL output without and with injection locking.

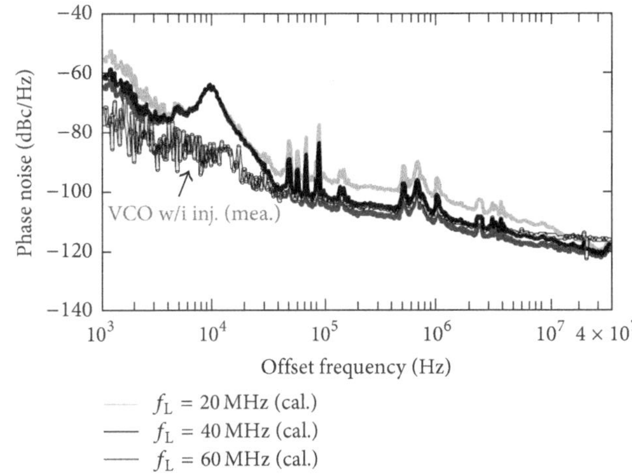

FIGURE 13: Calculated phase noise by using (4) and measure phase noise characteristics as shown in Figure 12.

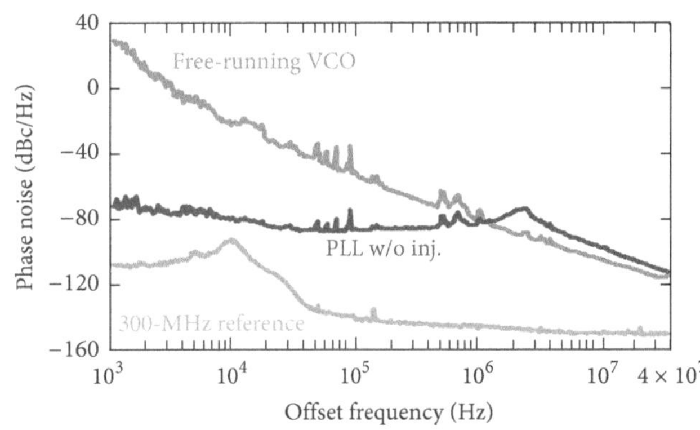

FIGURE 14: Measured phase noise characteristics at 10.8 GHz from the VCO and PLL output without injection locking.

output frequency could be tuned quite linearly versus the rail-to-rail control voltage (V_b) due to the bias-level-shift circuit. When the VCO output frequency (f_0) is 7.18 GHz, the total power consumption of the VCO (with the bias-level-shift circuit and pulser) was 8.4 mW.

Phase noise characteristics of the VCO and PLL at $f_0 \simeq f_{out}$ = 7.2 GHz without and with injection locking are shown in Figure 12 as measured by the signal source analyzer. In addition to them, phase noise characteristics of the 300 MHz reference signal are shown in Figures 12 and 14. A 1 MHz-offset phase noise of −75.8 dBc/Hz was generated in the free-running VCO. With injection locking, a 1 MHz offset phase noise of −108 dBc/Hz was generated, which was improved by 32 dB compared to the former. On the other hand, a 1 MHz-offset phase noise of −91.3 dBc/Hz was generated in the PLL when the PLL was only locked by the phase-locked loop. Due to the poor phase margin, gain peaking at the offset frequency of about 3 MHz was observed. With injection locking, a 1 MHz offset phase noise of −107 dBc/Hz was generated, which was improved by 16 dB compared to the former.

Figure 13 shows calculated phase noise characteristics by using (4) and the measurement phase noise of the free-running VCO and the reference signal as shown in Figure 12. The results show that wider lock range makes lower phase noise characteristics within the lock range. From the calculated result of $f_L = \omega_L/2\pi$ = 40 MHz, (4) is well matched to the measurement results except the offset-frequency region up to about 30 kHz. It is because that flicker noise model as expressed in [5] is not included for simplicity and certain spurs occurred at the offset frequency of about 10 KHz were measured.

Phase noise characteristics of the VCO and PLL at $f_0 \simeq f_{out}$ = 10.8 GHz are shown in Figure 14. A 1MHzoffset phase noise of −79.5 dBc/Hz and −83.7 dBc/Hz were generated in the free-running VCO and the PLL, respectively. Phase noise reduction with injection locking could not be achieved since it was difficult to generate effective injection pulses with sufficient power for achieving the injection-locked condition at that high output frequency.

4.2. Cascaded PLL. Figure 15 shows a chip micrograph of the proposed CPLL. It was fabricated by a 90 nm Si CMOS process. It includes both RPLL and MPLL that occupy an area of 0.11 mm^2. It was measured in 1.0 V supply condition. Also, the PLL circuit was measured using 100 μA current

FIGURE 15: A chip micrograph of the proposed cascaded PLL.

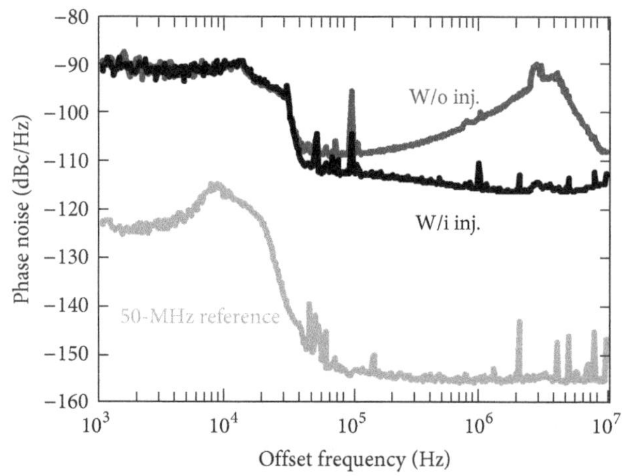

FIGURE 16: Measured phase noise characteristics of RPLL output at 1.6 GHz.

TABLE 3: Performance summary at f_{out} = 7.2 GHz.

This work	MPLL VCO	MPLL	CPLL	
f_{out} [GHz]		7.2		
f_{ref} [MHz]	—	300	50	
f_{inj} [MHz]	300	300	200	1600
f_{out}/f_{inj}	24	24	36	4.5
$L(\Delta f)\|_{\Delta f=1\,\text{MHz}}$ [dBc/Hz]	−108	−107	−99	−101
Power consumption [mW]	8.4	14	25	
Area [mm^2]	0.0014	0.080	0.11	

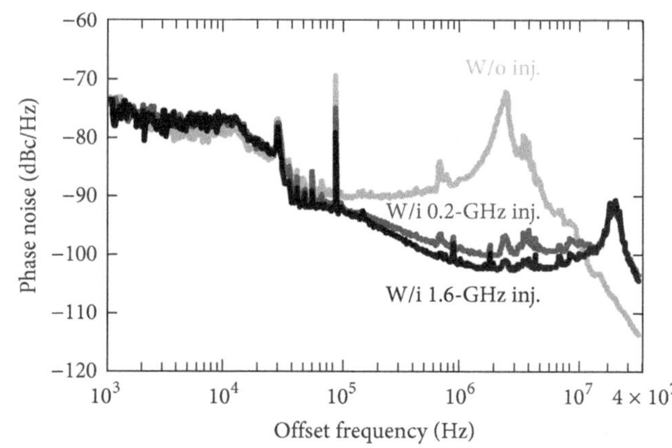

FIGURE 17: Measured phase noise characteristics of CPLL output at 7.2 GHz.

sources into RPLL charge pump and 20 μA current-sources into MPLL charge pump. RPLL was locked to reference signals of 50 MHz which were generated by the pulse pattern generator.

Figure 16 shows the phase noise characteristics at f_{rout} = 1.6 GHz (⊠ 32 × 50 MHz) as measured by an Agilent Technologies E5052A signal source analyzer. Without injection locking, a 1 MHz-offset phase noise of −100 dBc/Hz was generated in RPLL. Due to the poor phase margin, gain peaking at the offset frequency of about 4 MHz was observed. With injection locking, the measured phase noise was −116 dBc/Hz at an offset of 1 MHz. It shows a 16-dB phase-noise reduction with injection locking. Also, phase noise characteristics of the external reference signal are shown Figure 16. At 10 KHz and 1 MHz offset, the phase noise of the reference signal were −117 and −155 dBc/Hz, respectively.

Figure 17 shows the phase noise characteristics at f_{mout} = 7.2 GHz (⊠ 144 × 50 MHz). 0.2 GHz injection signals were injected when N_2, N_3, and N_4 are corresponding to 36, 8, and 8, respectively. Also, 1.6 GHz injection signals were injected when N_2, N_3, and N_4 are corresponding to 36, 1, and 8, respectively. Without injection into MPLL, a 1 MHz offset phase noise of −88 dBc/Hz was generated in the PLL. With integral subharmonic injection locking (f_{mout} = 36 × f_{inj2}, f_{inj2} = 0.2 GHz), the measured phase noise was −99 dBc/Hz at an offset of 1 MHz. With high-frequency half-integral subharmonic locking (f_{mout} = 4.5 × f_{inj2}, f_{inj2} = 1.6 GHz), we successfully achieved 2 dB lower phase noise at 1 MHz offset than the former. A 4 MHz offset phase noise was improved by 4 dB in the latter case, compared with the former. The results show that high-frequency reference injections can widen the injection lock range. However, there was a spur around the offset frequency of 25 MHz owing to the RPLL spur level, and the spur limited the lock range widening with high-frequency signal injections.

Usually, spurs are induced by periodic phase shift due to injection locking. The spur level can be expressed as follows:

$$\sqrt{\frac{P_{\text{spur},\omega_{\text{inj}}}}{P_0}} \approx \frac{\omega_0}{2Q\omega_{\text{inj}}} \cdot \sqrt{\frac{P_{\text{inj}N}}{P_{\text{osc}}}} = \frac{\omega_L}{\omega_{\text{inj}}}, \quad (8)$$

where $P_{\text{spur},\omega_{\text{inj}}}$ represents the spur levels occurred by the reference signal at $f_0 \pm f_{inj}$, and P_0 is the injection-locked output power of the oscillator [22]. As shown in this equation, the spur level would be reduced lowering the lock range with the same reference frequency, however, which is undesirable to reduce phase noise characteristics.

Calculated phase noise characteristics by using (6) and measure phase noise characteristics, as shown in Figures 12 and 16, are shown in Figure 18. In this case, the lock range was supposed to be proportional to the input frequency and the coefficient was 0.14, which was expected in Figure 13. In

TABLE 4: Performance comparison of PLLs.

Reference	CMOS Technology	f_0 [GHz]	$N = f_0/f_{ref}$	$L_{in\text{-}band}$ [dBc/Hz]	$L^*_{normalized}$ [dBc/Hz^2]	Power [mW]	Area [mm^2]	VCO
This work	90 nm	1.6	32	**−116**	**−223**	11 (sim.)	0.031	Ring
		7.2	144	**−101**	**−221**	25	0.11	
		9.6	192	**−93**	**−216**	27		
[11]	90 nm	20	20	−113	−229	38	0.46	LC
[18]	90 nm	9.24	35	−97	−212	56	0.09	Ring
[19]	0.18 μm	8.98	17	−110	−222	58	0.77	LC
[20]	0.18 μm	8.45	32	−98	−212	117	5.5	LC

*Normalized in-band phase noise ⊠ $L_{in\text{-}band} - 20\log N - 10\log f_{ref}$.

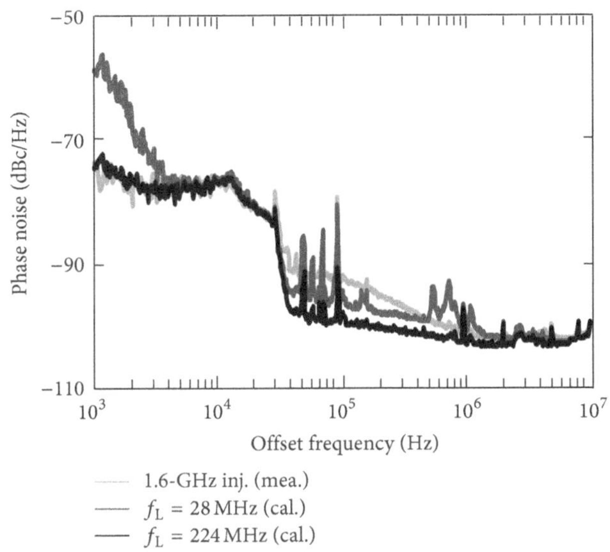

FIGURE 18: Calculated phase noise by using (6) and measuring phase noise characteristics as shown in Figures 12 and 16.

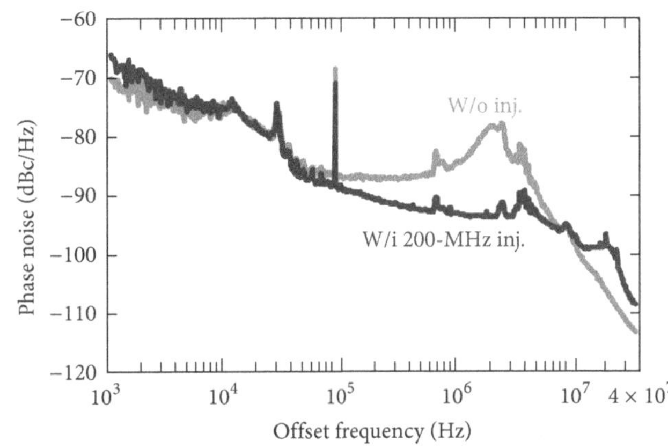

FIGURE 19: Measured phase noise characteristics of CPLL output at 9.6 GHz.

the results, phase noise characteristics especially at the offset frequency of 500 kHz, 700 kHz, and lower than 5 kHz due to the secondary VCO (VCO2) would be reduced by using high-frequency injection signals (f_{inj} = 1.6 GHz, f_L = 224 MHz). In Figure 18, measured phase noise characteristics from the offset frequency of 30 kHz to 1 MHz were degraded compared with calculated results due to induced noise from the MPLL loop.

Figure 19 shows the phase noise characteristics at f_{mout} = 9.6 GHz (⊠ 192 × 50 MHz). In these cases, N_2, N_3, and N_4 are corresponding to 24, 8, and 8, respectively. Without injection into MPLL, a 1 MHz offset phase noise of −85 dBc/Hz was generated in the PLL. With integral subharmonic injection locking (f_{mout} = 48 × f_{inj2}, f_{inj2} = 0.2 GHz), the measured phase noise was −93 dBc/Hz at an offset of 1 MHz. In Figure 19, spur levels around the offset frequency of 25 MHz were decreased, because the phase-locking effect of injection locking was decreased.

The PLL generated reference spurs of lower than −31 dBc at the output frequency of 7.2 GHz with 1.6 GHz injections, as shown in Figure 20(a). At the output frequency of 9.6 GHz with 0.2 GHz injections, reference spurs of lower than −27 dBc were measured as shown in Figure 20(b).

A performance summary at the output frequency of 7.2 GHz of the fabricated chips are given in Table 3, when injection locking was established. It shows that high-frequency injections are effective to reduce the phase noise because a wide injection lock range can be achieved.

A performance comparison of the PLL with other PLLs that were designed using various kinds of phase-locking methods is given in Table 4. Unfortunately, the proposed PLL cannot cover wide frequency range from 6 GHz to 12 GHz as shown in Table 1, due to the VCO tuning range and limitation of tunable divider ratio. To make a fair in-band phase noise comparison between various kinds of PLL designs, the dependency of in-band phase noise on f_{ref} and N should be normalized out [23]. Therefore, normalized in-band phase noise $L_{normalized}$ was applied for comparison. The proposed PLL shows a relatively good $L_{normalized}$ value. Also its area and power consumption are small and comparable to that of other circuits.

5. Conclusion

An inductorless PLL architecture, using the combination of a phase-locked loop, and injection locking with a ring VCO was proposed. The proposed CPLL that consists of two PLLs was designed in order to generate high-frequency output signals with low-frequency external reference signals. High-frequency half-integral subharmonic injection locking

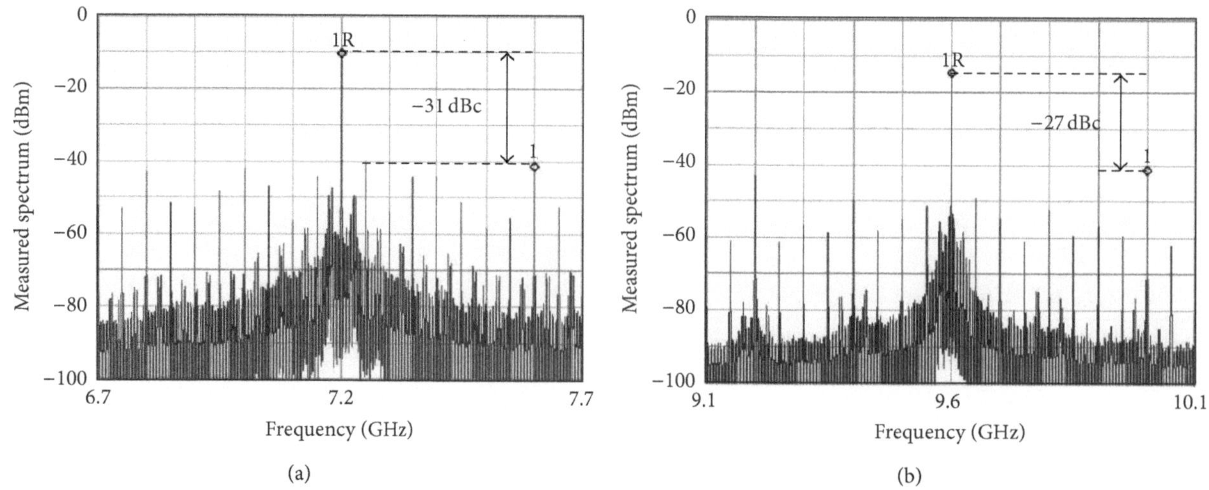

FIGURE 20: Measured frequency spectra of CPLL output (a) at 7.2 GHz and (b) at 9.6 GHz.

to improve the phase noise characteristics of the inductorless PLL was implemented.

The injection-locked PLL was fabricated by adopting 90 nm Si CMOS technology. A 1 MHz-offset phase noise of −101 dBc/Hz was achieved at an output frequency of 7.2 GHz, which was improved by 25 dB compared with that of the free-running VCO. The area of this inductorless PLL was as small as 0.11 mm^2 with low power consumption of 25 mW.

Acknowledgments

This work was partly supported by STARC, KAKENHI, MIC.SCOPE, and VDEC in collaboration with Agilent Technologies Japan, Ltd., Cadence Design Systems, Inc., and Mentor Graphics, Inc. The authors also acknowledge the JSPS Research Fellowship for Young Scientists from the Japan Society for the Promotion of Sciences.

References

[1] R. Adler, "A study of locking phenomena in oscillators," *Proceedings of the IRE*, vol. 34, pp. 351–357, 1946.

[2] Y. Ito, H. Sugawara, K. Okada, and K. Masu, "A 0.98 to 6.6 GHz tunable wideband VCO in a 180 nm CMOS technology for reconfigurable radio transceiver," in *Proceedings of the IEEE Asian Solid-State Circuits Conference (ASSCC '06)*, pp. 359–362, November 2006.

[3] B. Razavi, "Cognitive radio design challenges and techniques," *The IEEE Journal of Solid-State Circuits*, vol. 45, no. 8, pp. 1542–1553, 2010.

[4] F. M. Gardner, "Charge-pump phase-lock loops," *IEEE Transactions on Communications*, vol. 28, no. 11, pp. 1849–1858, 1980.

[5] ⊠. Zhang, ⊠. Zhou, and A. S. Daryoush, "A theoretical and experimental study of the noise behavior of subharmonically injection locked local oscillators," *IEEE Transactions on Microwave Theory and Techniques*, vol. 40, no. 5, pp. 895–902, 1992.

[6] B. Razavi, "A study of phase noise in CMOS oscillators," *IEEE Journal of Solid-State Circuits*, vol. 31, no. 3, pp. 331–343, 1996.

[7] S. Lee, N. Kanemaru, S. Ikeda et al., "A ring-VCO-based injection-locked frequency multiplier with novel pulse generation technique in 65 nm CMOS," *IEICE Transactions on Electronics*, vol. 95, no. 10, pp. 1589–1597, 2012.

[8] Y. Kobayashi, S. Amakawa, N. Ishihara, and K. Masu, "A low-phase-noise injection-locked differential ring-VCO with half-integral subharmonic locking in 0.18 μm CMOS," in *Proceedings of the 35th European Solid-State Circuits Conference (ESSCIRC '09)*, pp. 440–443, September 2009.

[9] T. Sugiura and S. Sugimoto, "FM noise reduction of Gunn-effect oscillators by injection locking," *Proceedings of the IEEE*, vol. 57, no. 1, pp. 77–78, 1969.

[10] M. C. Chen and C. Y. Wu, "Design and analysis of CMOS subharmonic injection-locked frequency triplers," *IEEE Transactions on Microwave Theory and Techniques*, vol. 56, no. 8, pp. 1869–1878, 2008.

[11] J. Lee and H. Wang, "Study of subharmonically injection-locked PLLs," *The IEEE Journal of Solid-State Circuits*, vol. 44, no. 5, pp. 1539–1553, 2009.

[12] S. Y. Lee, S. Amakawa, N. Ishihara, and K. Masu, "Low-phase-noise wide-frequency-range ring-VCO-based scalable PLL with subharmonic injection locking in 0.18 μm CMOS," in *Proceedings of the IEEE MTT-S International Microwave Symposium Digest (MTT '10)*, pp. 1178–1181, May 2010.

[13] S. Y. Lee, S. Amakawa, N. Ishihara, and K. Masu, "High-frequency half-integral subharmonic locked ring-VCO-based scalable PLL in 90 nm CMOS," in *Proceedings of the IEEE Asia-Pacific Microwave Conference (APMC '10)*, pp. 586–589, December 2010.

[14] B. Razavi, *Design of Analog CMOS Integrated Circuits*, McGraw-Hill, New York, NY, USA, 2001.

[15] M. Tiebout, "A CMOS direct injection-locked oscillator topology as high-frequency low-power frequency divider," *IEEE Journal of Solid-State Circuits*, vol. 39, no. 7, pp. 1170–1174, 2004.

[16] P. R. Gray, P. J. Hurst, S. H. Lewis, and R. G. Meyer, *Analysis and Design of Analog Integrated Circuits*, John Wiley ⊠ Sons, New York, NY, USA, 4th edition, 2001.

[17] T. Sekiguchi, S. Amakawa, N. Ishihara, and K. Masu, "Inductorless 8.9 mW 25 Gb/s 1:4 DEMU⊠ and 4 mW 13 Gb/s 4:1 MU⊠ in 90 nm CMOS," *Journal of Semiconductor Technology and Science*, vol. 10, no. 3, pp. 176–184, 2010.

[18] H. Kodama, H. Okada, H. Ishikawa, and A. Tanaka, "Wide lock-range, low phase-noise PLL using interpolative ring-VCO with coarse frequency tuning and frequency linearization," in *Proceedings of the IEEE Custom Integrated Circuits Conference (CICC '07)*, pp. 349–352, September 2007.

[19] G. Y. Tak, S. B. Hyun, T. Y. Kang, B. G. Choi, and S. S. Park, "A 6.3–9 GHz CMOS fast settling PLL for MB-OFDM UWB applications," *IEEE Journal of Solid-State Circuits*, vol. 40, no. 8, pp. 1671–1679, 2005.

[20] T. Y. Lu and W. Z. Chen, "A 3-to-10GHz 14-band CMOS frequency synthesizer with spurs reduction for MB-OFDM UWB system," in *Proceedings of the IEEE International Solid-State Circuits Conference—Digest of Technical Papers (ISSCC '08)*, pp. 126–601, February 2008.

[21] S. Ye, L. Jansson, and I. Galton, "A multiple-crystal interface PLL with VCO realignment to reduce phase noise," *IEEE Journal of Solid-State Circuits*, vol. 37, no. 12, pp. 1795–1803, 2002.

[22] M. M. Izad and C. Heng, "A pulse shaping technique for spur suppression in injection-locked synthesizers," *The IEEE Journal of Solid-State Circuits*, vol. 47, no. 3, pp. 652–664, 2012.

[23] ⊠. Gao, E. A. M. Klumperink, M. Bohsali, and B. Nauta, "A 2.2 GHz 7.6 mW sub-sampling PLL with −126 dBc/Hz in-band phase noise and 0.15 psrms jitter in 0.18 μm CMOS," in *Proceedings of the IEEE International Solid-State Circuits Conference—Digest of Technical Papers (ISSCC '09)*, pp. 392–393, 393a, February 2009.

19

Downconverting Module Architectures for High Performance Multipixel Cameras

Diego Palombini, Maciej Jankowski, and Ernesto Limiti

Department of Electronic Engineering, University of Roma Tor Vergata, Roma 00133, Italy

Correspondence should be addressed to Diego Palombini; diego.palombini@uniroma2.it

Academic Editor: Chang-Ho Lee

Multipixel cameras represent an emerging topology for arrays receivers, improving speed and accuracy of both security scanning systems and radioastronomical sky surveys by means of a matrix of phased elements. Difficulties in the generation and proper distribution to each pixel of the local oscillator signal still limit their use to frequency ranges below a few GHz or at least seriously affect the complexity of the implementable cameras. This work presents a full comparison between two possible system architectures, alternatively based on LO frequency multiplication or subharmonic mixing strategies, aiming to overcome the aforesaid limitations: design and performance of two compact test vehicles in MMIC technology, both operating in the Q-band frequency range with ultrabroadband IF section, are reported.

1. Introduction

Array receivers have been adopted since the beginning of the 90's in radioastronomical applications [1] and, more recently, in security scanning systems, improving active and passive millimetre-wave imaging capabilities [2].

In radioastronomical applications, this kind of receivers is usually placed at the focus of a large radiotelescope antenna and it is mainly used to survey large areas in a more efficient way than single-pixel receivers. Array receivers can be roughly classified into three different types: focal plane arrays (FPAs), which incorporate distinct superheterodyne receivers performing a spectroscopic analysis of the incoming signal; bolometric arrays (BAs), achieving an integral measure of the total channel power; phased array feeds (PAFs) consisting of a matrix of receiving elements fed by the same local oscillator and then digitally combined [3]. Major research efforts are currently devoted to the latter receiver architecture, exhibiting two key advantages over conventional multibeam horn feeds: firstly it enables the radiotelescope to support many simultaneous beams on the sky, thus increasing the instrument speed; secondly, it offers the opportunity to properly shape the pattern of the feed array in order to match the optical configuration of the telescope [4]. The major drawback of a PAF, often limiting its use for frequency ranges above a few GHz, is however the necessary requirement of distributing the same LO signal to each pixel, without loss of synchronism between different paths; the research for new circuit topologies allowing to overcome this limitation represents the main purpose of this work. Possible strategies have the common feature of reducing the main local oscillator operating frequency, thus alleviating the critical aspect of distributing a millimetre-wave signal among the different paths, and leaving the task of locally producing the latter to the front-end pixel circuitry.

Viable approaches to this goal can be roughly classified into two different categories: both the integration of $a \times N$ multiplier and the selection of a Nth order mixing product allow the scaling of the LO frequency by the same N factor. Clearly, the two architectures are featured by different pros and cons and undergo different criteria to be properly implemented. In the following sections, the two topologies are compared through the design and the resulting performance of the two subsystems implemented in monolithic technology, both operating at the same Q-band frequency range.

In particular, after the present introduction, in Section 2 the two more promising architectures are introduced and critically reviewed, while in Sections 3 and 4 the designs and performance of the two test vehicles are described in detail. Respective results are finally compared and summed up in Section 5.

2. Feasible Architectures Description

As already stated, the main factor limiting the use of multipixel cameras at high frequencies is related to the difficulty of correctly synchronizing the local oscillator among different paths. A real case will be adopted in the following to describe the features of the selected design approaches. In particular, a Q-Band receiver will be addressed, able to downconvert the incoming RF full-band signal (33–50 GHz) into the resulting low-frequency 1–18 GHz signal. In the present case, the frequency conversion would require the proper distribution, among the different pixel systems, of a 32 GHz LO signal. In order to circumvent this problem, two feasible strategies may be envisaged: the first one consists in multiplying the original LO signal while the second one in selecting a higher RF-LO mixing product; in the remainder of this work we will refer to the multiplier-based topology as MHM (Multiplied Harmonic Mixing) architecture and to the subharmonic mixing-based topology as SHM (Subharmonic Mixing) architecture. MHM architecture, depicted in Figure 1(a), offers the possibility of LO signal amplification and takes advantage of the higher conversion gain typical in fundamental LO mixing; unfortunately, the insertion of the multiplying-amplifying chain results in the twofold drawback of higher circuit complexity and DC power consumption. Conversely, SHM architecture, depicted in Figure 1(b), hardly allows LO signal amplification and suffers from higher conversion losses related to the use of a higher-order mixing product. On the other hand, circuit complexity and DC power consumption involved in the implementation of a subharmonic mixer are definitely lower. The higher the selected multiplication factor ($\times N$) is or the higher-order mixing product (Nth), the more pronounced the resulting effect is, thus emphasizing advantages and disadvantages of both implementations.

Further notes can be drawn considering the values of N that are actually selected in the present designs: ×8 for the multiplication factor and the selection of 2nd order mixing product. MHM architecture allows in this case the generation and distribution of a lower frequency LO (4 GHz) and maintains 1 GHz separation between LO and RF bands; on the other hand, some external hybrid circuitry is however needed to filter out unwanted LO harmonics. SHM architecture implies the distribution of a higher frequency LO (16 GHz) and an overlap between IF and LO bands (thus imposing a very high LO/IF isolation to the mixer) but, on the other hand, does not need any additional external filtering.

Both implementations have to be properly designed to handle the 180% relative bandwidth (1–18 GHz) at the IF port.

Two test vehicles have been designed and simulated in agilent ADS environment [5]. Given the inherent uniformity and volume production featured by multipixel camera applications, a monolithic approach has to be preferred. To this goal, a metamorphic GaAs 70 nm gate length foundry process by OMMIC (D007IH) has been selected. Such industrial-grade technology is featured by 1.0 dB minimum noise figure and 9.5 dB associated gain at 50 GHz (for a sample $4 \times 25\,\mu$m device, biased at $V_{DS} = 1$V and $I_{DS} = 20\%\ I_{DSS}$). The resulting design of the two architectures will be presented in the following.

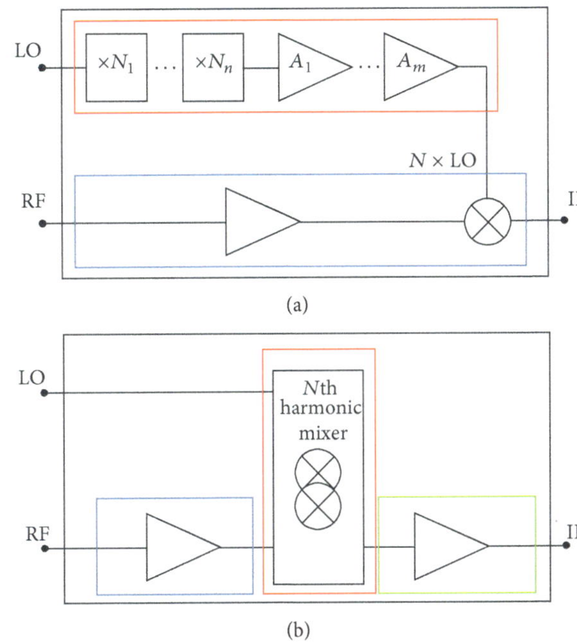

FIGURE 1: MHM (a) and SHM (b) architectures block diagrams.

3. MHM Architecture Test Vehicle Design

As stated above, the harmonic mixing topology (Figure 1(a)) mainly consists in the combination of two different functionalities, namely, an LO amplifying-multiplying chain and a RF mixing section, whose design will be fully described in the present section. The corresponding two subsystems, while intended to operate together, have been designed and implemented as separate ones onto the same MMIC die, for the purpose of testability and resulting design flexibility. The former feature actually allows the on-site measurement of the multiplier harmonic content, while the latter allows the insertion, if unacceptable harmonic levels result, of eventual external highly selective filtering structures.

3.1. Multiplying-Amplifying Chain. The aforementioned technology and the related process offer the possibility of using different nonlinear devices to perform the LO frequency multiplication: Schottky diodes (in varactor-like operation), drain-source-connected FETs (by using the resulting Schottky diodes), and mHEMT FETs (exploiting the pinch-off nonlinearity); however, manifold considerations about the lack of proper varactor diodes for mixing and multiplication purposes and the high losses introduced by the diode resistive multiplier [6] led to the selection of an active FET multiplier

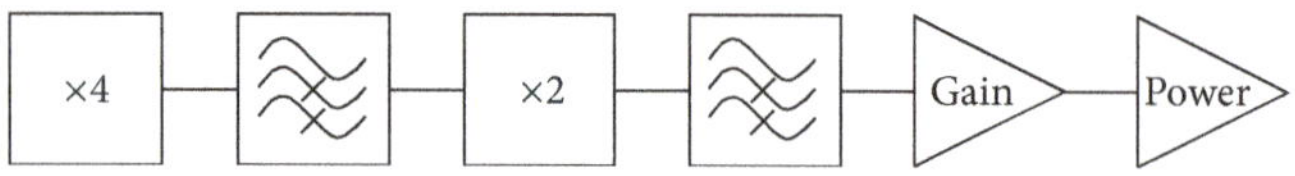

Figure 2: Multiplying-amplifying chain.

structure [7], providing at the same time a limited amount of gain.

Different chain architectures [8, 9] can be adopted to obtain the ×8 multiplication: cascading three doublers or using a quadrupler doubler topology, both in single-ended or balanced configurations. Multiplier architecture, made by self-biased ×4 and ×2 blocks, was selected as the best compromise between conversion gain, efficiency, and size requirements; a more accurate analysis allowed to select the single-ended configuration, given the poor performance improvement resulting from the balanced one, in conjunction with a major increase in circuit complexity, power consumption, and area occupation.

Fundamental frequency rejection can be therefore only guaranteed by the insertion, at the output of the two stages, of the proper high-pass filtering structures [10, 11] ensuring large-signal match at the desired harmonic frequency and optimum harmonic loading [12]. The proper conversion gain level and output power were finally achieved by further cascading a gain stage and an output buffer amplifier both maintaining independent gate/drain bias (see Figure 2), with the major consequence of separating the nonlinear operation of the two multiplying sections from the LO input of the mixer.

3.2. Amplifying-Mixing Section. The approach in the design of this subsystem was essentially driven by the wide operating band requested to the mixer and by the close proximity between RF and LO frequency bands; further, minimum area occupation and power consumption, dictated by the large number of pixels that are typically needed, are the driving constraints.

In light of the abovementioned severe specifications, a passive structure providing low noise operation, high linearity, and low DC power consumption was selected; in addition, the adoption of HEMTs in diode configuration as mixing elements typically requires less LO power than using actual diodes [13].

The choice of the most effective mixing topology has been the result of an intense investigation in which advantages and disadvantages of various implementations with different balancing levels and geometric arrangement have been compared: double balanced ring topology was found to guarantee an acceptable level of conversion gain and isolation for the concerned application [10, 14].

Different microstrip realizations of the balancing structures can be found in open literature, ranging from parallel coupled-line balun to Marchand-like baluns and planar transformers [15–17]; however, the need to comply with reduced area and wide bandwidth constraints suggested the use of spiral baluns, considering their extremely compact size and wideband performance [18, 19].

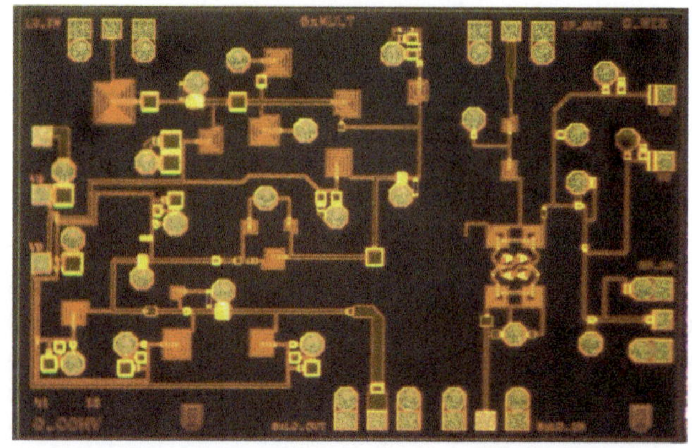

Figure 3: Microphotograph of the realized MHM architecture test vehicle MMIC.

The major drawback of the topology under discussion consists in the difficulties of a wideband IF extraction, mainly due to the inductive behaviour exhibited in the physical realization of the single-ended IF port [20], difficult to match over ultrawideband frequencies. To improve IF matching and isolations, an elliptic filter section, featuring low insertion loss in the 1–18 GHz band and high attenuation in the RF and LO frequencies was then inserted at IF port.

Finally, a two-stage buffer amplifier was added at RF port, thus improving input return loss and providing the entire system with a positive gain at the expense of a corresponding reduction of RF to IF isolation. Such an additional amplifier has the further beneficial effect of isolating the nonlinear RF mixer input from the output of the LNA that would likely be inserted just after each receiving pixel antenna; moreover, the resulting noise figure of the overall system is indeed improved.

3.3. MHM Architecture Test Vehicle Performance. The whole system implementation results in a really integrated and compact layout, with an area occupation of about $3 \times 2\,\mathrm{mm}^2$. The die microphotograph is shown in Figure 3, where both the multiplying-amplifying chain and the amplifying-mixing sections are shown, respectively on the left and right part of the chip: input port for the low-frequency local oscillator and output port for the multiplied LO signal are placed on opposite sides while, regarding the mixer, RF, LO, and IF ports are clockwise arranged on the right, lower, and upper borders of the chip, respectively. Multiplier output and LO mixer ports are juxtaposed specifically to be both directly interconnected by means of a simple wire bond connection or separated by an high-order microstrip hybrid filter selecting the desired 8th LO harmonic and rejecting an eventually higher unwanted spurious level.

In Figure 4, a comparison between measured and simulated multiplier port matching is reported, demonstrating better than 25 dB input return losses and 15 dB output matching.

The whole multiplying-amplifying chain exhibits a total DC power consumption of approximately 200 mW, with gain and output power performance, at 3 dBm input power from

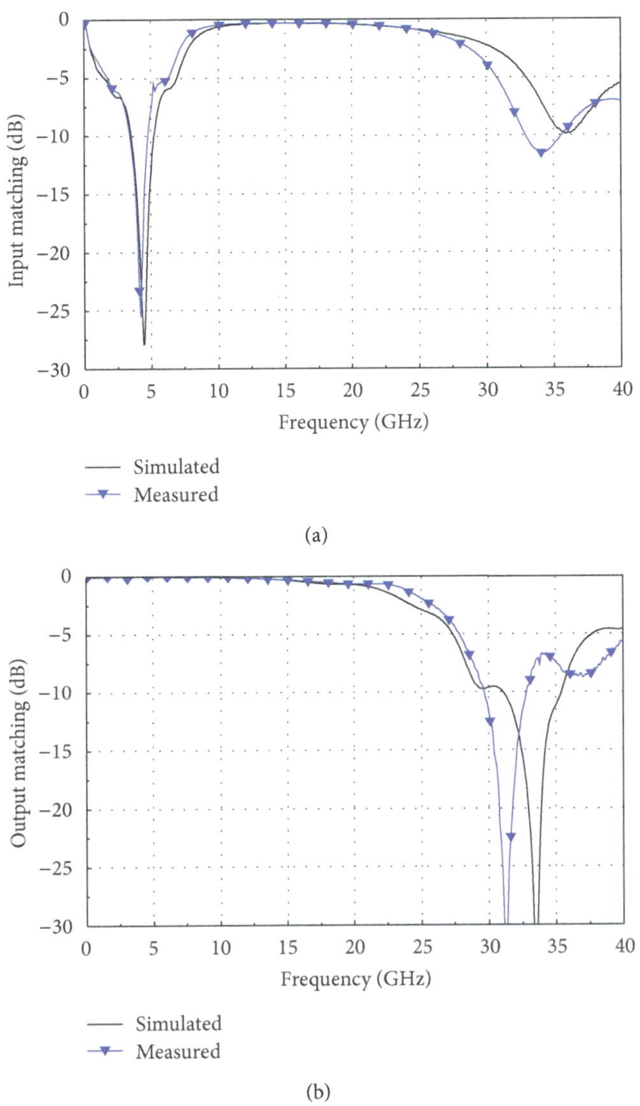

FIGURE 4: Simulated (black trace) and measured (blue trace) multiplying-amplifying chain input (a) and output (b) matching.

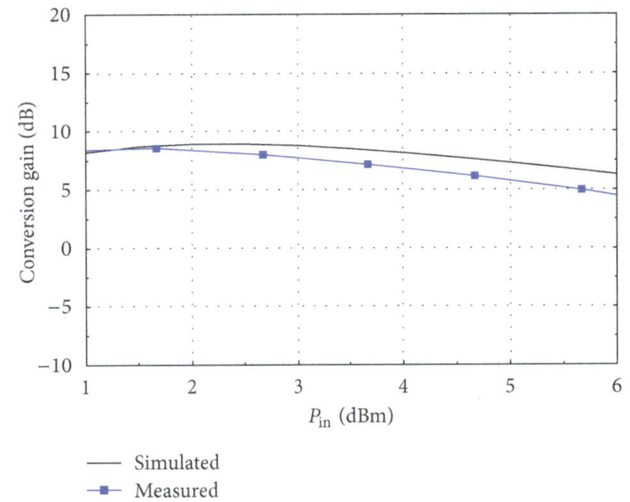

FIGURE 5: Simulated (black trace) and measured (blue trace) multiplying-amplifying chain conversion gain as a function of input LO power drive.

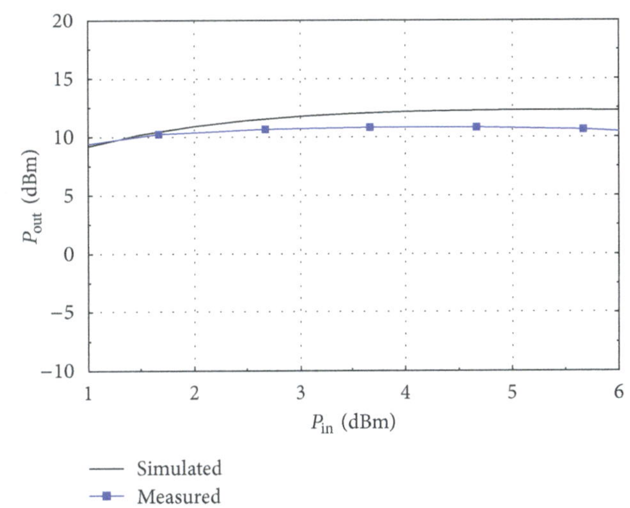

FIGURE 6: Simulated (black trace) and measured (blue trace) multiplying-amplifying chain output power as a function of input LO power drive.

the LO source, equal to 8 dB, and 11 dBm, respectively, in good agreement with simulated results (Figures 5 and 6 resp.). Under the same large-signal operating conditions, the ratio between the desired 8th LO harmonic and the highest spurious product is around 20 dBc, a rejection level that actually allows the direct connection between the amplifying/multiplying chain and the mixer LO input.

The amplifying-mixing section exhibits a DC power consumption of about 40 mW. Moreover, the structure is featured by a pretty good matching level (see Figure 7), especially remarkable if the relative bandwidths are considered: −15 dB at the RF port (40% bandwidth) and −9 dB at the IF port (180% bandwidth).

Conversion gain and RF/IF isolation at 7.5 dBm LO power are reported in Figure 8: the former is typically around 10 dB on the entire 33–50 GHz frequency range, while the latter varies from 10 dB to 35 dB as frequency increases; LO leakage toward RF and IF ports is lower than −43 dB and −33 dB, respectively.

4. SHM Architecture Test Vehicle Design

Such a topology can be split into three main functional blocks (Figure 1(b)), namely, sub-harmonic mixer, RF, and IF amplifiers, whose overall design provides sufficient degrees of freedom to meet different downconverter requirements while maintaining compact overall size.

4.1. Subharmonic Mixer. Considerations similar to the previous, about operating bandwidth, area occupation, and power consumption, make HEMTs in diode configuration the best

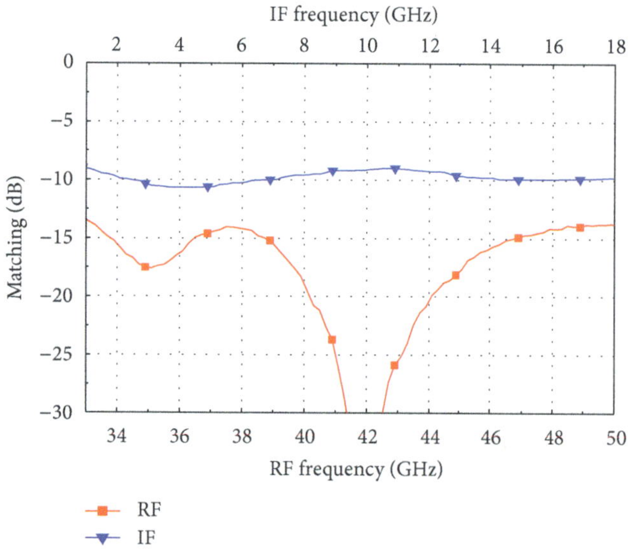

FIGURE 7: Measured amplifying-mixing section RF and IF matching.

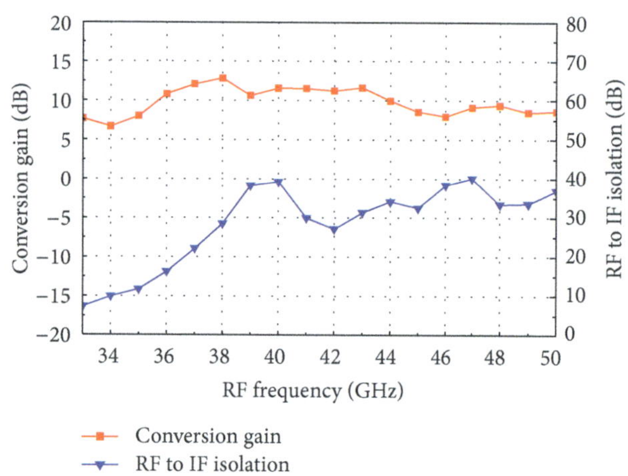

FIGURE 8: Measured amplifying-mixing section conversion gain and RF/IF isolation.

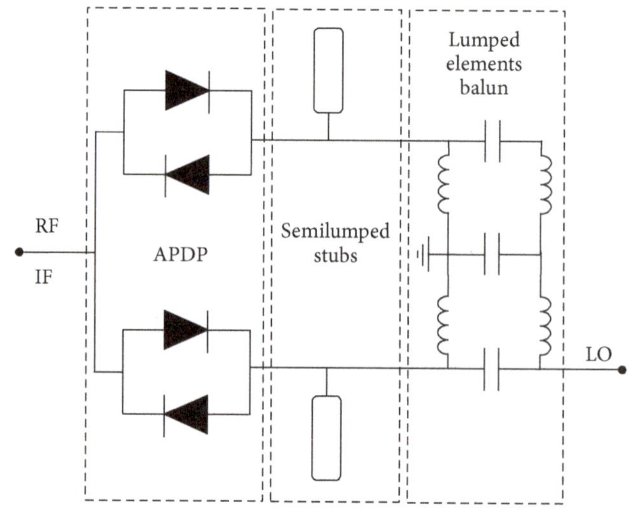

FIGURE 9: simplified schematic of sub-harmonic mixer.

starting point to develop the mixer structure; a particular care has to be posed to ensure structure symmetry and balancing, given the overlap between IF and LO bands. An anti-parallel diode pair (APDP) constitutes the basic core element with the aim to perform a 2nd order harmonic mixing, when an appropriate balancing level is selected to isolate LO and IF ports. In spite of LO and IF bands overlap, a single-balanced arrangement was found to guarantee an acceptable compromise between conversion gain (better than in doubly balanced ones), isolation (slightly worse), and area occupation (much lower) [21], provided the balancing structure is appropriately selected and sized. To this goal, open literature offers different topologies of balun-based networks that enable an effective trade off between operating bandwidth and area occupation; in this case lumped-element implementation [22], featuring compact size at the expense of narrowband performance, was found to represent the best choice. Finally, the proper application of the RF voltage across the balanced APDP was achieved by means of two stubs, implemented with semilumped elements, behaving as an open circuit at LO frequency and as short circuit at RF frequency. The resulting simplified subharmonic mixer schematic is reported in Figure 9.

4.2. RF Amplifier. The main purpose of this block is to provide the system with the adequate amount of conversion gain improving, at the same time, RF port matching and easing overall system noise properties optimization. For this reason, the amplifier was designed by cascading three single stages and accounting for the effective termination on the output port provided by the mixer input impedance. as it will be detailed in the following section, the IF amplifier was designed to exhibit a high impedance level at RF frequency. Three stages, although biased at the same drain voltage, maintain separate DC power supplies; active devices are then self-biased to 0 V regarding the gate supply, in order to reduce the complexity of the biasing circuitry. Furthermore, additional care was paid in controlling output impedance of the RF amplifier at IF frequencies; this impedance was designed to be as high as possible not to affect, with its load, the mixer input and thus allowing the downconverted signal to merely flow towards IF port.

4.3. IF Amplifier. Amplifying section on the IF port is responsible for meeting three fundamental figures of merit: it provides to the system, together with the RF amplifying section, the adequate amount of conversion gain, improves the wideband IF port matching, and guarantees an acceptable level of RF to IF isolation, otherwise unacceptably low. Although the last requirement can be achieved by providing the IF amplifier, in the RF band, with any reactive termination, it is fundamental to choose a high impedance level in order not to affect the RF signal power flow toward the APDP. In light of the aforementioned reasons, distributed amplifier (DA) topology actually represents the only one capable of satisfying the simultaneous requirements of adequate gain and

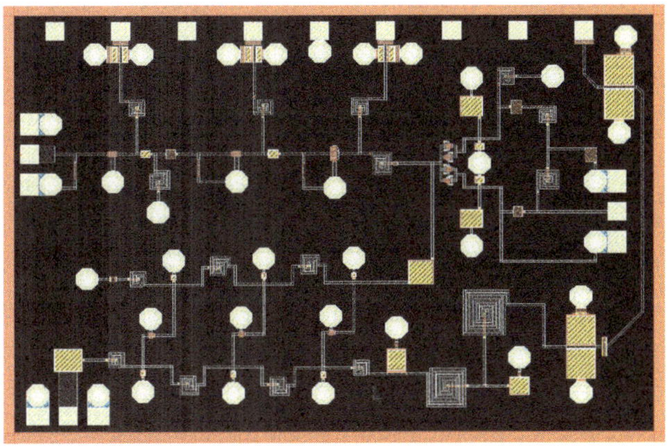

FIGURE 10: Layout of the designed SHM architecture test vehicle MMIC.

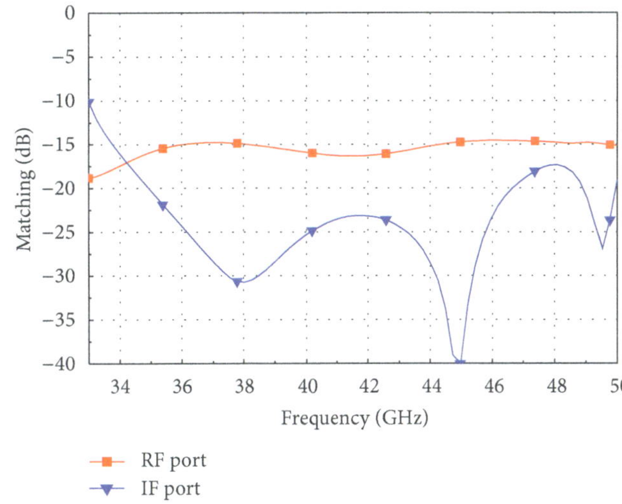

FIGURE 11: Simulated down-converting module RF and IF matching.

input/output matching in a wideband operation. DA design can be performed both on image impedance base [23], by identically replicating the elementary cells of an artificial transmission line, and on filter synthesis approaches [24], by properly selecting the resulting ladder network values. If requested by design constraints, further degrees of freedom may be achieved by tapering active devices gate periphery [25]. In the present case, the filter synthesis approach had to be preferred to appropriately control the stop-band frequency response and properly filter out the RF signal leakage. Three-cell structure provides a sufficient number of elements to synthesize a 7th order chebycheff low-pass filter, featuring 40 dB out-of-band rejection; again, choosing a series element as the first one in the synthesis results in an input impedance higher than 500 Ohm in most of the RF band.

4.4. SHM Architecture Test Vehicle Performance. Even this second implementation, currently under realization, demonstrates a high level of integration: SHM MMIC shows total area occupation of $3 \times 2\,\text{mm}^2$ and results completely comparable with the MHM one (again, $3 \times 2\,\text{mm}^2$). System layout, depicted in Figure 10, shows RF and IF amplifiers, respectively, in the upper and bottom left part of the chip, with a sub-harmonic mixer occupying the remaining right part. LO, IF, and RF ports are clockwise arranged on the right, lower, and left borders of the chip respectively, whereas amplifiers biasing pads can be recognized along the upper edge. The entire system exhibits a total DC power consumption of 120 mW.

In Figure 11 simulated input (RF) and output (IF) matching are reported: the former is better than −15 dB all over the bandwidth and better than −20 dB in most of the 1–18 GHz range, the latter is typically around −15 dB.

Simulated conversion gain and RF/IF isolation at 10 dBm LO power are reported in Figure 12: conversion gain results to vary between 17 dB and 19 dB in the whole operating bandwidth while RF/IF isolation ranges from 35 dB to 80 dB as frequency increases; LO leakage toward RF and IF ports is lower than −34 dB and −135 dB, respectively.

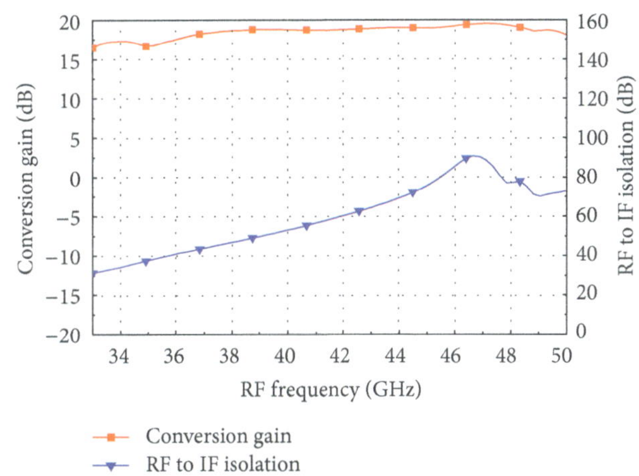

FIGURE 12: Simulated downconverting module conversion gain and RF to IF isolation.

5. Architectures Comparison

Architectures described in previous sections represent different approaches in the realization of downconverting module. Beyond the considerations already performed in Section 2, actual circuit realizations exhibit evident differences concerning the main figures of merit: in Table 1 a summary enabling to draw a direct comparison between two described implementations is reported.

As it can be easily noticed, MHM architecture suffers from a lower conversion gain, worse isolation, and higher DC power consumption but requires 3-4 dBm LO power only at 4 GHz to be effectively implemented. On the other hand, SHM architecture demonstrates better gain and isolation performance, featuring at the same time lower DC power consumption but requires the distribution of a higher LO power at a higher frequency (16 GHz).

Table 1: MHM architecture and SHM architecture main figures of merit.

Parameter	MHM architecture (measured)	SHM architecture (simulated)
Gain (dB)	≈10	≈18
RF/IF isolation (dB)	10 ÷ 40	35 ÷ 80
LO/IF isolation (dB)	33	34
LO/RF isolation (dB)	43	135
RF matching (dB)	<−15	<−15
IF matching (dB)	<−9	<−20
LO frequency (GHz)	4	16
LO power (dBm)	3 ÷ 4	10
DC power consumption (mW)	240	120

6. Conclusion

Two downconverting module architectures for high performance multipixel cameras, both operating in the Q-band frequency range (33–50 GHz) with really ultrabroadband IF section (1–18 GHz), have been illustrated and compared. MHM architecture, fabricated and successfully tested, is based on LO frequency multiplication, whereas SHM architecture, currently under realization, takes advantage of 2nd order harmonic mixing.

References

[1] J. J. Condon, J. J. Broderick, G. A. Seielstad, K. Douglas, and P. C. Gregory, "A 4.85 GHz sky survey. I—maps covering delta between 0 and +75 deg," *Astronomical Journal*, vol. 97, pp. 1064–1073, 1998.

[2] J. Richter, D. Nötel, F. Klöppelt, J. Huckt, H. Essent, and L. P. Schmidt, "A multi-channel radiometer with focal plane array antenna for W-band passive millimeterwave imaging," in *Proceedings of the 2006 IEEE MTT-S International Microwave Symposium Digest*, pp. 1592–1595, June 2006.

[3] D. T. Emerson and J. M. Payne, *Multi-Feed Systems for Radio Telescopes*, Conference Series, Astronomical Society of the Pacific, San Francisco, Calif, USA, 1995.

[4] B. Veidt and P. Dewdney, "Development of a phased-array feed demonstrator for radio telescopes," in *Proceedings of the 11th International Symposium on Antenna Technology and Applied Electromagnetics*, Saint-Malo, France, June 2005.

[5] http://www.home.agilent.com/.

[6] S. Maas, *The RF and Microwave Circuits Design Cookbook*, Artech House, Norwood, Mass, USA, 1998.

[7] P. Colantonio, F. Giannini, G. Leuzzi, and E. Limiti, "On the optimum design of microwave active frequency doublers," in *Proceedings of the IEEE MTT-S International Microwave Symposium Digest*, pp. 1423–1426, May 1995.

[8] C. Kärnfelt, R. Kozhuharov, H. Zirath, and I. Angelov, "High-purity 60-GHz-band single-chip ×8 multipliers in pHEMT and mHEMT technology," *IEEE Transactions on Microwave Theory and Techniques*, vol. 54, no. 6, pp. 2887–2898, 2006.

[9] H. Okazaki, T. Nakagawa, and K. Araki, "A low-power ku-band GaAs monolithic frequency quadrupler," in *Proceedings of the 29th European Microwave Conference*, vol. 2, pp. 353–356, Munich, Germany, October 1999.

[10] S. A. Maas, *Nonlinear Microwave and RF Circuits*, Artech House, Norwood, Mass, USA, 2nd edition, 2003.

[11] M. Ito, S. Kishimoto, T. Morimoto, Y. Hamada, and K. Maruhashi, "Highly integrated 60 GHz-band ×12 multiplier MMIC," in *Proceedings of the 2006 IEEE MTT-S International Microwave Symposium Digest*, pp. 1697–1700, June 2006.

[12] P. Colantonio, F. Giannini, G. Leuzzi, and E. Limiti, "Non-linear design of active frequency doublers," *International Journal of RF and Microwave Computer-Aided Engineering*, vol. 9, no. 2, pp. 117–128, 1999.

[13] T. H. Chen, K. W. Chang, S. B. T. Bui, L. C. T. Liu, G. S. Dow, and S. Pak, "Broadband single- and double-balanced resistive HEMT monolithic mixers," *IEEE Transactions on Microwave Theory and Techniques*, vol. 43, no. 3, pp. 477–484, 1995.

[14] S. Maas, M. Kintis, F. Fong, and M. Tan, "Broadband planar monolithic ring mixer," in *Proceedings of the IEEE Microwave and Millimeter-Wave Monolithic Circuits Symposium*, pp. 51–54, June 1996.

[15] D. Schneiderbanger, M. Nezadal, M. Sterns, R. Rehner, S. Martius, and L. P. Schmidt, "A double-balanced diode mixer with multi-decade bandwidth up to 119 GHz," in *Proceedings of the 1st European Wireless Technology Conference (EuWiT '08)*, pp. 5–8, October 2008.

[16] M. Yu, R. H. Walden, A. E. Schmitz, and M. Lui, "Ka/Q-band doubly balanced MMIC mixers with low LO power," *IEEE Microwave and Wireless Components Letters*, vol. 10, no. 10, pp. 424–426, 2000.

[17] G. Felic and E. Skafidas, "An integrated transformer balun for 60 GHz silicon RF IC design," in *Proceedings of the International Symposium on Signals, Systems and Electronics (ISSSE '07)*, pp. 541–542, August 2007.

[18] C. M. Lin, C. H. Lin, J. C. Chiu, and Y. H. Wang, "An ultra-broadband doubly balanced monolithic ring mixer for Ku- to Ka-band applications," *IEEE Microwave and Wireless Components Letters*, vol. 17, no. 10, pp. 733–735, 2007.

[19] C. J. Trantanella, "Ultra-small MMIC mixers for K- and Ka-band communications," in *Proceedings of the 1999 IEEE MTT-S International Microwave Symposium Digest*, vol. 2, pp. 647–650, June 2000.

[20] B. C. Henderson, "Mixers in microwave systems (part 2)," Tech Note of Watkins-Johnson Company, WJ Communications, San Jose,Calif, USA, 2001.

[21] M. Q. Lee, S. M. Moon, K. K. Ryu, D. P. Jang, and I. B. Yom, "Subharmonically pumped image rejection mixer for K-band applications," in *Proceedings of the Gallium Arsenide Applications Symposium*, Amsterdam, The Netherlands, October 2004.

[22] P. C. Yeh, W. C. Liu, and H. K. Chiou, "Compact 28-GHz subharmonically pumped resistive mixer MMIC using a lumped-element high-pass/band-pass balun," *IEEE Microwave and Wireless Components Letters*, vol. 15, no. 2, pp. 62–64, 2005.

[23] J. B. Beyer, S. N. Prasad, R. C. Becker, J. E. Nordman, and G. K. Hohenwarter, "MESFET distributed amplifier guidelines," *IEEE Transactions on Microwave Theory and Techniques*, vol. 32, no. 3, pp. 268–275, 1984.

[24] C. Paoloni and S. Kosslowski, "New approach on the design of travelling wave amplifier," in *Proceedings of the 34th Midwest Symposium on Circuits and Systems*, vol. 1, pp. 414–417, Monterey, Calif, USA, May 1991.

[25] K. Narendra, E. Limiti, C. Paoloni, J. M. Collantes, and R. H. Jansen, "Cascaded distributed power amplifier with non-identical transistors and inter-stage tapered impedance," in *Proceedings of the 40th European Microwave Conference (EuMC '10)*, pp. 990–993, Paris, France, September 2010.

Performance Comparison of UWB Pulse Modulation Schemes under White Gaussian Noise Channels

O. Abedi and M. C. E. Yagoub

School of Electrical Engineering and Computer Science, University of Ottawa, Ottawa, ON, Canada K1N 6N5

Correspondence should be addressed to M. C. E. Yagoub, myagoub@site.uottawa.ca

Academic Editor: Xianming Qing

In this paper, and for the first time, the performance of various ultra wideband pulse modulation schemes are discussed and compared in terms of narrowband interference robustness, symbol error rate, system complexity, data rate, and maximum transmit power with respect to transceiver distance and channel capacity. The channel model is the UWB IEEE 802.15.3a multipath indoor channel model under additive white Gaussian noise. The transmit power was evaluated by integrating the fifth derivative of the power spectrum density of the Gaussian pulse over the whole bandwidth.

1. Introduction

In ultra wideband (UWB) system design, there are always tradeoffs between maximum transmit distance, data rate, transmit power, and system complexity. Different data modulation schemes such as pulse-position modulation (PPM), pulse-amplitude modulation (PAM), phase-shift keying (PSK), or on-off keying (OOK) have been used in UWB communications with relative success depending on the targeted application [1, 2]. In fact, their performance can significantly vary according to which system parameters are considered such as narrowband interference (NBI) robustness, symbol error rate (SER), system complexity, data rate, or maximum transmit power with respect to transceiver distance and channel capacity. For instance, if minimum complexity is important, then OOK modulation would be the best choice. However, it is very sensitive to noise. On the other side, if interference robustness and power efficiency are the parameters to consider, binary PSK (BPSK) and M-ary PPM (M-PPM) can be the best candidates. Nevertheless, M-PPM deals with multidimensional signals, whereas M-PAM has only one dimension [3, 4].

This topic has been widely discussed in the last decade but most of published works [4] did not efficiently include all tradeoffs, such as comparison of bit error rate (BER), modulation level, channel capacity, and complexity. In fact, even if several technical papers have already discussed this issue, they mainly investigated the modulation performance for specific applications or system requirements and thus, to the best of our knowledge, did not cover the whole comparison spectrum [5].

The objective of this paper is indeed to evaluate all these parameters and their impact on each scheme performance. For this aim, we considered the PPM in time-hopping mode (TH-mode), the OOK in direct-sequence mode (DS-mode), and the binary PAM (BPAM) in both TH- and DS-modes. The channel model is the UWB IEEE 802.15.3a indoor channel model under additive white Gaussian noise (AWGN) [6]. Note that the white noise channel model is largely used in the literature due to its relative simplicity. Because the target here is to compare the performance of different UWB pulse modulation schemes under the same channel model, a relatively simple model can be the best candidate. Note also that this channel model can be specified for different propagation scenarios, such as extremely indoor non-line-of-sight (ENLOS) or outdoor line-of-sight (LOS). In this work, we focused on the indoor propagation channel model to take advantage of UWB frequency selective fading.

Also, the transmit power was evaluated by integrating the fifth derivative of the power spectrum density (PSD) of the Gaussian pulse over the whole bandwidth.

2. Modulation Schemes

2.1. M-PAM. In pulse-amplitude modulation, the same waveform is sent with different amplitudes corresponding to different data being transmitted. The BPAM can be presented using antipodal, monocyclic, or Gaussian pulses. The transmitted signal for M-PAM can be expressed as [7]

$$S_{\mathrm{PAM}}(t) = \sum_{j=-\infty}^{\infty} \sum_{m=1}^{M} \delta_m(j) d_m P(t - jT_s), \tag{1}$$

where $\delta(j) = [\delta_1(j), \delta_2(j) \ldots, \delta_M(j)]$ is an indicator transmitter pulse vector, that is, for each known unit impulse j, only one element of the vector is equal to unity otherwise zero; T_s is the pulse symbol period, M is the modulation level, and P is the pulse shape impulse response. For multiuser case, d_m is the amplitude of independent and identically distributed (IID) random signals according to an equiprobable distribution. The related PSD can be formulated as [4]

$$S_{\mathrm{PAM}}(f) = S_{\mathrm{cPAM}}(f) + S_{\mathrm{dPAM}}(f), \tag{2}$$

where $S_{\mathrm{cPAM}}(f)$ and $S_{\mathrm{dPAM}}(f)$ are, respectively, the desired continuous part (nonperiodic) and the undesired discrete part (periodic), which induces more spectral lines and lower mutual information [4]. For real data with equal probability of occurrence, $S_{\mathrm{PAM}}(f)$ becomes [6, 7]

$$S_{\mathrm{PAM}}(f) = \frac{\sigma_d^2}{T_s} |P(f)|^2 + \frac{\mu_d^2}{T_s^2} \sum_{j=-\infty}^{\infty} \left| P\left(\frac{j}{T_s}\right) \right|^2 \delta\left(f - \frac{j}{T_s}\right), \tag{3}$$

where σ_d^2 and μ_d are the variance and mean of the random data sequence, respectively. In case of repetition pulse coding [6, 8], each random data contain number of pulses with various amplitude (PAM case) that indicate just one bit of data. Let us consider the symbol period limited by trade of data rate and pulse numbers. For instance, for wider T_s, the data rate is indeed lower but does have better performance with higher pulse repetition coding (increasing pulse numbers), and inversely. On the other hand, if a time-hopping PAM (TH-PAM) is considered, they are dependent on frequency, which is not the case in this study.

In narrowband, using a filter bank receiver or a mixer bank receiver can mitigate the narrowband interference while splitting the UWB bandwidth to the narrowband. Thus, the average power or variance of the UWB signals can be calculated by integrating the PSD in frequency domain [9, 10]

$$\begin{aligned}\mathrm{Var}(S_{\mathrm{PAM}}(t)) &= \int_{-\infty}^{+\infty} S_{\mathrm{PAM}}(f)\,df \\ &= \int_{-\infty}^{+\infty} \left[\frac{\sigma_d^2}{T_s} |P(f)|^2 + \frac{\mu_d^2}{T_s^2} \sum_{j=-\infty}^{\infty} \left| P\left(\frac{j}{T_s}\right) \right|^2 \delta\left(f - \frac{j}{T_s}\right) \right] df \\ &= \frac{\sigma_d^2}{T_s} \int_{-\infty}^{+\infty} |P(f)|^2 df + \frac{\mu_d^2}{T_s^2} \sum_{j=-\infty}^{\infty} \left| P\left(\frac{j}{T_s}\right) \right|^2 .\end{aligned} \tag{4}$$

Assuming the carrier frequency of the receiver is f_{c0} and the bandwidth of the bandpass filter is B, the UWB interference power can be expressed as [10]

$$P_I = \frac{\sigma_d^2}{T_s} \int_{-\infty}^{+\infty} |P(f) H_T(f)|^2 df + \frac{\mu_d^2}{T_s^2} \sum_{(j/T) \pm f_{c0} \in [-B/2 \;\; B/2]} \left| P\left(\frac{j}{T_s}\right) H_T\left(\frac{j}{T_s}\right) \right|^2, \tag{5}$$

with $H_T(f) = H_{\mathrm{ch}}(f) H_B(f)$, where $H_B(f)$ and $H_{\mathrm{ch}}(f)$ are the Fourier transform for the filter and the channel impulse response, respectively. For an ideal bandpass filter, the above equation can be formulated as [10]

$$P_I = \frac{2\sigma_d^2}{T_s} \int_{f_{c0}-(B/2)}^{f_{c0}+(B/2)} |P(f)|^2 df + \frac{2\mu_d^2}{T_s^2} \sum_{(j/T) + f_{c0} \in [-B/2 \;\; B/2]} \left| P\left(\frac{j}{T_s}\right) \right|^2 . \tag{6}$$

Therefore, interference power can be affected by the following factors.

(i) The symbol period ($1/T_s$); in fact, increasing T_s will both increase the pulse repetition frequency and decrease the energy spectral density, while the corresponding continuous component of the power spectral density remains unchanged. So the pulse repetition leads to a discrete line spectrum. The line spectral density is composed of the delta function and separated by the pulse repetition frequency.

(ii) The energy spectral density (ESD) of the pulse waveform, $|P(f)|^2$.

(iii) The power spectral density (PSD) of the UWB chip sequence $S_{\mathrm{PAM}}(f)$ determined by σ_d^2 and μ_d for real and mutually uncorrelated sequences. Therefore, a proper pseudorandom (PR) code, directly related to the pulse repetition frequency and corresponding

PSD, could be selected to reduce the discrete spectral components. If the mean μ_d is not zero, discrete components (spectral lines) will appear in $S_{\text{PAM}}(f)$. However, for a sequence of ± 1, these components will disappear. On the other hand, if the UWB pulse train is modulated by an all "one" sequence ($\sigma_d^2 = 0$, $\mu_d = 1$), the PSD of $S_{\text{PAM}}(f)$ will be composed only of discrete components. Thus, if these components, which generally have a stronger PSD than continuous ones [2, 5], are in the band of narrowband systems, they will have a bigger effect on the performance of the victim receiver.

An alternative is to send alternating positive and negative pulses [2, 9]. This line coding scheme is often called alternate mark inversion (AMI). Following [10], it has been demonstrated that with equal probable binary elements, the PSD is given by

$$S_{\text{AMI}}(f) = \frac{1}{2T_s} |P(f)|^2 [1 - \cos(2\pi f T_s)], \tag{7}$$

which shows there are no spectral lines, and thus, the bipolar PAM signal has a lower bit error rate compared to the nonbipolar PAM signal, which suffers from the presence of more spectral lines (Figure 1).

2.2. OOK. Compared to others, OOK can be seen as the simplest scheme. In this technique, sending a pulse corresponds to bit "1" and no pulse means "0". Therefore, because of the equal probability of either symbols, $\sigma_d^2 = 0.25$, $\mu_d = 0.5$, and the PSD becomes [6, 10]

$$S_{\text{OOK}}(f) = \frac{1}{4T_s} |P(f)|^2 + \frac{1}{4T_s^2} \sum_{j=-\infty}^{\infty} \left| P\left(\frac{j}{T_s}\right) \right|^2 \delta\left(f - \frac{j}{T_s}\right). \tag{8}$$

In a variant (antipodal signals), the transmitter sends a pulse of duration T_p much higher than the bit time T_b to indicate bit "1" and the negative of this pulse for "0". Therefore, with equiprobable elements, we have $\sigma_d^2 = 1$, $\mu_d = 0$ and thus, a PSD of [10]

$$S_{\text{ant}}(f) = \frac{1}{T_s} |P(f)|^2, \tag{9}$$

showing that the PSD of antipodal does not have any spectral lines or discrete components. Note that OOK and binary-PPM are similar. Therefore, only 2-PPM performance was evaluated to avoid duplication.

2.3. M-PPM. In PPM, the chosen bit to be transmitted influences the position of the UWB pulses. The signal at the output of the transmitter cascade, $s(t)$, can be expressed as [1, 2, 6]:

$$S_{\text{PPM}}(t) = \sum_{j=-\infty}^{\infty} \sum_{m=1}^{M} P(t - jT_s - c_m T_C - a_m \varepsilon), \tag{10}$$

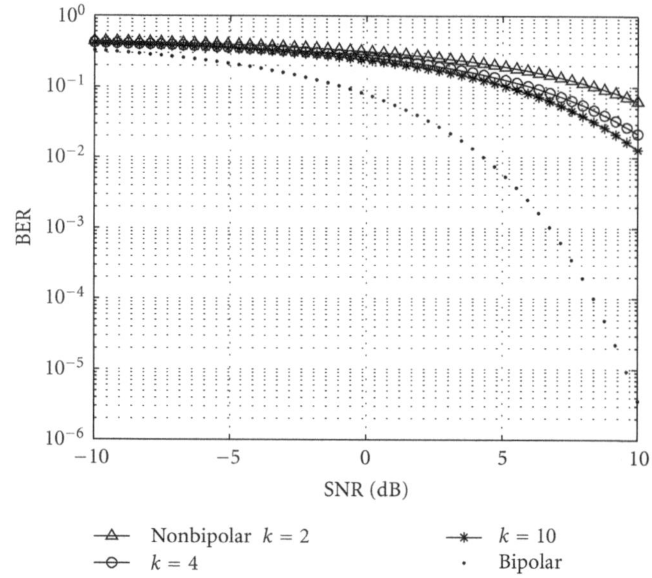

FIGURE 1: BER AWGN channel comparison with various polarization k.

where j is the number of transmitted bits, T_s is the symbol period, c_m is the time-hopping code, $c_m T_c$ is the pseudorandom code in chip duration T_C, a_m depends on j, and ε is the time shift associated with TH-PPM.

The chip time interval T_C represents only one pulse in the time frame period. Note that $a_m\varepsilon + T_p < T_c$ with T_p the pulse period.

The Fourier transform of the PSD of M-PAM can be expressed as [11]

$$S_{\text{PPM}}(f) = \int_{-\infty}^{+\infty} S_{\text{PPM}}(t) e^{-j2\pi f t} dt. \tag{11}$$

Leading to the following PSD, expressed in terms of continuous and discrete parts

$$PS_{\text{PPM}}(f) = PS_{\text{CPPM}}(f) + PS_{\text{dPPM}}(f) \tag{12}$$

with

$$PS_{\text{CPPM}}(f) = \frac{|P(f)|^2}{T_s} \left[1 - \left(\frac{1}{M} \frac{\sin(\pi f T)}{\sin(\pi f T/M)} \right)^2 \right]$$

$$PS_{\text{dPPM}}(f) = \frac{|P(f)|^2}{T_s^2} \left(\frac{1}{M} \frac{\sin(\pi f T)}{\sin(\pi f T/M)} \right)^2 \times \left[\sum_{j=-\infty}^{\infty} \delta\left(f - \frac{j}{T_s}\right) \right], \tag{13}$$

which can be simplified to [6, 11]

$$PS_{\text{PPM}}(t) = \frac{1}{T_s^2} \sum_{j=-\infty}^{\infty} \sum_{m=1}^{M} \left| P\left(\frac{jM}{T_s}\right) \right|^2 \delta\left(f - \frac{jM}{T_s}\right), \tag{14}$$

showing a smother PSD than of PAM, leading to less narrowband interference. The single pulse of the UWB signal is placed into one of $N = T_{sym}/T_h$ slots, where N is the number of slots per symbol, T_h is the pulse width and T_{sym} is the symbol time. The number of bits per one PPM symbol is equal to $\log_2^N$.

Note that by increasing M (modulation level) the data rate will increase as well. Furthermore, the average transmitted power will be increased by a factor of M.

In this case, the N slots are divided into L groups. Each group includes N/L slots with a single pulse. With this, the average bit number will be [11]

$$N_b' = \log_2^{(N/L)^L} = L\left(\log_2^N - \log_2^L\right) \quad N >> L, \qquad N_b' = L\log_2^N = N_b. \tag{15}$$

Thus, modifying the bit rate can improve the throughput of number of pulses per N slots, but it will raise other issues.

(i) A larger bandwidth requirement, as the number of pulses per symbol interval increases.

(ii) The pulse width relaxation will be limited.

(iii) All symbols can be orthogonal. To solve this for large N, one can use predefined pulse arrangement such that no pulse is put in the same position. In other words, to avoid pulse interpath interference (IPI) or pulse jamming, we should decrease the pulse period or increase the chip period.

Note that more information on PSK modulation can be found in [4, 6]. Because of its complexity, it has been introduced here just for comparison.

3. AWGN Channels

Under AWGN UWB channel models expressed by the IEEE 802.15.3a standard, symbol, or bit error rates can be easily expressed, leading to an optimum receiver [6, 7], consisting on an optimum demodulator (coherent demodulator). First, let us assume a zero system margin (no power loss between equipment connections), that is, $M_s = 0$.

For any binary modulation, the BER can be calculated using the distance between two symbols as [6]

$$P_b = Q\left(\sqrt{\frac{d}{2N_0}}\right). \tag{16}$$

This distance is different for each modulation scheme

$$d_{PPM} = \sqrt{2\gamma_b}; \qquad d_{PSK} = 2\sqrt{\gamma_b}; \qquad d_{OOK} = \sqrt{2\gamma_b}; \qquad d_{PAM} < \sqrt{2\gamma_b}, \tag{17}$$

$\gamma_b = E_{Rx}/N_0 = N_s\gamma_p$ is the average bit energy in the receiver and γ_p is its average pulse energy. Table 1 summarizes the BER and SER performance for different modulation schemes.

For simplicity, let us assume $N_s = 1$ with a normalized pulse energy. From theory, and as shown in Figure 2, binary PAM has 3 dB more power efficiency than PPM. As M is increasing, the performance of M-PPM also increases. This is due to space dimensionality that makes more noise orthogonal and causes the elements of the covariance noise matrix, Q_c, to be almost zero [9, 12]. Since the signal-to-noise ratio (SNR) is given by $SNR_r = S^T Q_c^{-1} S$, it will also increase. Also, M-PPM requires lower transmission power in high modulation levels. On the other hand, the performance of M-PAM becomes worse due to decreased symbol distance and as a result, $|Q_c|$ will increase leading to more interference at narrowband. Thus, by increasing the modulation level, the M-PPM will exhibit better performance.

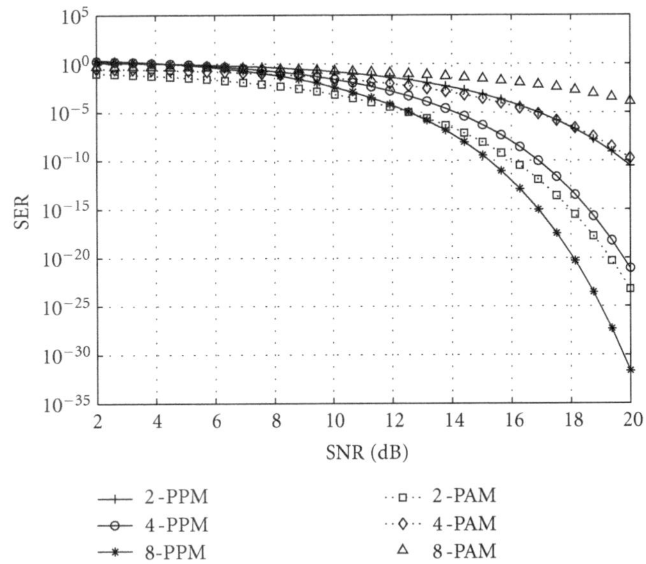

FIGURE 2: Symbol error rate for different M levels of M-PAM and M-PPM ($N_s = 1$).

Note that (17) confirms what was mentioned above, that is, simulating 2-PPM and OOK parameters leads to similar performance. Therefore, we just simulated and compared the M-PAM and the M-PPM in terms of BER versus SNR while a comparison between OOK and PPM is not necessary.

Table 2 summarizes the comparison between the different modulations schemes in terms of noise interference, spectral lines, transceiver complexity, pulse energy variance, and bandwidths.

As shown, the candidate pulse modulation is strongly related to the application. In fact, if simplicity is considered, the repeat pulse coding or OOK modulation will be the good option but in case of NBI the 4-PPM should be chosen to satisfy all spectrum requirements.

4. Data Rate and Maximum Distance

To further investigate the modulation scheme performance, let us assume that the signal propagation occurs over free space. Thus, the free-space attenuation A_m can be expressed as [6, 13]

$$A_m = \frac{4\pi D^2 L}{G_T S_R}, \tag{18}$$

TABLE 1: BER and SER in AWGN channel for different UWB modulations.

TH-BPPM	Orthogonal single pulse ($N_s = 1$) $P_{rb} = Q\left(\sqrt{N_s\gamma_b}\right), Q(x) = \frac{1}{2}\operatorname{erfc}\left(\frac{x}{\sqrt{2}}\right)$ $s_m = \begin{cases} s_0 = +\sqrt{p_{Tx}}[1 - R_0(m)], \\ s_1 = -\sqrt{p_{Tx}}[1 - R_0(m)] \end{cases}$ where $R_j(\varepsilon)$ is autocorrelation of the pulse S_m. orthogonal multipulse signal with TH coding: $P_{rb} = Q\left(\sqrt{N_s\gamma_b(1 - R_0(m))}\right)$
BPAM/BPSK	$P_b = Q(\sqrt{2\gamma_b})$
M-PPM	$P_M = \frac{1}{\sqrt{2\pi}}\int_{-\infty}^{\infty}\left[1 - \left(\frac{1}{\sqrt{2\pi}}\int_{-\infty}^{y} e^{-x^2/2}dx\right)^{M-1}\right]\cdots$ $\cdots\exp\left[\frac{-1}{2}\left(y - \sqrt{2\gamma_s}\right)^2\right]dy$ $P_b = \frac{2^{k-1}}{2^k - 1}P_M \approx \frac{1}{2}P_M$
M-PAM	$P_M = \frac{2(M-1)}{M}Q\left(\sqrt{\frac{6\log_2(M)\gamma_b}{M^2 - 1}}\right) \quad P_b = \frac{1}{k}P_M$ $\gamma_s = \log_2(M)\gamma_b \quad \gamma_b = N_s\frac{\xi_b}{N_0}$

TABLE 2: Comparison of various pulse modulation techniques in terms of bandwidth, spectral line, noise sensitivity, transceiver complexity, and pulse energy variance.

	Bandwidth requirement (normalized)	Power requirement (normalized)	Noise sensitively	Spectral line power distortion	Transceiver complexity	Pulse energy variance
PAM	1	1	Highest (1 dimension)	Yes	Low	$\left(\frac{M^2-1}{6}\right)\frac{1}{\gamma_s}$
PPM	$\left(\frac{M}{\log_2 M}\right)\left(\frac{\tau}{T_s}\right)$	$\sqrt{\frac{2}{M\log_2 M}}$	Lowest (M dimensions)	Yes (lower)	Highest (by M)	$\frac{1}{\gamma_s}$
OOK	1	1	Very high	Yes (high)	Lowest	$\frac{1}{\gamma_s}$
BPSK	$\frac{\tau}{T_s}$	1.12	Lower ($M/2$ dimensions)	No	High (by $M/2$)	$\frac{1}{\gamma_s}$

with D is the distance of propagation, S_R is the antenna effective area, L is the system hardware loss, and G_T is the transmitter antenna gain. In this work, we assumed $L = 1$ (no losses in the system hardware). The pulse waveform is the 5th derivative of a monocyclic Gaussian pulse and its PSD is given by [13]

$$P_s(f) = A_{\max}\frac{(2\pi f\delta)^{2n}e^{-(2\pi f\delta)^2}}{n^n e^{-n}};\qquad A_m = 10^{-13.125};\qquad \delta = 51\text{ ps}, \tag{19}$$

leading to the following receiving signal power

$$P_r = \frac{M_s \mathrm{SNR}_{\mathrm{spec}} N_0}{2}, \tag{20}$$

where N_0 is the thermal noise power, $\mathrm{SNR}_{\mathrm{spec}}$ is the system requirement SNR, and $H(f)$ is the frequency response of the indoor channel, which can be expressed as [13]

$$|H(f)| = \sqrt{\frac{1}{A_m(f)}R(f)}, \tag{21}$$

where $R(f)$ is the shadowing parameter that depends on the geography of the propagating environment (usually assumed to be 6 dB). The system margin M_s is assumed to be 1 dB. From this, the distance can be reformulated as [6, 7]

$$D^2 = \frac{\left(G_T G_R c^2/(4\pi)^2 R_b\right)2\int_{fL}^{fH}\left(P_s(f)/f^2\right)df}{M_s \mathrm{SNR}_{\mathrm{spec}}(1/2)N_0}, \qquad N_0 = kT_s(f) = k\left(T_A + (F(f)-1)T_0\right), \tag{22}$$

where T_s is the noise temperature at the receiver, T_A is the receiver antenna temperature, T_0 is the room temperature (usually $T_0 = 300°K$), k is the Boltzmann constant, $F(f)$ is the noise factor of the receiving device, and R_b is the data rate. Given a specific SNR (i.e., the SER of a specific modulation scheme), one can evaluate the maximum transmit distance. Figure 3 shows the maximum distance as function of data rate for PAM and PPM modulation schemes. The following assumptions have been made

(i) SER is lower than $10^{-5}(N_s = 1,\ T_s = 3ns)$.

(ii) No fading margin required and PSD of noise is constant (E_b/N_0).

It is observed that BPAM (BPSK) has the best data rate over short distances with respect to complexity. However, by increasing the modulation level M, the M-PPM data rate can be improved ($\mathrm{SER}_{8\,\mathrm{PPM}} < \mathrm{SER}_{4\,\mathrm{PPM}} < \mathrm{SER}_{2\,\mathrm{PPM}}$) while the opposite occurs for M-PAM.

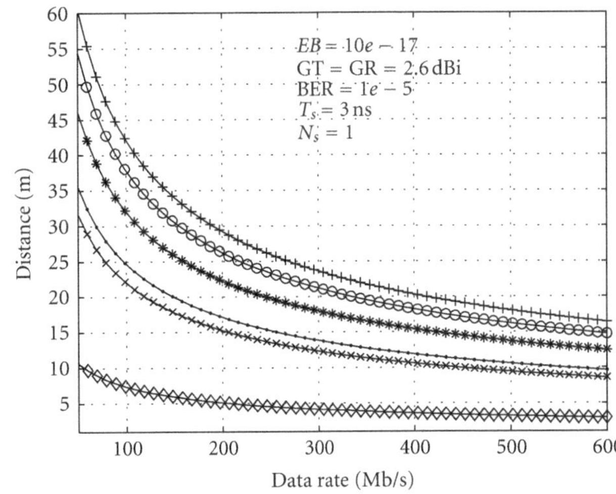

FIGURE 3: M-PAM and M-PPM signals data rate for short distances and uncoded systems.

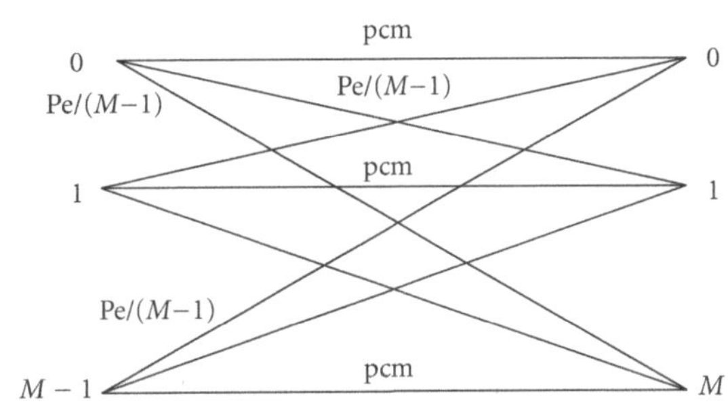

FIGURE 4: Channel capacity model for M-ary.

5. Channel Capacity

A channel with M-ary PPM/PAM modulation has discrete-valued inputs and continuous-valued outputs, which imposes an additional capacity calculation. Let s_m be the encoded M-ary PPM/PAM signal vector input to the channel and r_i the ith diversity channel output vector corrupted by the AWGN noise vector. Let w_i be an M-dimensional Gaussian vector with zero mean and variance $\sigma_w^2 = N_0/2$ along each real direction (multipath). This output vector is given by:

$$r_i = s_m + w_i \quad 1 \le i \le L, \tag{23}$$

where L is the number of multipaths. The channel capacity (bits/channel), with input signals restricted to an equiprobable M-ary signal constellation and with no restrictions on the channel output, can be written as [14]

$$C = \log_2(M) - \frac{1}{M}\sum_{m=1}^{M}\int P\left(\frac{r}{s_m}\right)\log_2\left(\frac{\sum_{k=1}^{M}P(r/s_k)}{P(r/s_m)}\right)dy,$$
$$P\left(\frac{r}{s_m}\right) = \prod_{i=1}^{L}\left[\left(\frac{1}{\pi N_0}\right)e^{(-(ri-\sqrt{E})^2/N_0)}\right], \tag{24}$$

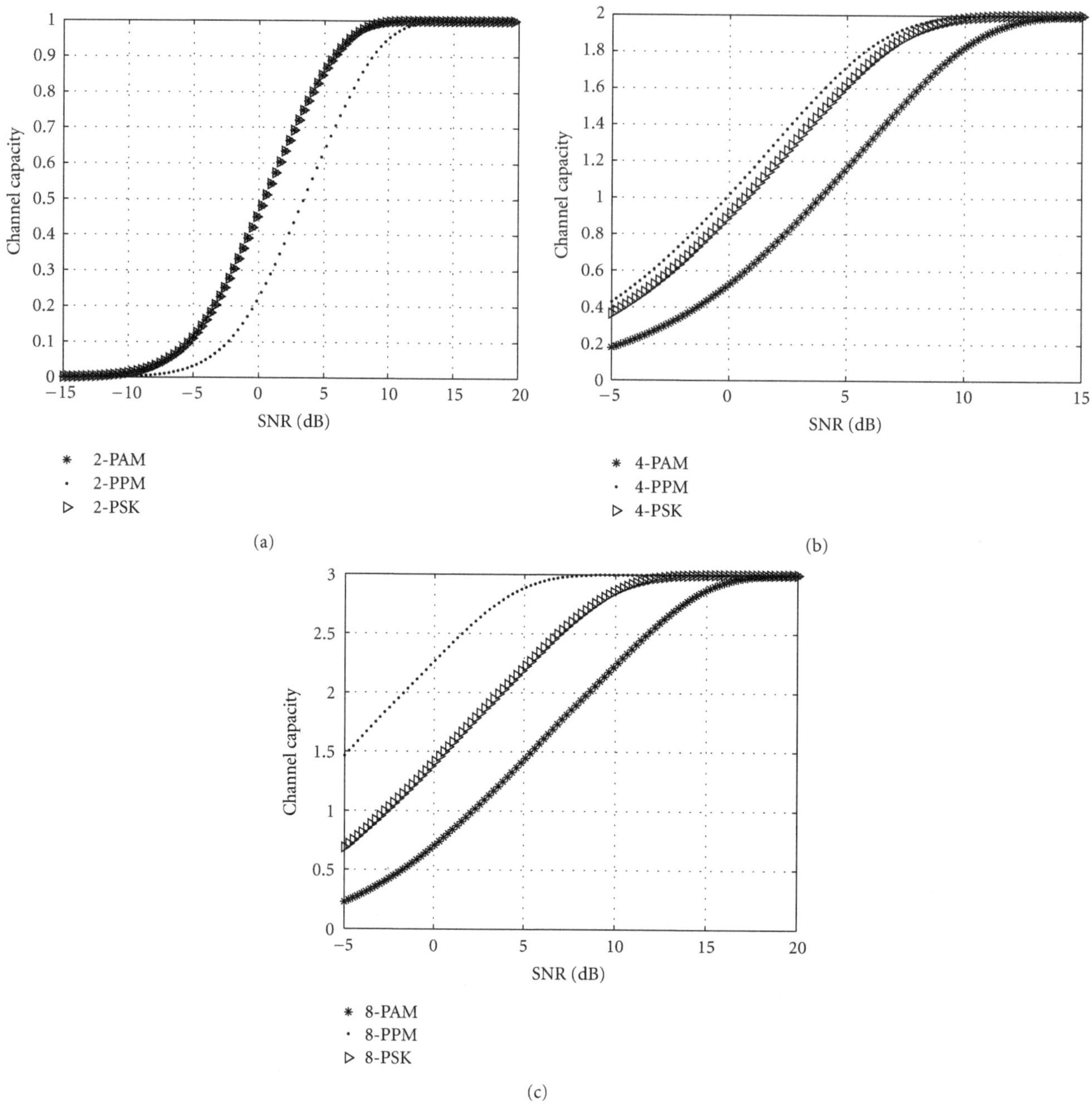

FIGURE 5: Channel capacity comparison of PAM, PPM, and PSK in various modulation levels.

with P is the probability to receive a signal from the transmitter and N_0 is the PSD of white noise. For M-PAM over an AWGN channel, we obtain

$$C = \log_2(M) - \frac{1}{M}\sum_{k=0}^{M-1} \times \left\langle \left\{ \log_2 \sum_{p=0}^{M-1} \exp\left(\sum_{i=1}^{L} \frac{|w_i|^2 - \left| s_k + w_i - s_p \right|^2}{N_0} \right) \right\} \right\rangle, \quad (25)$$

where s_p is one of the M-ary PAM signals. For M-PPM, the probability of receiving a signal is [7, 9, 14]

$$P\left(\frac{r}{s_m}\right) = \prod_{i=1}^{L}\left[\left(\frac{1}{\pi N_0}\right)^{M/2} \left(\prod_{j=1}^{M} e^{-r_{ij}^2/2\sigma_T^2} \right) e^{-(r_{ij}-\sqrt{E_g})^2/N_0} \right]$$

$$r_{ij} = \sqrt{E_g} + w_{ij},$$

$$\sigma_T^2(n) = \sigma_{\text{pre}}^2(n) + \sigma_{\text{post}}^2(n) + \sigma_w^2,$$

$$\sigma^2_{\text{pre}}(n) = \frac{1}{N}\sum_{k=-\infty}^{-1} h^2(n-k),$$

$$\sigma^2_{\text{post}}(n) = \frac{1}{N}\sum_{k=1}^{n} h^2(n-k) \quad \text{for } n > 0, \tag{26}$$

where σ_T^2 is the total noise variance, E_g is the signal energy, M is the signal dimension, $\sigma^2_{\text{pre}}(n)$ is the variance of the noise pulses preceding the present signal pulse at time n, $\sigma^2_{\text{post}}(n)$ is the variance of the noise of the pulses following the present signal, and $h(n)$ is the discrete channel response.

Giving the following channel capacity model for symmetric channel (Figure 4), and [9, 12, 15]

$$C_{M\text{-PPM}} = \frac{1}{M}\sum_{i=0}^{M-1}\sum_{j=0, j\neq i}^{M-1}\frac{P_e}{M-1}\log\left[\frac{(P_e/(M-1))}{(1/M)}\right] + \frac{1}{M}\sum_{i=0}^{M-1}\sum_{j=0, j\neq 0}^{M-1}P_{\text{cm}}\log\left[\frac{P_{cm}}{(1/M)}\right], \tag{27}$$

$$C_{M\text{-PPM}} = \log_2(M) + P_{\text{cm}}\log_2(P_{\text{cm}}) + (1-P_{\text{cm}})\log_2\left(\frac{1-P_{\text{cm}}}{M-1}\right).$$

Figure 5 shows the channel capacity of various modulation schemes. In this figure, the capacity can be improved by increasing the modulation level for M-PPM, M-PAM, and M-PSK. Note that the theory details about this comparison are expressed in Table 2.

6. Maximum Transfer Power

From a communication theory perspective, the most important characteristic of UWB systems is the power-limited regime operation. Following standards and regulations, there is a maximum allowed transmit power P_t. This limitation has a direct impact on the maximum transmit distance and data rate [8, 16]

$$\frac{P_t}{P_r}(\text{dB}) = C_0 + 10\chi\log_{10}\left(\frac{\lambda}{4\pi d}\right) + X_R(\text{dB}), \tag{28}$$

where

$$P_t(\text{dB}) = 10\log_{10}\left(\frac{M_s\, R_b\, E_b^0}{G_T\, G_R}\left(\frac{4\pi}{\lambda}\right)d^2\right). \tag{29}$$

In the above equations, X_R is the average shadowing system, χ is the path-loss component, and C_0 is the maximum UWB pulse power, assumed to be −41.42 dBm [8]. E_b^0 is the energy per bit ($1.82.\ 10^{-19}$ W) for a specified BER of 5.10^{-5}.

In Figure 6, we compared the transfer power for different modulation levels at a fixed data rate (200 MHz) and as function of maximum transmitter power and distance. It can be seen that BPSK soft decision has a higher maximum distance compared to other modulation schemes.

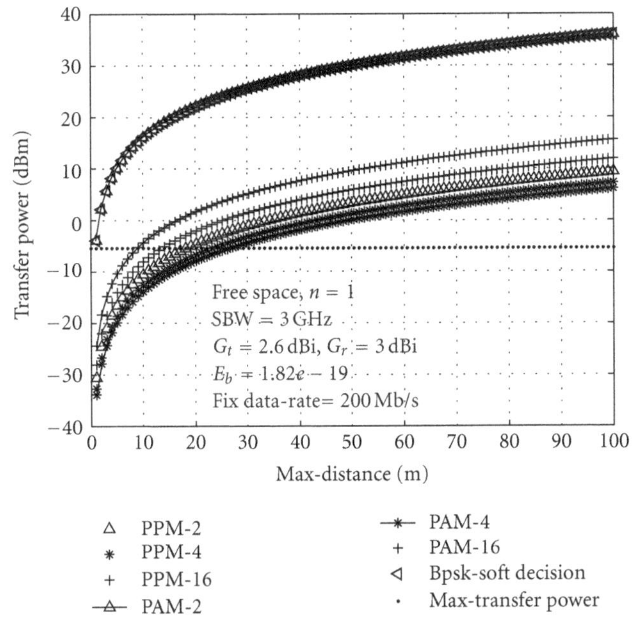

Figure 6: Maximum transfer power for M-PAM-M-PPM, BPSK for a fixed data rate of 200 Mb/s. The horizontal line (close to −5 dBm) represents the maximum transmit power specified by FCC regulations.

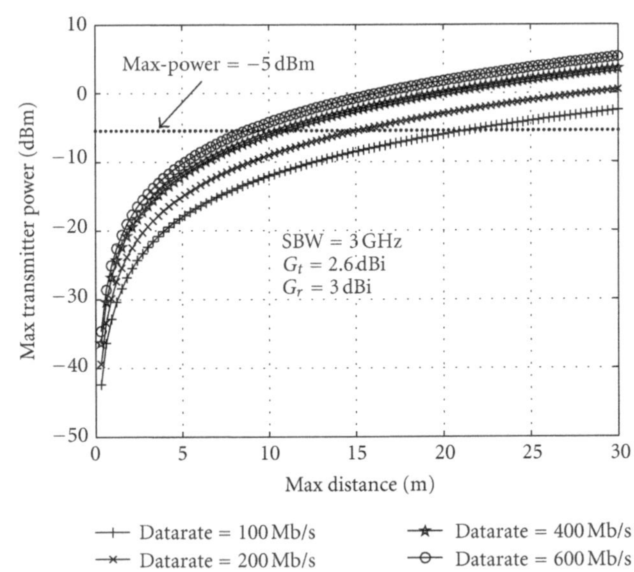

Figure 7: Maximum transfer power for 4-PPM with various data rate. The horizontal line (close to −5 dBm) represents the maximum transmit power specified by FCC regulations.

By increasing the modulation level M, M-PPM will be improved but M-PAM and M-PSK will exhibit a lower performance, due to more noise interference (noise dimensionality). Then again M-PPM has better narrowband interference robustness in terms of required less transmit power with respect to system complexity. Furthermore, the 4-PPM has an optimum modulation level which covers all tradeoff as sorted by Table 2.

In Figure 7, we compared the 4-PPM in various data rates using the same transmitter and channel propagation scenario as in Figure 5 for different modulation levels. As expected, by increasing distance, the data rate will decrease, requiring higher transmitter power. For short distance applications (<10 m), the data rate cannot exceed 200 Mb/s even with an LOS scenario and AWGN channel assumption. This demonstrated that UWB technology cannot practically provide the expected high data rate (>400–600 Mb/s) unless a pulse coding is considered. In fact, with more complexity and higher modulation levels, the data rate can be significantly improved but with the cost of higher transmitter power.

7. Conclusion

Various UWB pulse modulation schemes were evaluated in this paper and their performance was compared in terms of narrowband interference robustness, symbol error rate, system complexity, data rate, and maximum transmit power with respect to transceiver distance and channel capacity.

It appears that BPAM has the highest data rate in terms of maximum distance and transmit power. However, M-PPM has better power efficiency and channel capacity, at the expense of increased complexity and bandwidth.

If bandwidth is not an issue, the M-PPM is the best option, due to lower spectral lines (smoother PSD), higher dimensionality (orthogonally of noise), and narrowband interference robustness. However, if the UWB receiver is single band, the M-PPM is not the optimum option in terms of bandwidth.

If the system complexity and bandwidth are the limiting factors, OOK is the best choice, but has more spectral lines and higher probability of error which leads to higher risk of interference. Finally, because of signal orthogonally and lower spectral lines (smoother PSD), the 4-PPM (optimum modulation level) can be the best candidate for NBI robustness and can be used for unique UWB rake filter-bank receiver design.

References

[1] A. Naanaa and S. Belghith, "Performance enhancement of a time hopping—pulse position modulation ultra-wideband system using guided local search," *IET Journal of Communications*, vol. 5, no. 15, pp. 2212–2220, 2011.

[2] Q. Zhu, Z. Jia, and C. Zou, "On the performance of different modulation schemes for UWB systems in a multipath channel," in *Proceedings of the IET International Conference on Wireless, Mobile and Multimedia Networks*, pp. 1–4, November 2006.

[3] R. Hidayat and Y. Miyanaga, "IR-UWB pulse position modulation and pulse shape modulation through S-V channel model," in *Proceedings of the 2nd International Conference on Communication Software and Networks (ICCSN '10)*, pp. 214–217, February 2010.

[4] A. Al Zaman and N. Islam, "Modulation schemes and pulse shaping in ultra wideband," in *Proceedings of the IEEE Southeast conference*, pp. 142–146, April 2008.

[5] C. Ngo and D. Birru, "Analysis of UWB modulation techniques," in *Proceedings of the International Conference on Wireless Networks (ICWN '03)*, pp. 646–652, June 2003.

[6] M. G. Di Benedetto and G. Giancola, *Understanding Ultra Wide Band Radio Fundamentals*, Prentice Hall, New York, NY, USA, 2004.

[7] M. Ghavami, L. B. Michael, R. Kohno, and U. W. B. Signals a, *Systems in Communication Engineering*, John Wiley & Sons, New York, NY, USA, 2nd edition, 2007.

[8] H. Sheng, "On the spectral and power requirements for ultra-wideband transmission," in *Proceedings of the International Conference on Communications (ICC '03)*, pp. 738–742, May 2003.

[9] L. Ma and T. C. Wang, "Performance analysis and simulations of UWB-PAM communications in AWGN channel," in *Proceedings of the 4th International Conference on Microwave and Millimeter Wave Technology (ICMMT '04)*, pp. 830–833, August 2004.

[10] Z. Ye, "Power spectral density and in-band interference power of UWB signals at narrowband systems," in *Proceedings of the IEEE International Conference on Communications*, pp. 3561–3565, June 2004.

[11] J. Nielsen and S. Zwierzchowski, "Power spectral density of a UWB signal with discrete quantized pulse positions," *Canadian Journal of Electrical and Computer Engineering*, vol. 28, no. 3-4, pp. 145–154, 2003.

[12] P. E. McIllree, "Calculation of channel capacity for M-ary digital modulation signal sets," in *Proceedings of the IEEE Singapore International Conference on Networks*, pp. 639–643, Singapore, September 1993.

[13] S. S. Mo, "On the power spectral density of UWB signals in IEEE 802.15.3a," in *Proceedings of the IEEE Wireless Communications and Networking Conference (WCNC '04)*, pp. 999–1003, March 2004.

[14] L. Zhao and A. M. Haimovich, "Capacity of M-ary PPM ultra-wideband communications over AWGN channels," in *Proceedings of the IEEE 54th Vehicular Technology Conference (VTC FALL '01)*, pp. 1191–1195, October 2001.

[15] R. Pasand, "Capacity of PPM ultra-wideband communications with inter pulse interference," in *Proceedings of the Canadian Conference on Electrical and Computer Engineering*, pp. 2355–2358, May 2004.

[16] V. Somayazulu, J. R. Foerster, and S. Roy, "Design challenges for very high data rate UWB systems," in *Proceedings of the 36th Asilomar Conference on Signals Systems and Computers*, pp. 717–721, November 2002.

The Impact of Blue Inorganic Pigments on the Microwave Electrical Properties of Polymer Composites

L. C. Costa[1] and F. Henry[2]

[1] I3N and Physics Department, University of Aveiro, 3810-193 Aveiro, Portugal
[2] LRP, CNRS, 2 rue Henry Dunand, 94320 Thiais, France

Correspondence should be addressed to L. C. Costa, kady@ua.pt

Academic Editor: Paolo Colantonio

We present the results of the measurement of complex dielectric permittivity, in the microwave frequency region, on glass reinforced polybutylene terephthalate (PBT) with blue inorganic pigments. The cavity resonant method had been used in order to measure the shift in the resonant frequency of the cavity, caused by the insertion of a sample, which can be related to the real part of the complex permittivity. Also, the quality factor of the cavity decreases with the insertion of a sample. The changes in the inverse of this quality factor give the imaginary part. In order to predict the dielectric behavior of this composite, we had developed a program of numerical simulation to calculate the complex permittivity of the inclusion. By using some of dielectric mixture laws (Maxwell-Wagner-Sillars, Hanai, Looyenga, inverse and direct Wiener, and Bruggemann), we can predict the dielectric behavior of the composite in a large range of volume fraction of inclusions.

1. Introduction

Ultramarine blue is a nonhazardous pigment with a lot of variety of industrial applications. Its manufacturing process and possibility for close control over its physical, chemical, and particular color characteristics enable the production of several types of this blue pigment, which are readily accepted by polymers, printing ink, paint, cosmetic, and many other industries, due to advantages over other organic pigments.

The electrical conductivity of an insulating polymer, such as polybutylene terephtalate (PBT), can be altered by adding different conducting particles, like carbon [1–3], iron [4], nickel [5], or another conducting polymer [6–8]. The conductivity of the composite material can thus be controlled by properly choosing the components, their shape, and their relative volume fractions.

Small quantities of conducting particles can increase the dielectric constant, without exceeding the critical concentration of percolation [9], that is, avoiding high conductivity. If electrical losses become high, heating of the plastic will occur, resulting in melting or even carbonization.

For high frequencies, as in the microwave range, interfacial polarization mechanisms are not present, because such polarization occurs at frequencies lower than the times scales typical of dipolar polarizations [10].

In order to predict the electrical properties of the composite, different mixture laws can be used. Numerical simulations of mixtures can lead not only to a better understanding of the physics of dielectrics but also to improvement of designing of tailored materials without going to expensive attempts.

Numerous authors have studied the dielectric behavior of nonhomogeneous materials [11–13], and several theories can be applied, depending on the difference of the electrical properties of the host and the inclusion materials [14, 15].

2. Experimental

The blue inorganic pigment, with chemical formula $Na_6Al_6Si_6O_{24}S_4$, in the form of powder with mean particle size 3.8 μm, was obtained from Kremer Pigments Inc., USA. Glass fibers of 10 to 20 μm reinforced the polybutylene terephthalate matrix, which was purchased from DuPont,

USA. The powder samples were pressed at room temperature and 4 MPa to obtain cylinders of length 10 mm and diameter 4 mm.

The resonant cavity method was used to calculate the complex permittivity [16, 17], $\varepsilon^* = \varepsilon' - i\varepsilon''$. Two different rectangular cavities were used, operating in $TE_{3,5,5}$ and $TE_{0,1,11}$ modes, and resonant frequencies of 2.7 and 5 GHz, respectively. In this technique, we measured the shift in the cavity resonant frequency, Δf, caused by the insertion of the sample inside the cavity, which can be related to the real part of the complex permittivity, ε', and the change in the inverse of the quality factor of the cavity, $\Delta(1/Q)$, which is related with the imaginary part, ε''. The relations are simple when we consider only the first-order perturbation in the electric field caused by the sample [18],

$$\frac{\Delta f}{f_0} + i\Delta\left(\frac{1}{2Q}\right) = \varepsilon_0(\varepsilon^* - 1)\frac{\int_{V_s} E_i^* E_0 dV}{\int_V \varepsilon_0 E_0^2 dV}, \quad (1)$$

where f_0 is the resonance frequency of the cavity, ε^* the complex permittivity of the material, and E_i and E_0 the electric fields inside and outside the material. The integration is made in the volume of the sample, V_s, and in the volume of the cavity, V.

Splitting the real and imaginary parts, we can obtain the expressions for ε' and ε'',

$$\varepsilon' = K\frac{\Delta f}{f_0}\frac{V}{V_s} + 1, \qquad \varepsilon'' = \frac{K}{2}\Delta\left(\frac{1}{Q}\right)\frac{V}{V_s}, \quad (2)$$

where K is a constant related to the depolarization factor, which depends upon the geometric parameters. In fact, this factor introduces the effect of the shape and dimensions of the sample in the electromagnetic field perturbation. Maxwell equations and the boundary conditions were used to deduce the previous equations [19], taking into account the depolarization field, which appears outside the dielectric. To calculate the factor K, we used a sample of known dielectric constant of polytetrafluorethylene (PTFE) ($\varepsilon' = 2.1$ at microwave frequencies), with the same size and shape of the studied samples.

In order to couple the microwave to the cavity, we used the quarter-wavelength flange joints and small circular irises (10 mm in diameter).

For the measurements, we used an HP 8753D Network Analyzer with the excitation power of 1 mW. All the measurements were carried out at constant temperature. The measured transmissions are the S_{12} parameters that quantify how the microwave energy propagates through a multiport network. The subscripts "12" refers to the ratio of signal from port 1 (input) to port 2 (output).

The samples, ultramarine blue pigments, in concentrations up to 4% in a glass reinforced polybutylene terephtalate matrix, were introduced into the cavity through circular holes milled in the centre of the cavity. There, the electric field is in its strongest value and the insertion of the samples causes the maximal frequency shift. That is, a coupling between the samples and the electric field is produced

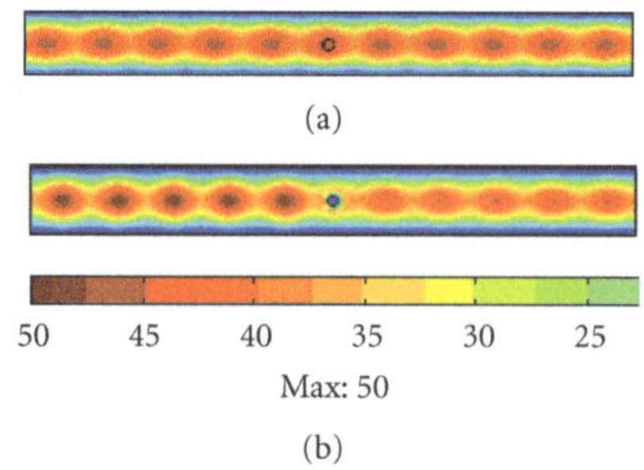

Figure 1: Calculated electric field in the 5 GHz cavity, without (a) and with a sample in the center (b). The vertical bar shows that the highest electric field is about 50 kVm^{-1}.

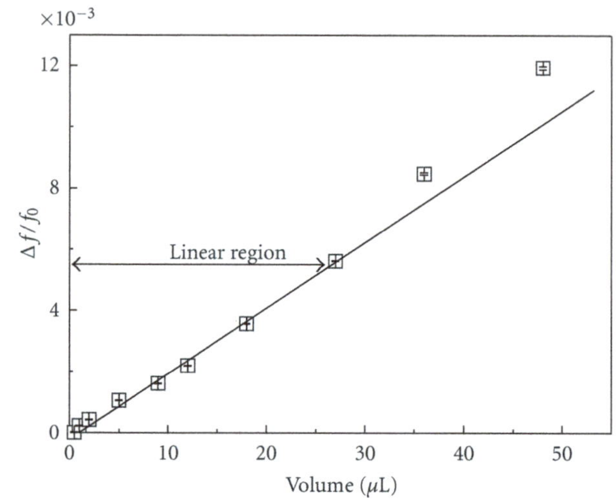

Figure 2: $\Delta f/f_0$ versus volume of water. f_0 is the resonance frequency of the cavity and Δf the variation of resonant frequency by the insertion of the samples.

in this region [19], resulting in a small perturbation of that field. The interest lies in the cases in which the samples are homogeneous and thier volumes are very small compared with the volume of the cavity. Figure 1 shows the electric field distribution in the 5 GHz cavity, without and with a sample, in a simulation made using the COMSOL software. The changes in the electric field in the center of the cavity, due to by the insertion of a sample in this region, are clearly visible, when comparing Figures 1(a) and 1(b).

3. Results

To study the linearity of the cavity, and consequently to infer the possibility to use the small perturbation theory [19], we carried out measurements using glass microtubes filled with distilled water. In Figures 2 and 3 we present, as an example, $\Delta f/f_0$ and $\Delta(1/Q)$ for the different volume of distilled water (5 GHz resonant cavity), and the linear fit parameters that confirm the possibility of using (2).

The shift in the resonant frequency of the cavity remains in the linear regime for measurements up to 27 μL of water, which corresponds to $\Delta f/f_0$ of 0.56%. The linearity in the

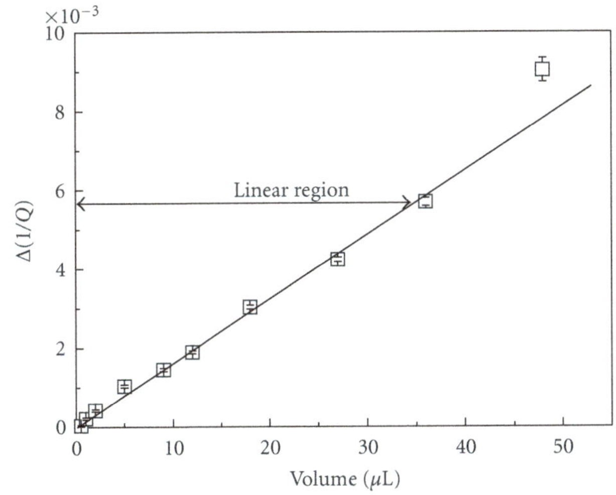

Figure 3: $\Delta(1/Q)$ versus volume of water. Q is quality factor of the cavity, which is degraded by the insertion of the samples.

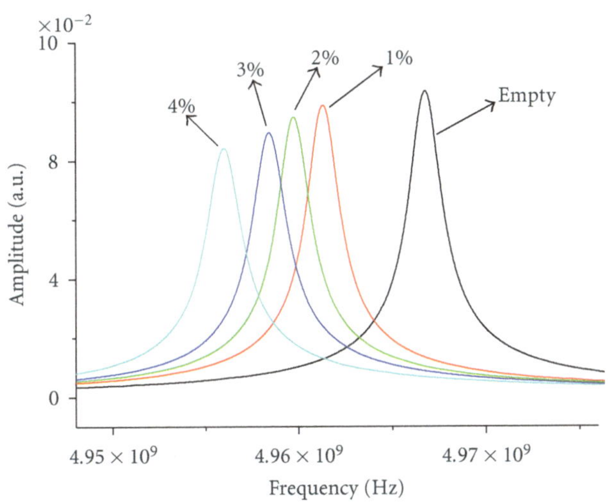

Figure 4: Transmission for the empty and loaded cavity with samples of different concentrations of blue pigments.

inverse of the quality factor is observed up 36 μL that is up to 5.7×10^{-3}. The conjugation of both conditions permits us to conclude that the cavity can be used up to variations of $\Delta f/f_0$ of 0.56% and $\Delta(1/Q)$ of about 4.2×10^{-3}. If the measurements overcome this limit, the sample volume should be reduced until the results falls into the linear regime [20].

In Figure 4 we show the transmissions of the 5 GHz cavity, for the empty and loaded cavity with samples of different concentrations of blue pigments. The most perturbating sample was referred to that with highest concentration of blue inorganic pigments, corresponding to the highest complex permittivity. This means that the real and imaginary parts of the complex permittivity of the filler has considerable values.

4. Discussion

A very common mixture law was proposed by Hanai [21]. If the filler particles, with a complex permittivity $\varepsilon_f^*(\omega)$ and volume fraction φ_f are dispersed in a matrix material with complex permittivity $\varepsilon_m^*(\omega)$ and volume fraction φ_m, then, according to this theory,

$$\frac{\varepsilon_f^* - \varepsilon^*}{\varepsilon_f^* - \varepsilon_m^*}\left(\frac{\varepsilon_m^*}{\varepsilon^*}\right)^{1/3} = 1 - \varphi_f. \qquad (3)$$

Another theory, developed by Maxwell-Wagner-Sillars predicts the polarization process due to differences in conductivity and permittivity of the constituents [22]. The complex permittivity of the mixture can be calculated from

$$\varepsilon^* = \varepsilon_m^* \frac{\left(n\varepsilon_f^* * +(1-n)\varepsilon_m^*\right) + (1-n)\left(\varepsilon_f^* - \varepsilon_m^*\right)\varphi_f}{\left(n\varepsilon_f^* + (1-n)\,\varepsilon_m^*\right) - n\left(\varepsilon_f^* - \varepsilon_m^*\right)\varphi_f}, \qquad (4)$$

in which n, with $0 \le n \le 1$, is the shape factor of the dispersed particles in the direction of the electrical field lines. For spherical particles, $n = 1/3$ [23].

The Bruggeman effective medium theory has the virtue of simplicity. It can be used with the assumption that the inclusions do not interact each other and that are randomly distributed in the matrix [24]. In the most commonly used form, that is, the implicit form,

$$\varphi_f \frac{\varepsilon_f^* - \varepsilon^*}{\varepsilon_f^* + A\varepsilon^*} + \left(1 - \varphi_f\right)\frac{\varepsilon_m^* - \varepsilon^*}{\varepsilon_m^* + A\varepsilon^*} = 0, \qquad (5)$$

where, for spherical particles embedded in a host matrix, $A = 2$. This law predicts a percolation threshold of conduction for $\varphi_f = 1/(A+3)$.

Looyenga introduced a new mixture formula [25], usually known as generalized Looyenga law,

$$\varepsilon^*(\omega)^{1/t} = \varphi_f \varepsilon_f^*(\omega)^{1/t} + \varphi_m \varepsilon_m^*(\omega)^{1/t}, \qquad (6)$$

where for spherical inclusions $t = 3$. For $t = \pm 1$, we arrive to the Wiener laws (direct and inverse) [15].

Table 1 summarizes the obtained results for the fit of the data with the four previous laws. The values for the real and imaginary parts of the complex permittivity of the charge are indicative that generalized Looyenga model is more accurate, which is confirmed by the obtained values of the standard deviation χ^2.

Figure 5 shows the measured real and imaginary parts of the complex permittivity for different concentrations of the blue inorganic pigment, at 2.7 GHz, and the fit with the different mixture laws, confirming the good accuracy in the fit using generalized Looeynga. Similar results, from the point of view of quality of the fit, are obtained for the 5 GHz measurements frequency.

For these small quantities of blue inorganic particles, the percolation threshold is not observed and the composite is maintained as an insulator.

Finally, the small polarity present in PBT explains the measured complex permittivity. The obtained values make

Table 1: Calculated values for ε_f' and ε_f'' of the filler, and standard deviation χ^2, using the inversion of Wiener, Hanai, Looyenga, Maxwell-Wagner-Sillars, and Bruggeman laws. A is a parameter of Bruggeman and t the exponent of generalized Looyenga laws.

	Parameters	Direct Wiener	Invere Wienr	Hanai	Maxwell-Wagner-Sillars	Generalized Looyenga	Bruggeman
2.7 GHz	ε_f'	23.63	6.13E5	2.75	9.87E5	108.3	52.1
	ε_f''	3.72	—	0.42	1.24E5	39.58	19.18
	t	—	—	—	—	3.50	—
	A	—	—	—	—	—	7.97
	χ^2	0.0218	0.2455	0.2901	0.1627	0.0226	0.0248
5 GHz	ε_f'	22.36	7.3E5	1.1E5	1.0	106.9	1.0
	ε_f''	0.214	8.0E5	354.6	355.0	102.3	155.1
	t	—	—	—	—	3.94	—
	A	—	—	—	—	—	4.86
	χ^2	0.0446	0.2289	0.1381	0.1402	0.0415	0.0780

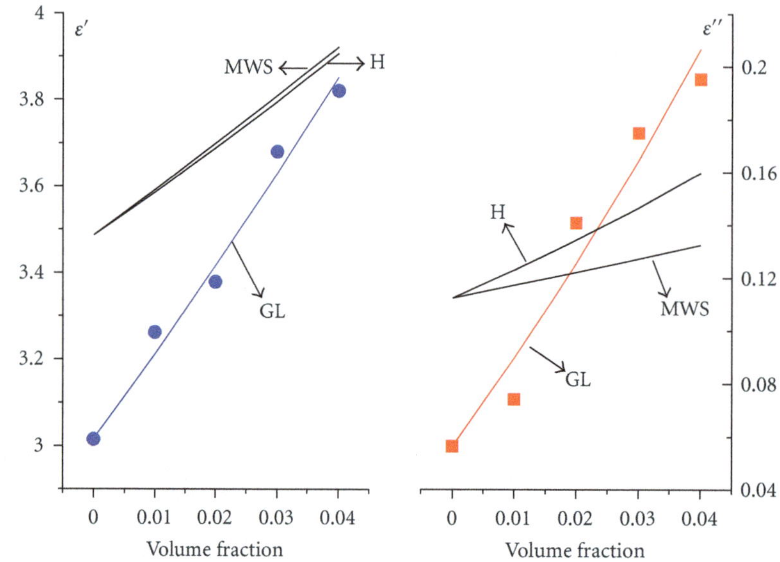

Figure 5: Measured real and imaginary parts of the complex permittivity for different concentrations of the blue inorganic pigment, at 2.7 GHz, and the fit with the generalized Looyenga (GL), Hanai (H), and Maxwell-Wagner-Sillars (MWS) mixture laws.

this material a good choice for applications in telecommunications products. The higher ε' permits to reduce the dimensions of the rods in dielectric antennas. In espite of the higher ε'', the values are still quite low, avoiding the heating of the material.

5. Conclusions

The cavity perturbation method presents a good accuracy to evaluate the dielectric permittivity of low loss materials.

The dielectric function of two phase materials can be accurately deduced using the generalized Looyenga model, in particular when the inclusion is conductive, and in low filler volume fractions. In this case, the percolation critical concentration is not reached.

With this law, we can choose the adequate doping concentrations, and then to control the electrical properties, in order to obtain the desired behavior, for a particular application.

References

[1] C. Brosseau, F. Boulic, P. Queffelec et al., "Dielectric and microstructure properties of polymer carbon black composites," *Journal of Applied Physics*, vol. 81, no. 2, pp. 882–891, 1997.

[2] I. Krupa and I. Chodak, "Physical properties of thermoplastic/graphite composites," *European Polymer Journal*, vol. 37, no. 11, pp. 2159–2168, 2001.

[3] G. Chen, D. Wu, W. Weng, H. Bin, and W. Yan, "Preparation of polystyrene-graphite conducting nanocompositesvia intercalation polymerization," *Polymer International*, vol. 50, no. 9, pp. 980–985, 2001.

[4] L. Costa, M. Valente, F. Henry, and L. Ramanitra, "Electrical properties of the polymer composite polystyrene/iron particles," *Journal de Chimie Physique et de Physico-Chimie Biologique*, vol. 95, no. 6, pp. 1453–1456, 1998.

[5] A. A. Khoroshilov, I. P. Koroleva, and Y. Y. Volodin, "Nickel/polystyrene composite electrode materials," *Russian Journal of Applied Chemistry*, vol. 73, no. 11, pp. 1918–1921, 2000.

[6] L. C. Costa, F. Henry, and A. André, "Propriétés électrique et mécanique de matériaux nanocomposites à polymère conducteur," in *Proceedings of the Journées Polymères Conducteurs*, Collonges la Rouge, France, 1996.

[7] G. Street, *Handbook of Conducting Polymers*, Dekker, New York, NY, USA, 1986.

[8] Y. Roichman, M. S. Silverstein, A. Siegmann, and M. Narkis, "Percolation of electrical conductivity in solution-cast blends containing polyaniline," *Journal of Macromolecular Science*, vol. B38, no. 1-2, pp. 145–161, 1999.

[9] M. A. Valente, L. C. Costa, S. K. Mendiratta, F. Henry, and L. Ramanitra, "Structural and electrical properties of polystyrene-carbon composites," *Solid State Communications*, vol. 112, no. 2, pp. 67–72, 1999.

[10] A. Schonhals and F. Kremer, "Analysis of dielectric spectra," in *Broadband Dielectric Spectroscopy*, F. Kremer and A. Schonhals, Eds., Springer, Berlin, Germany, 2003.

[11] G. Bánhegyi, "Comparison of electrical mixture rules for composites," *Colloid & Polymer Science*, vol. 264, no. 12, pp. 1030–1050, 1986.

[12] D. K. Hale, "The physical properties of composite materials," *Journal of Materials Science*, vol. 11, no. 11, pp. 2105–2141, 1976.

[13] F. Henry, F. Cansell, J. L. Guillaume, and C. Pichot, "Study of the polymer latex coalescence by dielectric measurements at microwave frequency-feasibility of the method," *Colloid & Polymer Science*, vol. 267, no. 2, pp. 167–178, 1989.

[14] Y. Roichman, M. S. Silverstein, A. Siegmann, and M. Narkis, "Percolation of electrical conductivity in solution-cast blends containing polyaniline," *Journal of Macromolecular Science*, vol. B38, no. 1-2, pp. 145–161, 1999.

[15] F. Henry and L. C. Costa, "Percolation and order-disorder transition," *Physica B*, vol. 387, no. 1-2, pp. 250–258, 2007.

[16] A. Rejou-Michel, F. Henry, M. De Villardi, and M. Delmotte, "Protein and ion hydration variation in fixed aqueous solutions: measurement by dielectric decrement," *Physics in Medicine and Biology*, vol. 30, no. 8, pp. 831–837, 1985.

[17] C. P. L. Rubinger and L. C. Costa, "Building a resonant cavity for the measurement of microwave dielectric permittivity of high loss materials," *Microwave and Optical Technology Letters*, vol. 49, no. 7, pp. 1687–1690, 2007.

[18] D. S. McLachlan, A. Priou, I. Chennerie, E. Isaac, and F. Henry, "Modeling the permittivity of composite materials with a general effective medium equation," *Journal of Electromagnetic Waves and Applications*, vol. 6, no. 9, pp. 1099–1131, 1992.

[19] D. X. Gouveia, L. C. Costa, and M. A. Valente, "Resonant cavity for the measurement of microwave magnetic permeability using the small perturbation theory," *Microwave and Optical Technology Letters*, vol. 50, no. 2, pp. 399–402, 2008.

[20] L. C. Costa, S. Devesa, P. André, and F. Henry, "Microwave dielectric properties of polybutylene terephtalate (PBT) with carbon black particles," *Microwave and Optical Technology Letters*, vol. 46, no. 1, pp. 61–63, 2005.

[21] T. Hanai, "Theory of the dielectric dispersion due to interfacial polarisation and its application to emulsions," *Kolloid-Zeitschrift*, vol. 171, pp. 23–31, 1960.

[22] R. W. Sillars, "The behaviour of polar molecules in solid paraffin wax. Ibid," *Proceedings of Real Society of London*, vol. 169, p. 66, 1939.

[23] V. Beek, "Dielectric behaviour of heterogeneous systems," in *Progress in Dielectrics*, J. B. Birks, Ed., pp. 69–87, Heywood Books, London, UK, 1967.

[24] C. Brosseau, P. Quéffélec, and P. Talbot, "Microwave characterization of filled polymers," *Journal of Applied Physics*, vol. 89, no. 8, pp. 4532–4540, 2001.

[25] H. Looyenga, "Dielectric constants of heterogeneous mixtures," *Physica*, vol. 31, no. 3, pp. 401–406, 1965.

22

Buried Object Detection by an Inexact Newton Method Applied to Nonlinear Inverse Scattering

Matteo Pastorino and Andrea Randazzo

Department of Naval, Electrical, Electronic, and Telecommunication Engineering, University of Genoa, Via Opera Pia 11A, 16145 Genova, Italy

Correspondence should be addressed to Matteo Pastorino, matteo.pastorino@unige.it

Academic Editor: Kristen M. Donnell

An approach to reconstruct buried objects is proposed. It is based on the integral equations of the electromagnetic inverse scattering problem, written in terms of the Green's function for half-space geometries. The full nonlinearity of the problem is exploited in order to inspect strong scatterers. After discretization of the continuous model, the resulting equations are solved in a regularization sense by means of a two-step inexact Newton algorithm. The capabilities and limitations of the method are evaluated by means of some numerical simulations.

1. Introduction

Subsurface imaging is an important subject in several applicative areas, including seismic and geophysical prospecting, nondestructive, testing and medical diagnostics [1–39]. Electromagnetic techniques are widely applied to face this problem. Ground penetrating radar [40] is the basic instrumentation able to retrieve discontinuities in the lower subspace (the ground). Processing of GPR data can allow an improvement in the detection and localization capabilities of the system. In some cases, neural network-based approaches have been adopted [41]. Moreover, inverse-scattering-based techniques have been proposed, too. By using these methods, the measured values of the field scattered by the discontinuities present in the lower half space are "inverted" in order to obtain the spatial distributions of dielectric parameters (e.g., the dielectric permittivity and the electric conductivity) inside a fixed inspection domain. Approaches of this kind have been proposed, for example, for the detection of water leaking from pipes [42]. If an accurate model of the buried scatterers is available, it would be possible to retrieve information about the buried objects in terms of their dielectric parameters.

The present authors developed in [43] an inverse-scattering-based method for buried object imaging under the second-order Born approximation (SOBA) [21, 44–48]. By using the SOBA, the reconstruction of buried objects has been obtained with a better accuracy with respect to linearized approximations. In particular, due to the nonlinear nature of the scattering equations, in which the electric field inside the inspected region is an unknown quantity as well as the spatial distributions of the dielectric properties of the medium, the SOBA allows to approximately retrieve the field distribution, too.

However, in order to detect strong discontinuities, the exact nonlinear inverse scattering formulation must be considered. This approach has been faced in [49, 50], in which strong scatterers have been reconstructed in free space under tomographic imaging conditions. The approach proposed in [49], which is based on the use an inexact Newton method, has been experimentally tested in [51]. The same method has been successfully applied in [52] within the contrast source formulation (under free-space conditions).

In the present paper, the approach developed in [43] for buried object detection under the SOBA is extended to treat strong scatterers. In this case, the exact nonlinear inverse

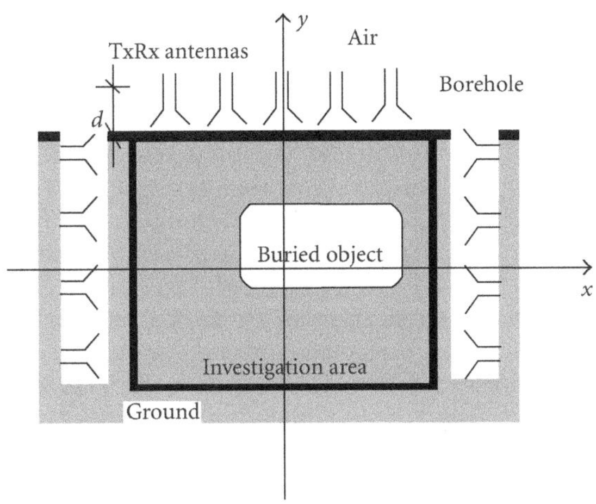

FIGURE 1: Geometrical configuration of the problem.

scattering formulation, in terms of the half-space Green's function, is taken into account. In particular, starting from the integral equations for the scattered field for cylindrical objects, slides of the spatial distributions of the dielectric parameters are reconstructed by numerically solving these equations in a regularization sense.

The paper is organized as follows. In Section 2 the mathematical formulation of the approach is reported. In Section 3, the application of the inexact Newton method is described. Numerical results are reported in Section 4, and conclusions are drawn in Section 5.

2. Mathematical Formulation

Let us consider a cylindrical object buried in a homogeneous-half-space medium (Figure 1). A 2D transverse-magnetic (TM) configuration is assumed. The dielectric properties of the buried object are assumed to be independent of the z coordinate. The transmitting and receiving antennas are located in the transverse plane and generate a z-polarized electric field in both the half spaces (this field is independent of the axial coordinate, too).

The dielectric characteristics of the investigation domain S, in the lower half space, are described by the so-called *object function*, which is defined as

$$\tau(\mathbf{r}) = j\omega[\epsilon(\mathbf{r}) - \epsilon_b], \tag{1}$$

where $\epsilon(\mathbf{r})$ is the complex dielectric permittivity at point $\mathbf{r}$, and ϵ_b denotes the dielectric permittivity of the soil.

The investigation area is illuminated by V line-current sources, with angular frequency ω (in the following, the $e^{j\omega t}$ time dependence is assumed and omitted). A mixed illumination configuration (partially borehole and partially from the top), which allows direct comparisons with the results reported in [43], is exploited. Under such assumptions, also the scattered electric field is z-polarized and independent of z. Consequently, only the transverse plane can be considered for the imaging procedure.

The scattered electric field, $u_{\text{scatt}}^{(v)}$, $v = 1, \ldots, V$, is collected for any incident wave in M measurement points $\mathbf{s}_m^{(v)}$, which constitute the so-called observation domain O. The dielectric properties of S can be related to the scattered field by means of the following set of nonlinear integral equations:

$$\begin{aligned} u_{\text{scatt}}^{(v)}(\mathbf{r}) &= G_{\text{lower}}^{\text{upper}} \tau u_{\text{tot}}^{(v)}(\mathbf{r}), \quad \mathbf{r} \in O, \\ u_{\text{tot}}^{(v)}(\mathbf{r}) &= u_{\text{inc}}^{(v)}(\mathbf{r}) + G_{\text{lower}}^{\text{lower}} \tau u_{\text{tot}}^{(v)}(\mathbf{r}), \quad \mathbf{r} \in S, \end{aligned} \tag{2}$$

where $G_{\text{lower}}^{\text{upper}}(\cdot)(\mathbf{r}) = j\omega\mu_0 \int_S(\cdot) g_{\text{HS}}(\mathbf{r}, \mathbf{r}')d\mathbf{r}'$, $\mathbf{r} \in O$ and $G_{\text{lower}}^{\text{lower}}(\cdot)(\mathbf{r}) = j\omega\mu_0 \int_S(\cdot) g_{\text{HS}}(\mathbf{r}, \mathbf{r}')d\mathbf{r}'$, $\mathbf{r} \in S$ are the data and state operators, where g_{HS} is the half-space Green's function [43] (superscripts in the operators recall where they are calculated, whereas subscripts recall where they are integrated).

Equations (2) can be rewritten in a compact operator form as

$$\text{HS}(\mathbf{x}) = \mathbf{y}, \tag{3}$$

where $\mathbf{x} = [\tau \;\; u_{\text{tot}}^{(1)} \;\; \cdots \;\; u_{\text{tot}}^{(V)}]^t$ is an array containing the values of the object function and of the total electric field inside the investigation area for any illumination. In (3), $\mathbf{y} = [u_{\text{scatt}}^{(1)} \;\; \cdots \;\; u_{\text{scatt}}^{(V)} \;\; u_{\text{inc}}^{(1)} \;\; \cdots \;\; u_{\text{inc}}^{(V)}]^t$ is an array containing the values of the scattered electric field acquired by the measurement probes in the upper half space and the values of the incident electric field inside the investigation area. The half-space operator HS can be expressed, according to (2), as

$$\text{HS}(\mathbf{x}) = \begin{bmatrix} G_{\text{lower}}^{\text{upper}} \tau u_{\text{tot}}^{(1)} \\ \vdots \\ G_{\text{lower}}^{\text{upper}} \tau u_{\text{tot}}^{(V)} \\ u_{\text{tot}}^{(1)} - G_{\text{lower}}^{\text{lower}} \tau u_{\text{tot}}^{(1)} \\ \vdots \\ u_{\text{tot}}^{(V)} - G_{\text{lower}}^{\text{lower}} \tau u_{\text{tot}}^{(V)} \end{bmatrix}. \tag{4}$$

As it is well known, the inverse problem (4) is non-linear and ill-posed. Consequently, nonlinear regularizing methods must be used.

3. Regularized Solution of (3)

Due to the mentioned ill posedness, (3) must be solved in a regularized sense. The inexact Newton regularizing scheme developed in [49, 50] is applied here. This method includes two nested loops, in which a linearization (with respect to the current solution guess $\mathbf{x}_k$) of the operator HS is constructed (*outer loop*) and solved by means of the truncated Landweber method (*inner loop*). Exactly as in [49], the algorithm performs the following steps.

(1) For $k = 0$ (k denotes the outer iteration index), a starting guess $\mathbf{x}_0$ is chosen (initialization phase).

(2) The following linearized version of equation is constructed

$$\text{HS}'_{\mathbf{x}_k}\mathbf{h} = \mathbf{y}_k, \tag{5}$$

where $\mathbf{h} = [h_\tau \ h_u^{(1)} \cdots \ h_u^{(V)}]^t$, $\mathbf{y}_k = \mathbf{y} - \mathrm{HS}(\mathbf{x}_k)$, and $\mathrm{HS}'_{\mathbf{x}_k}$ is the Fréchet derivative of HS at point $\mathbf{x}_k$. A regularized solution $\widehat{\mathbf{h}}_k$ to (5) is obtained by applying an iterative truncated Landweber algorithm, for which we have

$$\mathbf{h}_{k,0} = 0,$$
$$\mathbf{h}_{k,l+1} = \mathbf{h}_{k,l} - \gamma \mathrm{HS}'^{\,*}_{\mathbf{x}_k} \left(HS'_{\mathbf{x}_k} \mathbf{h}_{k,l} - \mathbf{y}_k \right), \quad (6)$$

where $\mathrm{HS}'^{\,*}_{\mathbf{x}_k}$ is the adjoint operator of $\mathrm{HS}'_{\mathbf{x}_k}$.

(3) The current solution is updated by using the following relation:

$$\mathbf{x}_{k+1} = \mathbf{x}_k + \widehat{\mathbf{h}}_k. \quad (7)$$

(4) A predefined stopping rule is finally applied. If this criterion is not satisfied, a fixed maximum number of outer iterations, $k_{\max}$, is performed. In both cases, at any iteration, the procedure can be stopped, otherwise steps (2)–(4) are repeated.

It should be noted that step (2) is actually the inner loop of the algorithm, whereas steps (1)–(3) constitute the outer loop.

4. Numerical Results

The developed method has been tested by means of numerical simulations. The working frequency has been set equal to 300 MHz (i.e., similar to that used in ground penetrating radars). A homogeneous lossy soil characterized by a relative dielectric permittivity $\epsilon_{r_b} = 4$ and an electric conductivity $\sigma_b = 0.01$ S/m has been considered (in order to simulate a dry sandy soil [53]). Similar to [43], a mixed measurement configuration is assumed, where reflection and cross-borehole operating modes are mixed together in order to acquire as much information as possible. Correspondingly, the antennas are located at points (Cartesian coordinates)

$$\mathbf{r}_n^{\mathrm{ant}} = \begin{cases} \left(-\dfrac{2}{3}\lambda_0, -\dfrac{4}{3}\lambda_0 + (n-1)\dfrac{2\lambda_0}{21}\right), & n = 1,\ldots,15, \\ \left(-\dfrac{2}{3}\lambda_0 + (n-15)\dfrac{2\lambda_0}{21}, 0\right), & n = 16,\ldots,28, \\ \left(\dfrac{2}{3}\lambda_0, -(n-29)\dfrac{2\lambda_0}{21}\right), & n = 29,\ldots,43, \end{cases} \quad (8)$$

where λ_0 is the free-space wavelength in the upper medium. A subset of these antennas act both as transmitter (TX) and receiver (RX). When an antenna operates in TX mode, all the remaining ones collect the scattered electric field (i.e., they work in RX mode). Consequently, for every illumination, $M = 42$ field samples are collected. In particular, in the cases reported in the following, $V = 7$ antennas are used. They are located at positions denoted by the indexes given by $n = 7(v-1) + 1$, $v = 1,\ldots,V$.

The investigation area is a square domain of dimensions $\lambda_0 \times \lambda_0$, whose upper side coincides with the air-ground interface. In order to numerically solve the equations involved in the inverse-scattering problem, such domain has been discretized into $N = 20 \times 20$ square subdomains.

The scattered field values used as input data for the considered simulations have been numerically obtained by using a computational code based on the method of moments [54]. In order to avoid the so-called *inverse crime*, a finer discretization (such as the ratio of the cell dimensions in the direct and inverse problems is not an integer number) has been used for the computation of the synthetic data. Moreover, in order to simulate a more realistic scenario, a Gaussian noise with zero mean value and variance corresponding to a signal-to-noise ratio SNR = 25 dB has also been added to the computed data.

In all the simulations reported in this section, the initialization phase has been always performed by assuming an empty investigation domain in the lower half space. Consequently, in this phase, the object function is equal to zero, and the total electric field is assumed equal to the one transmitted in the lower half space.

The first reported case concerns the reconstruction of a single circular cylinder of radius $a = 0.2\lambda_0$ centered at $\mathbf{r}^{\mathrm{obj}} = (0.2\lambda_0, -0.5\lambda_0)$. The dielectric properties of the cylinder are $\epsilon_{r_{\mathrm{obj}}} = 5.6$ and $\sigma_{\mathrm{obj}} = 0.01$ S/m. The parameters of the inexact Newton algorithm are the following: maximum number of inner iterations, $k_{\mathrm{LW}_{\max}} = 30$; maximum number of outer iterations, $k_{\mathrm{IN}_{\max}} = 100$. Figure 2 reports the behavior of two error parameters for different values of the number of outer iterations of the iterative procedure. In particular, the two parameters are defined as

$$r_{\mathrm{data}} = \frac{\left\| u^{\mathrm{actual}} - u^{\mathrm{calc}} \right\|_2}{\left\| u^{\mathrm{actual}} \right\|_2},$$
$$r_{\mathrm{solution}} = \frac{\left\| \tau^{\mathrm{actual}} - \tau^{\mathrm{rec}} \right\|_2}{\left\| \tau^{\mathrm{actual}} \right\|_2}, \quad (9)$$

with τ^{actual} and τ^{rec} being the actual and reconstructed permittivity distributions (at the current iteration) and u^{actual} and u^{calc} the measured and computed values of the electric field data (at the current iteration, too). As can be seen from Figure 2, the plot of the "residual" on the data (the error parameter r_{data}) decreases monotonically as expected. Moreover, the plot of the solution "residual" (the error parameter r_{solution}) shows a typical *semiconvergence* behavior, that is, after a certain value of the iteration number the reconstruction quality degrades [55, 56].

The final image of the distribution of the relative dielectric permittivity inside the investigation region (corresponding to the optimum iteration number) is shown in Figure 3. As can be seen, the method is able to correctly retrieve the dielectric profile of the investigation area. In particular, both the shape of the buried inclusion and the value of its dielectric permittivity are correctly identified.

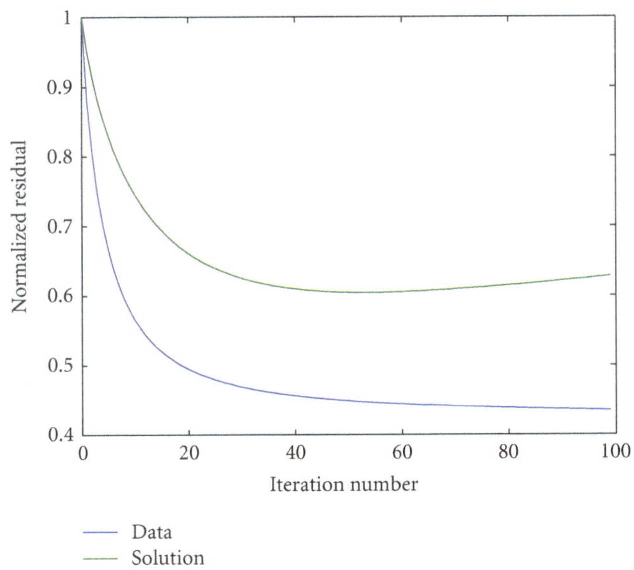

FIGURE 2: Data and solution residuals versus the outer iteration number (error parameters of (9)). Single circular cylinder buried in a lossy half space.

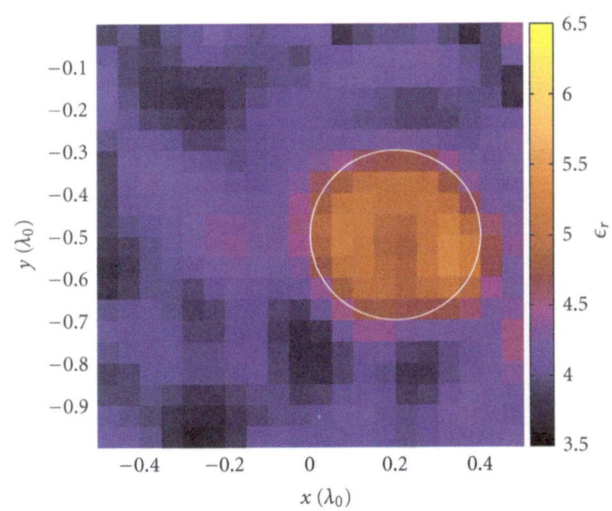

FIGURE 3: Reconstructed distribution of the relative dielectric permittivity. Single circular cylinder buried in a lossy half space.

For the present case, an analysis of the effects of the noise has also been performed. In particular, the mean relative error on the reconstruction, defined as the L_1-norm

$$e = \frac{1}{N}\sum_{n=1}^{N}\frac{|\epsilon_n^{\text{actual}} - \epsilon_n^{\text{rec}}|}{|\epsilon_n^{\text{actual}}|} \quad (10)$$

has been calculated. In (10), $\epsilon_n^{\text{actual}}$ and ϵ_n^{rec} denote the values of the original and reconstructed complex permittivity inside the nth discretization subdomain of the investigation region. Moreover, the optimal value of the outer iteration have been evaluated, too. The obtained results are reported in Table 1.

As expected, as the noise level increases, the reconstruction "quality" decreases. However, even in presence of strong noise levels, the method is able to yield rather good reconstructions (e.g., the error parameter e is equal to about 8% error for SNR = 15 dB).

TABLE 1: Mean relative errors and optimal iteration numbers versus the signal-to-noise ratio. Single circular cylinder buried in a lossy half space.

SNR	25	20	15	10
Mean relative error	0.065	0.078	0.085	0.12
Optimum outer iteration number	53	34	14	5

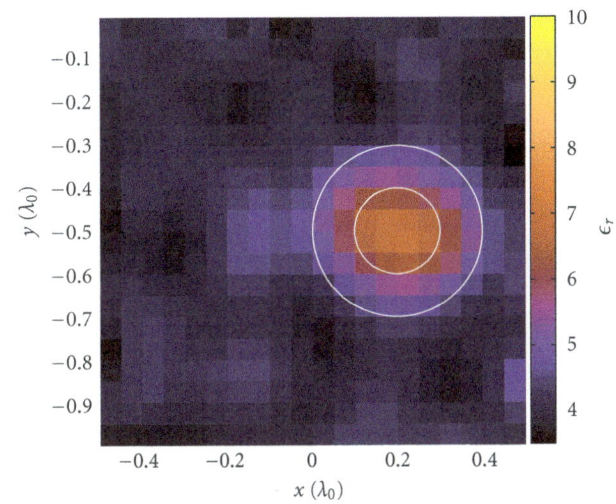

FIGURE 4: Reconstructed distribution of the relative dielectric permittivity. Double-layer circular cylinder buried in a lossy half space.

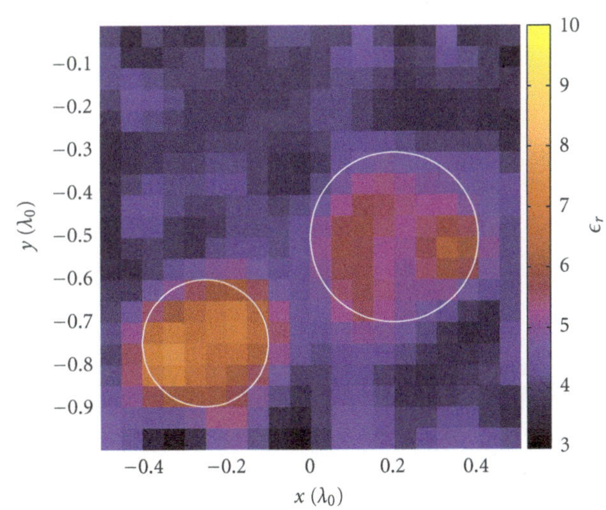

FIGURE 5: Reconstructed distribution of the relative dielectric permittivity. Two separate circular cylinders buried in a lossy half space.

In the second set of simulations, the reconstruction of a more complex configuration has been considered. In particular, the presence of a double-layer circular cylinder has been assumed. The outer layer has radius $a_1 = 0.2\lambda_0$ and is characterized by a relative dielectric permittivity $\epsilon_{r_1} = 5.6$ and an electric conductivity $\sigma_1 = 0.01$ S/m, whereas

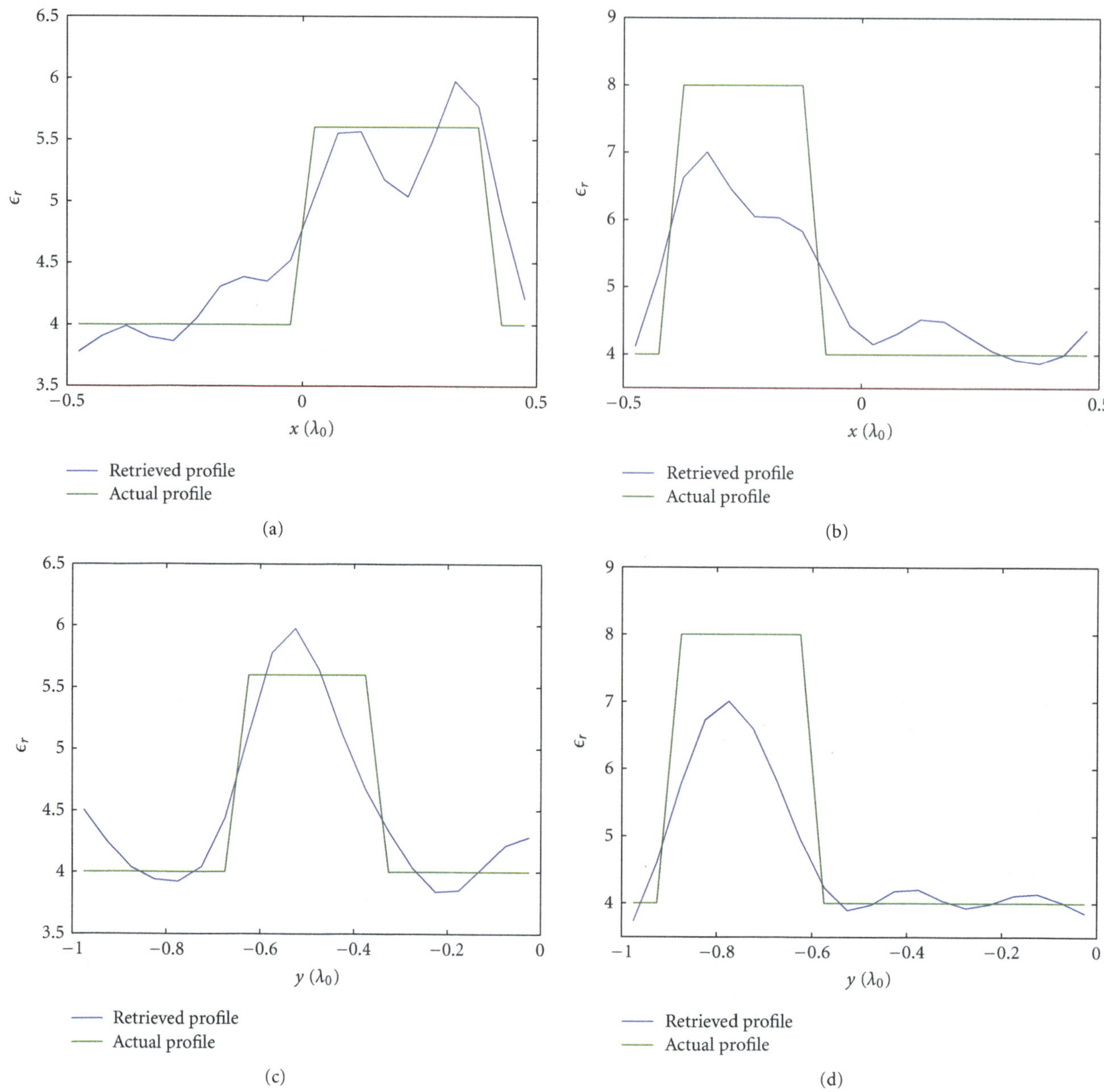

FIGURE 6: Horizontal cuts ((a) $y = -0.525\lambda_0$ and (b) $y = -0.775\lambda_0$)) and vertical cuts ((c) $x = 0.325\lambda_0$ and (d) $x = -0.325\lambda_0$)) of the reconstructed distribution of the relative dielectric permittivity (Figure 4). Two circular cylinders buried in a lossy half space.

the inner one has a radius $a_2 = 0.1\lambda_0$ and is characterized by a relative dielectric permittivity $\epsilon_{r_2} = 8$ and an electric conductivity $\sigma_2 = 0.01$ S/m. The object is centered at $\mathbf{r}^{\text{obj}} = (0.2\lambda_0, -0.5\lambda_0)$. The parameters of the inexact Newton method are the same as the ones adopted in the first case. The reconstructed distribution of the relative dielectric permittivity (final image) is shown in Figure 4. As can be seen, the object is correctly located, and the values of the dielectric properties are identified with a quite good accuracy.

Finally, the reconstruction of two separate circular cylinders have been considered. The radii of the two cylinders are $a_1 = 0.2\lambda_0$ and $a_2 = 0.15\lambda_0$. They are buried with centers located at points $\mathbf{r}_1^{\text{obj}} = (0.2\lambda_0, -0.5\lambda_0)$ and $\mathbf{r}_2^{\text{obj}} = (-0.25\lambda_0, -0.75\lambda_0)$. The dielectric properties of the two scatterers are the following: $\epsilon_{r_1} = 5.6$, $\epsilon_{r_2} = 8$, and $\sigma_1 = \sigma_2 = 0.01$ S/m. The signal-to-noise ratio is SNR $= 25\,dB$. The parameters of the iterative inversion algorithm are the same as in the previous cases. The reconstructed distribution of the relative dielectric permittivity is shown in Figure 5. This image corresponds to the optimal solution. As can be seen, also in this quite complex case, the method is able to correctly reconstruct the investigated scene. In particular, the two objects are clearly identified and separated. Moreover, both shapes and dielectric properties of the two targets are estimated with quite good accuracies (although for the second cylinder the dielectric permittivity is slightly underestimated). In order to better assess the reconstruction accuracy, Figure 6 gives the profiles of the retrieved distribution of Figure 5 along some cuts parallel to the coordinate axes.

5. Conclusion

In this paper, a previously proposed approach for the imaging of buried objects has been extended to inspect strong scatterers, for which the exact nonlinear equations of the inverse scattering problem must be considered. The developed method is based on a two-step inexact Newton method, which allows us to obtain a regularized solution of the (discretized) integral equations relating the samples of scattered electric field collected in the upper half space to the distributions of the dielectric parameters inside the test domain located in the ground. The reported preliminary results of numerical simulations seem to indicate the validity of the proposed approach, which needs further evaluation assessments, including an experimental validation, which will be the subject of future research activity in this area.

References

[1] T. J. Cui and W.C. Chew, "Inverse-scattering methods for three-dimensional targets buried in the lossy earth," in *Proceedings of the IEEE International Symposium on Antennas and Propagation*, vol. 3, pp. 1776–1779, 2000.

[2] C. J. Lin and C. C. Chiu, "Inverse scattering of buried inhomogeneous dielectric material coated on a conductor," in *Proceedings of the 5th International Kharkov Symposium on Physics and Engineering of Microwaves, Millimeter, and Submillimeter Waves (MSMW '04)*, pp. 281–283, June 2004.

[3] R. Persico, R. Bernini, and F. Soldovieri, "The role of the measurement configuration in inverse scattering from buried objects under the born approximation," *IEEE Transactions on Antennas and Propagation*, vol. 53, no. 6, pp. 1875–1887, 2005.

[4] C. J. Lin and C. C. Chiu, "Inverse scattering of buried complex object by TE wave illumination," in *Proceedings of the IEEE International Symposium on Microwave, Antenna, Propagation and EMC Technologies for Wireless Communications (MAPE '05)*, pp. 784–787, August 2005.

[5] A. N. Askarpour and R. Faraji-Dana, "Inverse scattering of 2-D dielectric objects buried in multilayer media using complex images green's function in born iteration method," in *Proceedings of the IEEE Antennas and Propagation Society International Symposium (APS '06)*, pp. 1085–1088, July 2006.

[6] C. C. Chiu and Y. W. Kiang, "Electromagnetic inverse scattering of a conducting cylinder buried in a lossy half-space," *IEEE Transactions on Antennas and Propagation*, vol. 40, no. 12, pp. 1562–1563, 1992.

[7] X. Chen, K. R. Shao, Y. Guo, J. Zhu, and J. D. Lavers, "A inverse scattering technique for objects buried in planar layered media based on an estimation of distribution algorithm," in *Proceedings of the 14th Biennial IEEE Conference on Electromagnetic Field Computation (CEFC '10)*, May 2010.

[8] L. Crocco, M. D'Urso, and T. Isernia, "New models and tools for forward and inverse scattering from buried dielectric targets," in *Proceedings of the 10th International Conference Ground Penetrating Radar (GPR '04)*, pp. 59–62, June 2004.

[9] K. Belkebir, R. E. Kleinman, and C. Pichot, "Microwave imaging—location and shape reconstruction from multifrequency scattering data," *IEEE Transactions on Microwave Theory and Techniques*, vol. 45, no. 4, pp. 469–476, 1997.

[10] L. Xu, S. K. Davis, S. C. Hagness, D. W. van der Weide, and B. D. Van Veen, "Microwave imaging via space-time beamforming: experimental investigation of tumor detection in multilayer breast phantoms," *IEEE Transactions on Microwave Theory and Techniques*, vol. 52, no. 8, pp. 1856–1865, 2004.

[11] T. J. Cui, A. A. Aydiner, W. C. Chew, D. L. Wright, and D. V. Smith, "Three-dimensional imaging of buried objects in very lossy earth by inversion of VETEM data," *IEEE Transactions on Geoscience and Remote Sensing*, vol. 41, no. 10, pp. 2197–2210, 2003.

[12] M. Lambert and D. Lesselier, "Binary-constrained inversion of a buried cylindrical obstacle from complete and phaseless magnetic fields," *Inverse Problems*, vol. 16, no. 3, pp. 563–576, 2000.

[13] G. C. Giakos, M. Pastorino, F. Russo, S. Chiwdhury, N. Shah, and W. Davros, "Noninvasive imaging for the new century," *IEEE Instrumentation & Measurement Magazine*, vol. 2, pp. 32–35, 1999.

[14] R. Zoughi, *Microwave Nondestructive Testing and Evaluation*, Kluwer Academic, Amsterdam, The Netherlands, 2000.

[15] D. Lesselier and J. Bowler, "Foreword to the special section on electromagnetic and ultrasonic nondestructive evaluation," *Inverse Problems*, vol. 18, no. 6, pp. 1–2, 2002.

[16] S. Kharkovsky and R. Zoughi, "Microwave and millimeter wave nondestructive testing and evaluation," *IEEE Instrumentation and Measurement Magazine*, vol. 10, no. 2, pp. 26–38, 2007.

[17] C. C. Chen, J. T. Johnson, M. Sato, and A. G. Yarovoy, "Guest editorial foreword to the special issue on subsurface sensing using Ground-Penetrating Radar (GPR)," *IEEE Transactions on Geoscience and Remote Sensing*, vol. 45, no. 8, p. 2419, 2007.

[18] O. Dorn and D. Lesselier, "Special issue on electromagnetic inverse problems: emerging methods and novel applications," *Inverse Problems*, vol. 26, no. 7, 2010.

[19] A. Baussard, E. L. Miller, and D. Lesselier, "Adaptive multiscale reconstruction of buried objects," *Inverse Problems*, vol. 20, no. 6, supplement, pp. S1–S15, 2004.

[20] A. Massa, M. Pastorino, and A. Randazzo, "Reconstruction of two-dimensional buried objects by a differential evolution method," *Inverse Problems*, vol. 20, no. 6, supplement, pp. S135–S150, 2004.

[21] M. Pastorino, *Microwave Imaging*, John Wiley & Sons, Hoboken, NJ, USA, 2010.

[22] C. H. Kuo and M. Moghaddam, "Electromagnetic scattering from a buried cylinder in layered media with rough interfaces," *IEEE Transactions on Antennas and Propagation*, vol. 54, no. 8, pp. 2392–2401, 2006.

[23] A. Bréard, G. Perrusson, and D. Lesselier, "Hybrid differential evolution and retrieval of buried spheres in subsoil," *IEEE Geoscience and Remote Sensing Letters*, vol. 5, pp. 788–792, 2008.

[24] A. Tabatabaeenejad and M. Moghaddam, "Inversion of subsurface properties of layered dielectric structures with random slightly rough interfaces using the method of simulated annealing," *IEEE Transactions on Geoscience and Remote Sensing*, vol. 47, no. 7, pp. 2035–2046, 2009.

[25] M. El-Shenawee, O. Dorn, and M. Moscoso, "An adjoint-field technique for shape reconstruction of 3-D penetrable object immersed in lossy medium," *IEEE Transactions on Antennas and Propagation*, vol. 57, no. 2, pp. 520–534, 2009.

[26] P. M. Meaney, M. W. Fanning, D. Li, S. P. Poplack, and K. D. Paulsen, "A clinical prototype for active microwave imaging of the breast," *IEEE Transactions on Microwave Theory and Techniques*, vol. 48, no. 1, pp. 1841–1853, 2000.

[27] G. Bozza, M. Brignone, and M. Pastorino, "Application of the no-sampling linear sampling method to breast cancer detection," *IEEE Transactions on Biomedical Engineering*, vol. 57, no. 10, pp. 2525–2534, 2010.

[28] G. Bellizzi, O. Bucci, and I. Catapano, "Microwave cancer imaging exploiting magnetic nanoparticles as contrast agent," *IEEE Transactions on Biomedical Engineering*, vol. 58, pp. 2528–2536, 2011.

[29] H. Harada, D. J. N. Wall, T. Takenaka, and M. Tanaka, "Conjugate gradient method applied to inverse scattering problem," *IEEE Transactions on Antennas and Propagation*, vol. 43, no. 8, pp. 784–792, 1995.

[30] P. M. van den Berg and A. Abubakar, "Contrast source inversion method: state of art," *Journal of Electromagnetic Waves and Applications*, vol. 15, no. 11, pp. 1503–1505, 2001.

[31] I. T. Rekanos, "Shape reconstruction of a perfectly conducting scatterer using differential evolution and particle swarm optimization," *IEEE Transactions on Geoscience and Remote Sensing*, vol. 46, no. 7, pp. 1967–1974, 2008.

[32] A. Litman, D. Lesselier, and F. Santosa, "Reconstruction of a two-dimensional binary obstacle by controlled evolution of a level-set," *Inverse Problems*, vol. 14, no. 3, pp. 685–706, 1998.

[33] R. Ferrayé, J. Y. Dauvignac, and C. Pichot, "Reconstruction of complex and multiple shape object contours using a level set method," *Journal of Electromagnetic Waves and Applications*, vol. 17, no. 2, pp. 153–181, 2003.

[34] A. Abubakar and T. M. Habashy, "Constrained parametric minimization for the inversion of electromagnetic measurements in the earth," in *Proceedings of the Progress in Electromagnetics Research Symposium (PIERS '04)*, pp. 49–52, March 2004.

[35] D. Colton, H. Haddar, and M. Piana, "The linear sampling method in inverse electromagnetic scattering theory," *Inverse Problems*, vol. 19, no. 6, supplement, pp. S105–S137, 2003.

[36] L. Pan, X. Chen, Y. Zhong, and S. P. Yeo, "Comparison among the variants of subspace-based optimization method for addressing inverse scattering problems: transverse electric case," *Journal of the Optical Society of America A*, vol. 27, no. 10, pp. 2208–2215, 2010.

[37] Y. Zhong, X. Chen, and K. Agarwal, "An improved subspace-based optimization method and its implementation in solving three-dimensional inverse problems," *IEEE Transactions on Geoscience and Remote Sensing*, vol. 48, no. 10, pp. 3763–3768, 2010.

[38] J. De Zaeytijd, A. Franchois, C. Eyraud, and J. M. Geffrin, "Full-wave three-dimensional microwave imaging with a regularized Gauss-Newton method-theory and experiment," *IEEE Transactions on Antennas and Propagation*, vol. 55, no. 11, pp. 3279–3292, 2007.

[39] C. Gilmore, P. Mojabi, and J. LoVetri, "Comparison of an enhanced distorted born iterative method and the multiplicative-regularized contrast source inversion method," *IEEE Transactions on Antennas and Propagation*, vol. 57, no. 8, pp. 2341–2351, 2009.

[40] D. J. Daniels, *Ground Penetrating Radars*, IEE, London, UK, 2nd edition, 2004.

[41] R. Mittra, M. Ji-Fu, and Y. Wenhua, "Identifying buried objects using the neural network approach," in *Proceedings of the IEEE Antennas and Propagation Society International Symposium*, pp. 2596–2599, 1999.

[42] L. Crocco, G. Prisco, F. Soldovieri, and N. J. Cassidy, "Advanced forward modeling and tomographic inversion for leaking water pipes monitoring," in *Proceedings of the 4th International Workshop on Advanced Ground Penetrating Radar (IWAGPR '07)*, pp. 127–131, June 2007.

[43] G. Bozza, C. Estatico, M. Pastorino, and A. Randazzo, "Application of an inexact-Newton method within the second-order Born approximation to buried objects," *IEEE Geoscience and Remote Sensing Letters*, vol. 4, no. 1, pp. 51–55, 2007.

[44] R. Pierri and G. Leone, "Inverse scattering of dielectric cylinders by a second-order born approximation," *IEEE Transactions on Geoscience and Remote Sensing*, vol. 37, no. 1, pp. 374–382, 1999.

[45] A. Brancaccio, V. Pascazio, and R. Pierri, "Quadratic model for inverse profiling: the one-dimensional case," *Journal of Electromagnetic Waves and Applications*, vol. 9, no. 5-6, pp. 673–696, 1995.

[46] C. Estatico, M. Pastorino, and A. Randazzo, "An inexact-Newton method for short-range microwave imaging within the second-order born approximation," *IEEE Transactions on Geoscience and Remote Sensing*, vol. 43, no. 11, pp. 2593–2605, 2005.

[47] S. Caorsi, A. Costa, and M. Pastorino, "Microwave imaging within the second-order born approximation: stochastic optimization by a genetic algorithm," *IEEE Transactions on Antennas and Propagation*, vol. 49, no. 1, pp. 22–31, 2001.

[48] G. Bozza, C. Estatico, M. Pastorino, and A. Randazzo, "A regularization approach to microwave imaging under the second-order Born approximation with real data," in *Proceedings of the IEEE International Workshop on Imaging Systems and Techniques, Niagara Falls*, pp. 14–19, Ontario, Canada, May 2005.

[49] G. Bozza, C. Estatico, M. Pastorino, and A. Randazzo, "An inexact Newton method for microwave reconstruction of strong scatterers," *IEEE Antennas and Wireless Propagation Letters*, vol. 5, no. 1, pp. 61–64, 2006.

[50] G. Bozza, C. Estatico, A. Massa, M. Pastorino, and A. Randazzo, "Short-range image-based method for the inspection of strong scatterers using microwaves," *IEEE Transactions on Instrumentation and Measurement*, vol. 56, no. 4, pp. 1181–1188, 2007.

[51] C. Estatico, G. Bozza, A. Massa, M. Pastorino, and A. Randazzo, "A two-step iterative inexact-Newton method for electromagnetic imaging of dielectric structures from real data," *Inverse Problems*, vol. 21, no. 6, supplement, pp. S81–S94, 2005.

[52] G. Bozza and M. Pastorino, "An inexact Newton-based approach to microwave imaging within the contrast source formulation," *IEEE Transactions on Antennas and Propagation*, vol. 57, no. 4, pp. 1122–1132, 2009.

[53] H. M. Jol, *Ground Penetrating Radar: Theory and Applications*, Elsevier Science, Oxford, UK, 2009.

[54] J. H. Richmond, "Scattering by a dielectric cylinder of arbitrary cross shape," *IEEE Transactions on Antennas and Propagation*, vol. 13, pp. 334–341, 1965.

[55] M. Bertero and P. Boccacci, *Introduction to Inverse Problems in Imaging*, IOP, Bristol, UK, 1998.

[56] G. Bozza, C. Estatico, M. Pastorino, and A. Randazzo, "A numerical assessment of the semiconvergence behavior in an inverse-scattering approach to electromagnetic imaging," in *Proceedings of the IEEE Instrumentation and Measurement Technology Synergy of Science and Technology in Instrumentation and Measurement (IMTC '07)*, Warsaw, Poland, May 2007.

A UWB Bidirectional Rectangular Ring Antenna Fed by CDM with a Rod and Ridges for Constant Beam Direction

S. Vongsack,[1] S. Lamultree,[2] P. Osklang,[3] C. Phongcharoenpanich,[3] S. Kosulvit,[3] K. Hamamoto,[4] and T. Wakabayashi[5]

[1] *International College, King Mongkut's Institute of Technology Ladkrabang, Bangkok 10520, Thailand*
[2] *Faculty of Engineering and Technology, Asian University, Chonburi 20260, Thailand*
[3] *Faculty of Engineering, King Mongkut's Institute of Technology Ladkrabang, Bangkok 10520, Thailand*
[4] *School of Information and Telecommunication Engineering, Tokai University, Kanagawa 259-1292, Japan*
[5] *Malaysia-Japan International Institute of Technology (MJIIT), Universiti Technologi Malaysia, Kuala Lumpur 54100, Malaysia*

Correspondence should be addressed to S. Lamultree, suthasineel@asianust.ac.th

Academic Editor: Beatriz Ortega

This paper presents an ultra-wideband (UWB) rectangular ring antenna excited by a circular disc monopole (CDM) with a conducting rod and two double ridges to radiate bidirectional pattern with constant beam direction along the entire UWB frequency range of 3.1–10.6 GHz. The conducting rod and double ridges at the upper wall of the ring are added to solve the tilted beam problem at the higher edge frequency whereas the double ridges at the lower wall are used to enhance the impedance bandwidth. The dimensions of the rectangular ring and the CDM are initially considered to achieve the bidirectional pattern with the suitable resonant frequencies and bandwidth. Then, the parameters of copper rod and two double ridges are determined by parametric study using CST Microwave Studio simulation software. The prototype antenna was fabricated, and the measured results show good agreement with the simulated ones. The obtained bandwidth of $|S_{11}| < -10$ dB can cover the UWB frequency range as well as the bidirectional beam radiation with constant beam direction ($\theta = 0^\circ, 180^\circ$ and $\phi = 90^\circ$). The minimum and maximum measured gains are 3.1 dBi to 5.3 dBi, respectively. The proposed antenna possesses compact size with good radiation performance that can be a promising candidate for UWB applications.

1. Introduction

Nowadays, the wireless communication systems become the significant part of daily-life essence. The ultra-wideband (UWB) is a key technology for future short-range, high-speed, and indoor-data wireless communication. UWB technology is extensively and continuously spread since the regulation of Federal Communication Commission (FCC) defined the frequency spectrum from 3.1 GHz to 10.6 GHz for this system [1]. The demand for the multipurpose and multifunction of this system imposes inevitably the use of antenna to cover very wideband frequency range. For the typical cellular system, the service area is decomposed into small cells for the urban area. The cell shape is approximated by a circle where omnidirectional antenna plays a vital role to employ at the base station [2]. However, for some specific application that requires the long-range service area such as point-to-point communication between the buildings, when the mobiles move along the confined paths such as street, highways, tunnels, and corridors, for these environments, a unidirectional or bidirectional antennas are more suitable than the omnidirectional one to serve these demands [3–6]. The omnidirectional monopole antenna is modified by adding different shapes of reflectors such as planar reflector [7], corner reflectors [8, 9], curved reflectors [10], ring reflectors [11], and others to obtain the unidirectional or bidirectional beams. To enhance the bandwidth, the linear monopole is extended to surface monopole [12] or three-dimensional (3D) monopole [13–15]. To realize the bidirectional pattern along the UWB frequency range, this paper presents the circular disc monopole (CDM) excited rectangular ring antenna. However, the beam direction at

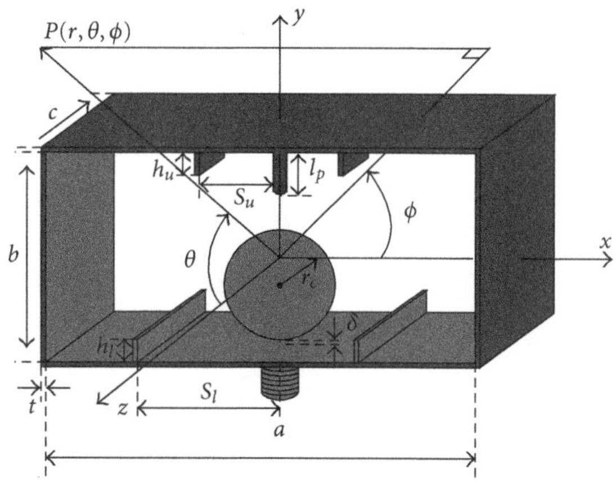

FIGURE 1: Antenna structure.

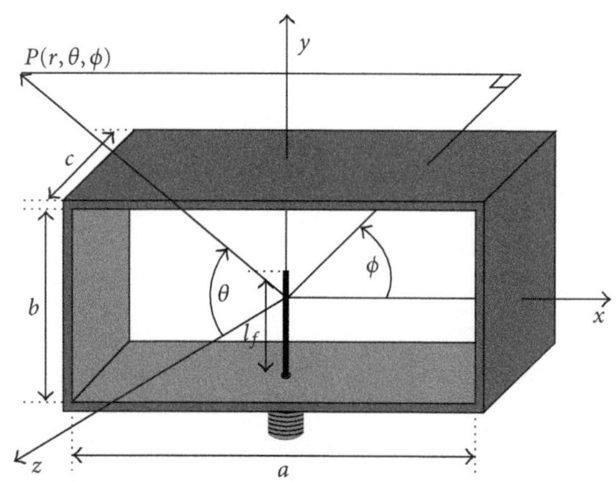

FIGURE 2: An initial antenna structure.

the higher edge frequency is tilted due to unsymmetrical structure of the antenna along the upper and lower parts of the ring. The conducting rod and two double ridges are added at the upper wall of the ring to induce the current along the upper part of the ring that makes the constant beam direction at the higher edge frequency. This tilted beam problem can be solved at the expense of slightly narrower impedance bandwidth. To enhance the impedance bandwidth to meet the frequency requirement of UWB system along 3.1–10.6 GHz, the double ridges at the lower wall of the ring are added. The design of this antenna is easy and straightforward. The ring dimension and radius of CDM are firstly considered to operate at the desired resonant frequency with very wide bandwidth. Subsequently, the appropriate parameters of conducting rod and two double ridges are determined from the parametric study. The antenna characteristics such as $|S_{11}|$, radiation pattern, and gain along the UWB frequency are reported.

The content of this paper is organized as follows. The proposed antenna structure and its parameters are illustrated in Section 2. Section 3 describes the design principle and parametric study. To verify the simulated results, Section 4 presents the measured results. The conclusions are drawn in Section 5.

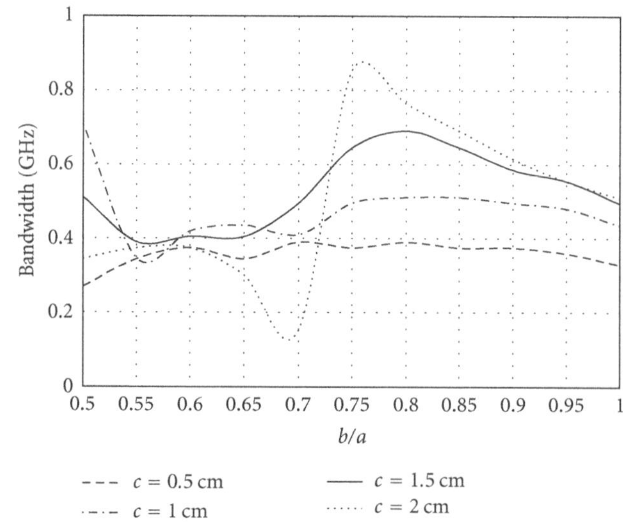

FIGURE 3: Bandwidth of the lower resonant frequency as the function of b/a (a = 5 cm and l_f = 2.5 cm).

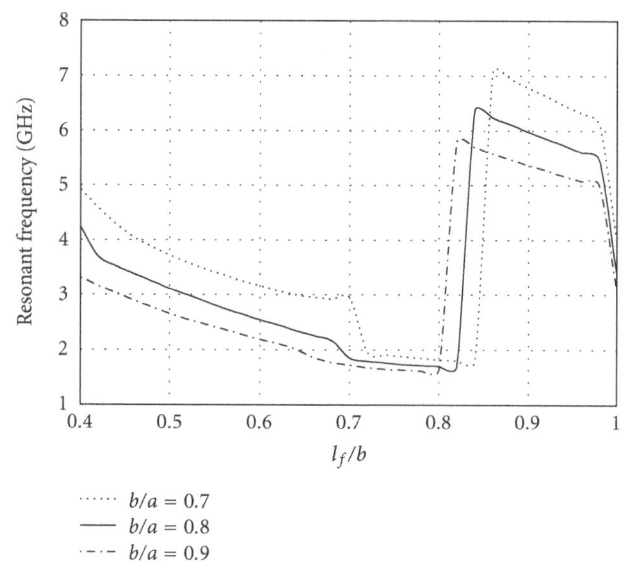

FIGURE 4: The lower resonant frequency for various l_f/b (a = 5 cm and c = 1.5 cm).

2. Antenna Description

The proposed antenna structure consists of the rectangular ring excited by CDM as shown in Figure 1. The ring is used to confine the CDM radiating omnidirectional pattern to become bidirectional beam. The ring is made of metallic material with the width of a, the height of b, and the length of c. The thickness of the ring is denoted by t. This structure is excited by the CDM of the radius r_c via 50 Ω SMA connector with delta gap δ. The CDM is located at the center of the lower wall of the ring. The copper rod of the length l_p is located at the center of the upper wall of the ring. The double ridge of the height h_u is located with s_u apart from the ring

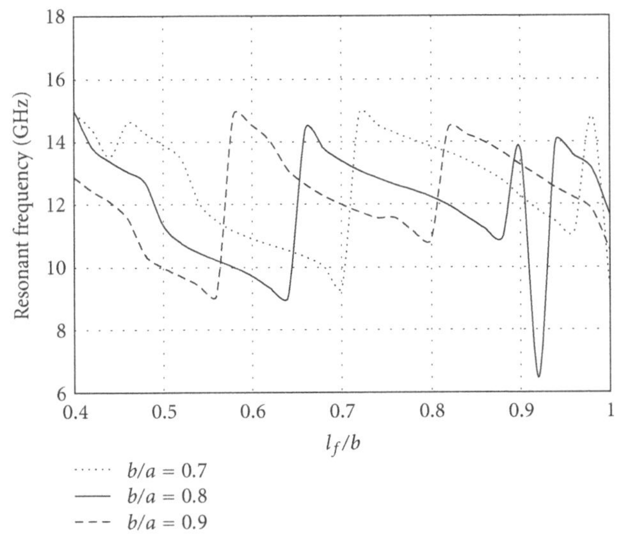

FIGURE 5: The upper resonant frequency for various l_f/b ($a = 5$ cm and $c = 1.5$ cm).

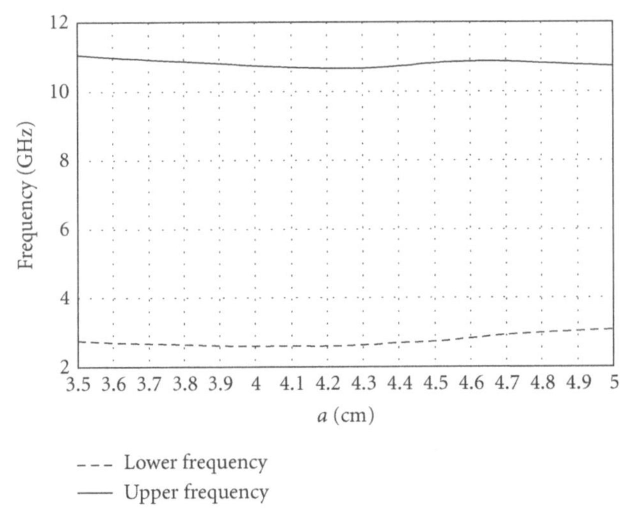

FIGURE 6: Resonant frequency for various a ($b/a = 0.7$, $l_f = 0.62b$).

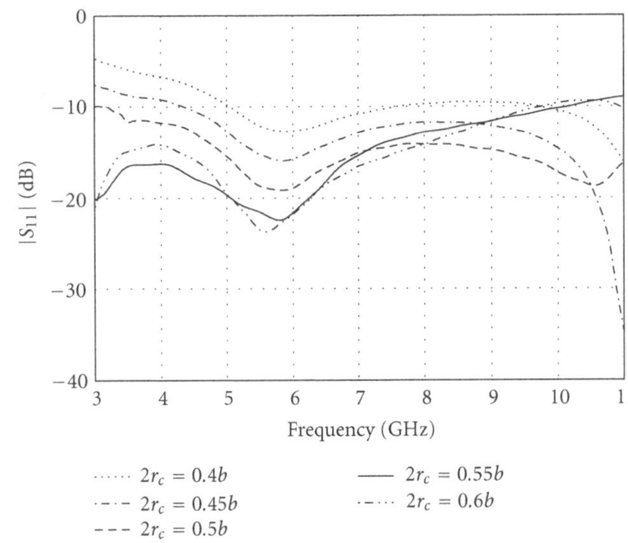

FIGURE 7: $|S_{11}|$ versus frequency as a function of $2r_c$.

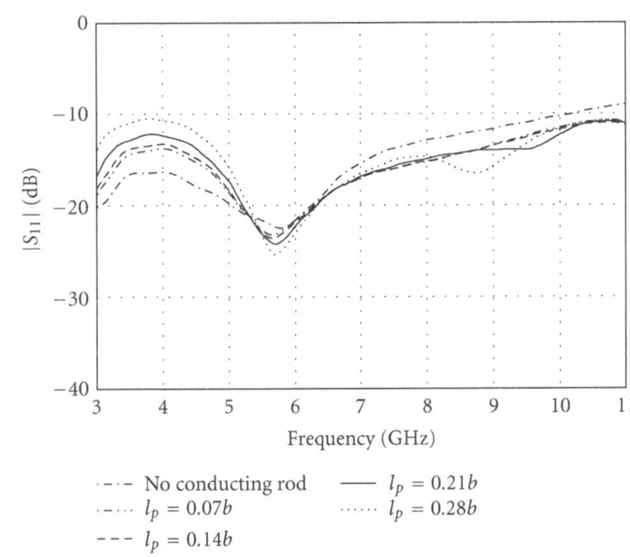

FIGURE 8: $|S_{11}|$ versus frequency as a function of l_p.

center on the upper wall of the ring. At the lower wall of the ring, there are double ridges of the height h_l located at the distance s_l from the center. It is noted that the length of the ridges is identical to the length of the ring (c).

3. Design Principle and Parametric Study

The principle of the antenna design starts with consideration of the conventional linear monopole antenna of the length l_f. This structure radiates the omnidirectional beam. To obtain the bidirectional pattern, this structure is surrounded by the rectangular ring to focus the main beam into two opposite directions along $+z$ and $-z$ axes as shown in Figure 2. The initial parameter of the ring width a of 5 cm ($\lambda/2$) is chosen with TE_{10} mode operation at 3 GHz that is around the lower edge frequency of UWB. It is noted that the ring in this paper is made of aluminum with thickness t of 8 mm for the reason of available fabricated material.

The length l_f of linear monopole is $\lambda/4$ at 3 GHz (l_f = 2.5 cm). Figure 3 shows the bandwidth of the resonant frequency closest to the frequency of 3 GHz as a function of the ratio between the height and the width (b/a) of the ring for different ring lengths (c). The CST Microwave Studio [16] is used as the simulation tool.

From Figure 3, the ring dimensions of a, b, and c can be selected from the compact size with the widest bandwidth. Therefore, the ring width (a) of 5 cm, ring height (b) of 3.5 cm ($b/a = 0.7$), and ring length (c) of 1.5 cm are chosen. The next step is to find the suitable feeding structure and its dimension to cover the UWB frequency range of 3.1–10.6 GHz. The linear monopole is the simplest structure

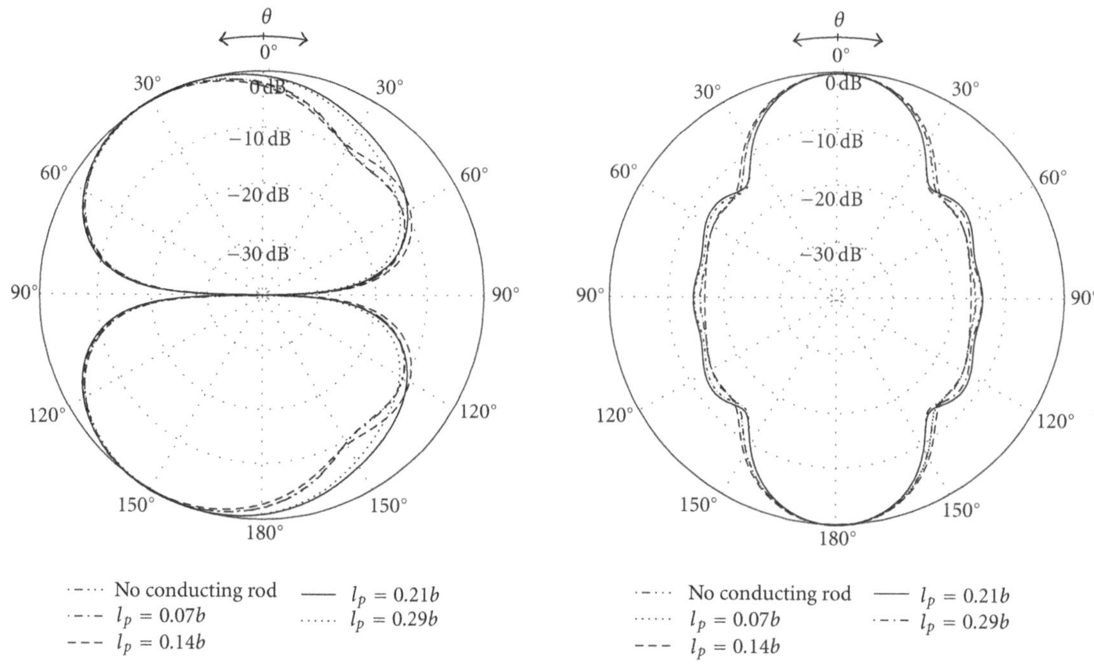

FIGURE 9: The radiation pattern in yz and xz planes as a function of l_p at 10.6 GHz.

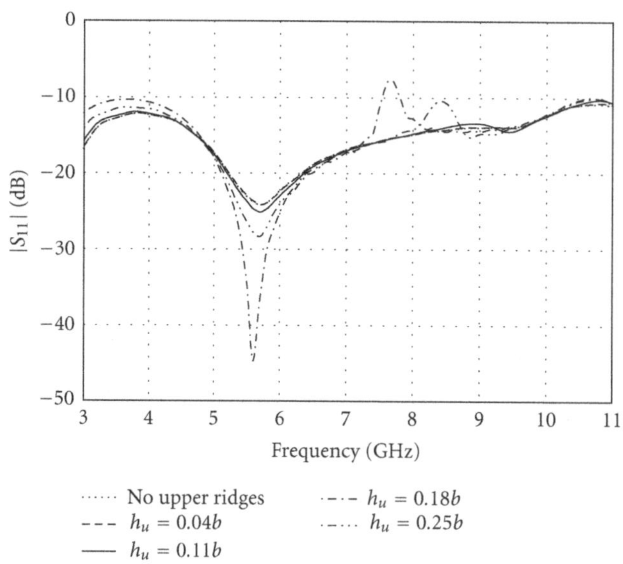

FIGURE 10: $|S_{11}|$ versus frequency as a function of h_u.

for consideration. Figures 4 and 5 show the lower and upper resonant frequencies of the rectangular ring with the dimension mentioned above, and it is excited by the linear electric probe of the length l_f. The dimension of the probe length is expressed in term of l_f/b ratio. To achieve the lower and upper resonant frequencies of 3.1 and 10.6 GHz, respectively, the length of the probe can be selected from different b/a. For example, if b/a is 0.7, the probe length should be $0.62b$. In the same fashion, the suitable l_f/b for $b/a = 0.8$ and $b/a = 0.9$ are 0.52 and 1.0, respectively. For compact size, $l_f = 0.62b$ for $b = 0.7a$ is chosen in this paper. Then, the effect of the ring width is studied to determine the possible smallest size. Figure 6 shows the lower and upper resonant frequencies as a function of the ring width. It is apparent that the suitable ring width is 4 cm. Therefore, the dimension of the ring of $a = 4$ cm, $b = 0.7a = 2.8$ cm, $c = 1.5$ cm is used hereafter.

To extend the bandwidth of the antenna, the one-dimensional feeding structure of linear monopole will be modified to two-dimensional monopole, that is CDM. The CDM is introduced to increase the bandwidth to 7.5 GHz (UWB: 3.1–10.6 GHz). The radius r_c of CDM is varied to determine the appropriate dimension. Figure 7 shows the $|S_{11}|$ as the function of the frequency for various diameters

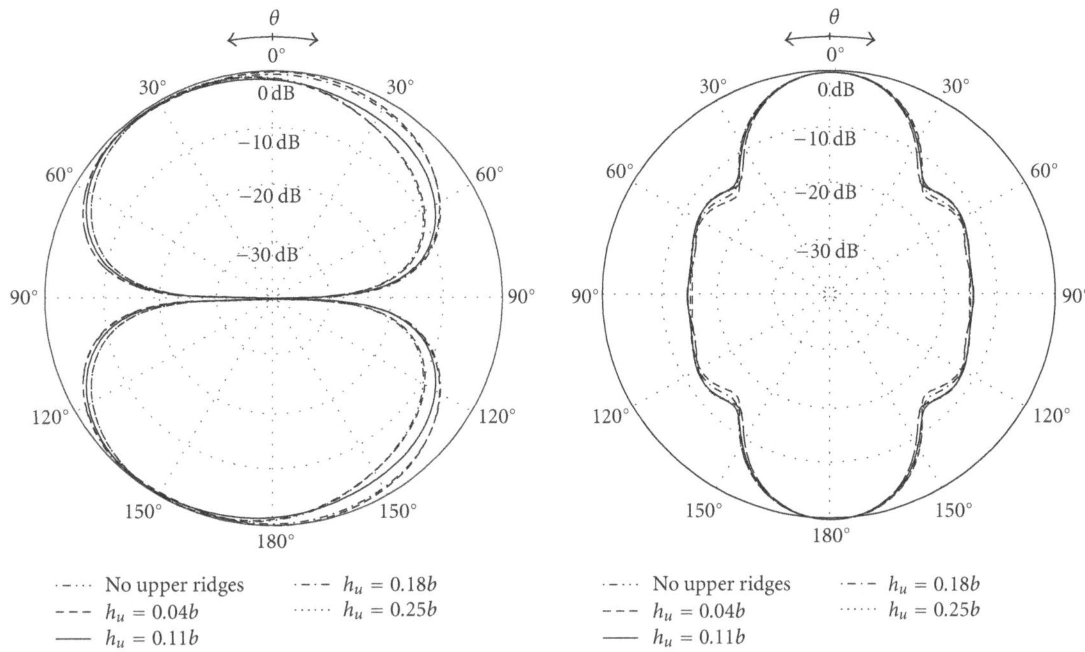

FIGURE 11: The radiation pattern in yz and xz planes as a function of h_u at 10.6 GHz.

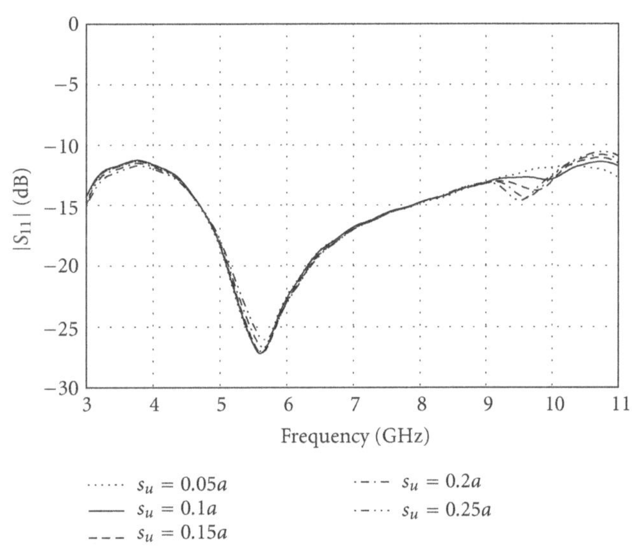

FIGURE 12: $|S_{11}|$ versus frequency as a function of s_u.

$(2r_c)$ of CDM. It is obvious that $2r_c$ of $0.55b$ or 1.5 cm is chosen to cover UWB frequency range. At this step, the bidirectional pattern along the frequency range of 3.0–10.2 GHz is obtained.

Figures 8 and 9 show the $|S_{11}|$ and the radiation pattern for different rod lengths compared with case of without conducting rod. The bandwidth can be further enhanced by adding conduction rod. It can be seen from Figure 9 that the beam direction in yz-plane without conducting rod at the higher edge frequency is shifted from the desired direction ($\theta = 0°, 180°$ and $\phi = 90°$). Therefore, the conducting rod of the length l_p is inserted at the upper wall of the ring to solve the tilted-beam problem. It is found that l_p of $0.21b$ (6 mm) is appropriately chosen to obtain the widest bandwidth with minimized beam tilt in yz-plane at 10.6 GHz.

Although the conducting rod can mitigate the beam tilt problem at the higher edge frequency, the radiation pattern can be further improved. From the simulated results, it is found that the radiation pattern is still slightly tilted. Thus, the double ridges at the upper wall of the ring with the height h_u and the position s_u from the center are added to improve the radiation patterns at the higher edge frequency. To study the influence of the double ridges, the height h_u is varied. From the results of parametric study, it is found that

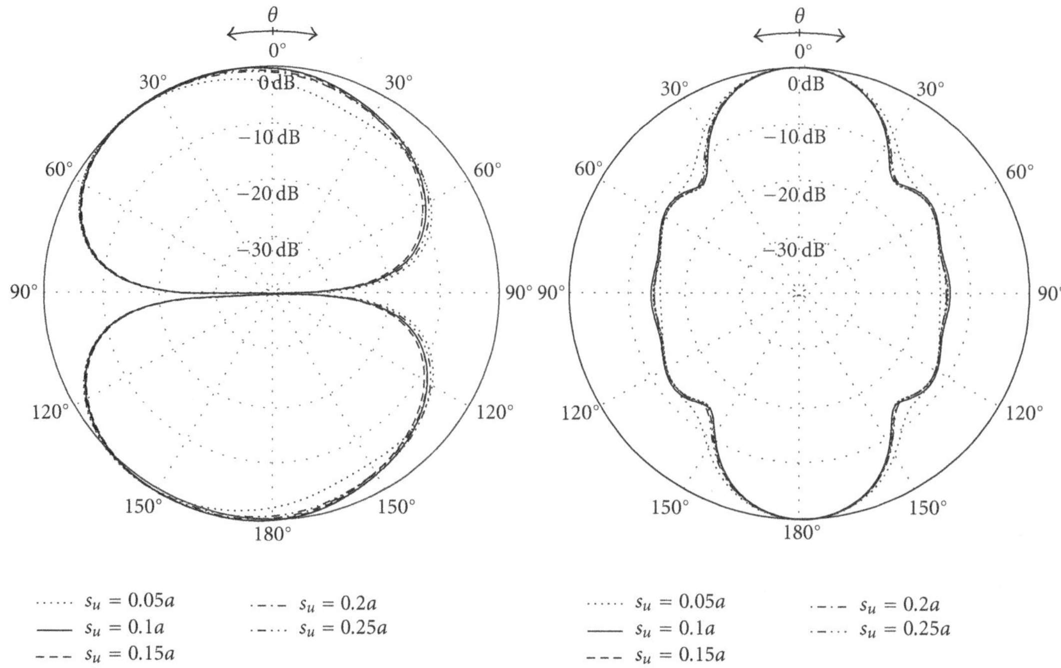

Figure 13: The radiation pattern in *yz* and *xz* planes as a function of s_u at 10.6 GHz.

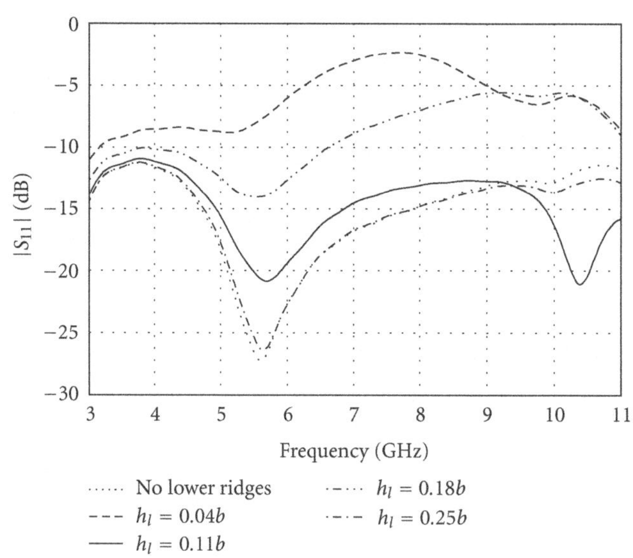

Figure 14: $|S_{11}|$ versus frequency as a function of h_l.

the height of double ridges equal or higher than 0.18b can improve level of $|S_{11}|$ but the frequency band cannot cover UWB as shown in Figure 10. Instead, the double ridges of the length 0.11b (3 mm) can cover the UWB. From Figure 11, it is observed that the radiation pattern is slightly tilted when the frequency become higher. Moreover, to change the main beam to the desired direction, the positions s_u are varied from 0.05a to 0.25a. From the results, it is obvious that variation of s_u does not affect to $|S_{11}|$ at the lower edge frequency as shown in Figure 12. When the frequency becomes higher than 9 GHz, the beam possesses minimum beam tilt when s_u equals to 0.1a (0.4 cm) as shown in Figure 13. As a result, s_u of 0.1a is selected as the design parameter.

To further determine the suitable parameter for the widest bandwidth and compact size, the height of double ridges h_l and the position of ridge s_l at the lower wall of the ring are varied. The height h_l and the position of the ridge s_l are varied at 0–0.25b and 0.175a–0.375a, respectively. From Figures 14 and 15, the impedance bandwidth and radiation pattern are depended on the height h_l at the lower wall of the ring. From the results, it is found that $|S_{11}|$ is higher than −10 dB making the bandwidth cannot cover UWB when the height of double ridges h_l is increased higher than 0.11b. The

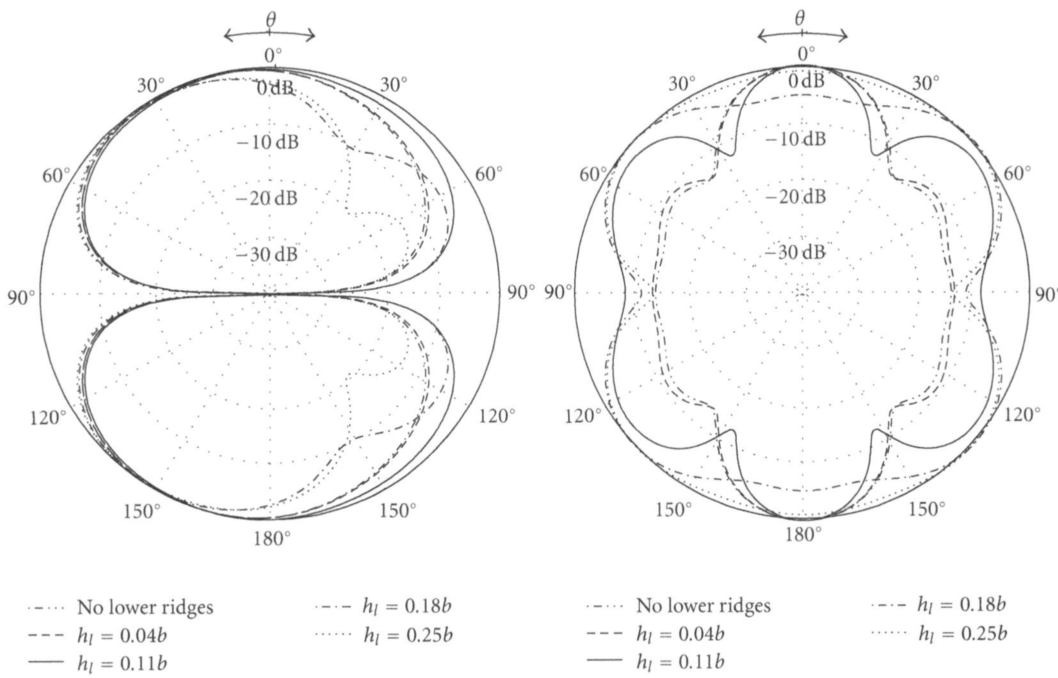

FIGURE 15: The radiation pattern in *yz* and *xz* planes as a function of h_l at 10.6 GHz.

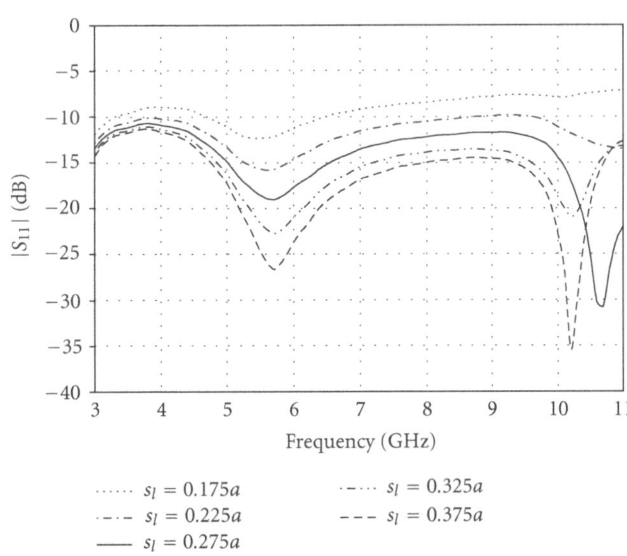

FIGURE 16: $|S_{11}|$ versus frequency as a function of s_l.

radiation patterns at the higher edge frequency will be worse with the increased height of double ridges at the lower wall of ring as illustrated in Figure 15. Moreover, the $|S_{11}|$ and radiation pattern at the higher edge frequency can be improved by varying the position of ridge s_l at the lower wall of the ring. From the results, it is obvious that $|S_{11}|$ can be reduced to lower than −10 dB with minimized beam tilt in *yz*-plane at 10.6 GHz when the position of ridge is increased as illustrated in Figures 16 and 17, respectively. Hence, the height of double ridges h_l of 0.11b (3 mm) and the position of ridges s_l of 0.275a (1.1 cm) are chosen as the design parameter.

Figures 18 and 19 show the compared $|S_{11}|$ of the proposed antenna by using CDM excited rectangular ring with rod and double ridges and without rod and double ridges. It is found that $|S_{11}|$ of the rectangular ring antenna fed by CDM without rod and double ridges is higher than −10 dB at the higher edge frequency that makes this structure cannot cover the frequency band from 3.1 to 10.6 GHz. The bandwidth can be extended to cover the UWB frequency range by adding the rod and double ridges. Moreover, the radiation pattern at the higher edge frequency can also be improved with the rod and double ridges.

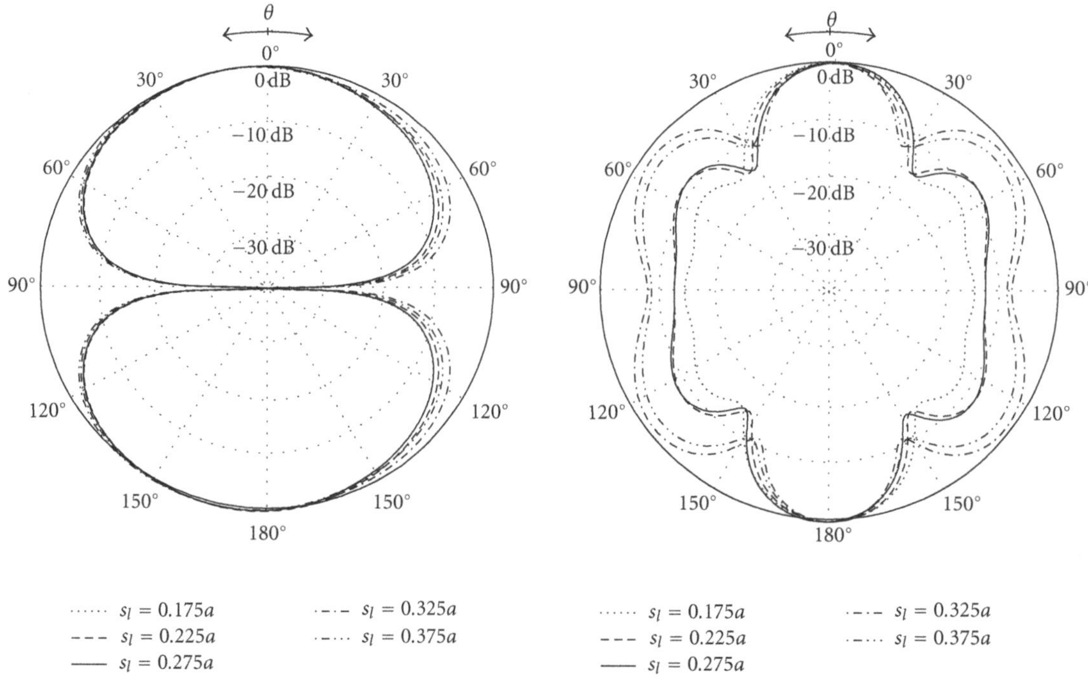

FIGURE 17: The radiation pattern in yz and xz planes as a function of s_l at 10.6 GHz.

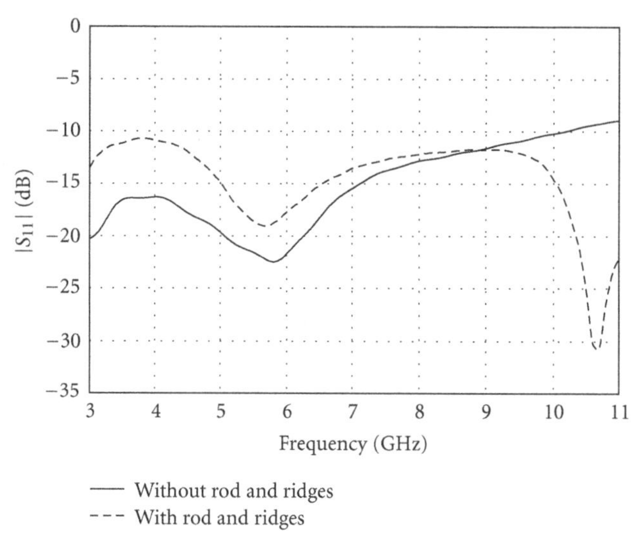

FIGURE 18: Comparison of $|S_{11}|$ between with and without rod and double ridges.

Ultimately, from the simulated results of the proposed antenna, the suitable parameters are obtained as tabulated in Table 1.

4. Measured Results

To confirm the simulated results, the prototype antenna with suitable parameters as tabulated in Table 1 was fabricated as depicted in Figure 20.

4.1. Impedance Bandwidth. The impedance characteristics and radiation pattern are measured by using an HP8720C Network Analyzer inside anechoic chamber. The compared $|S_{11}|$ between the simulated and measured impedance bandwidth is plotted in Figure 21. Apparently, the simulated and measured impedance bandwidths of $|S_{11}| < -10$ dB can cover the frequency range of 3–11 GHz. In addition, Figure 22 shows the simulated and measured gain at the desired direction ($\theta = 0°$ and $\phi = 90°$) along the frequency

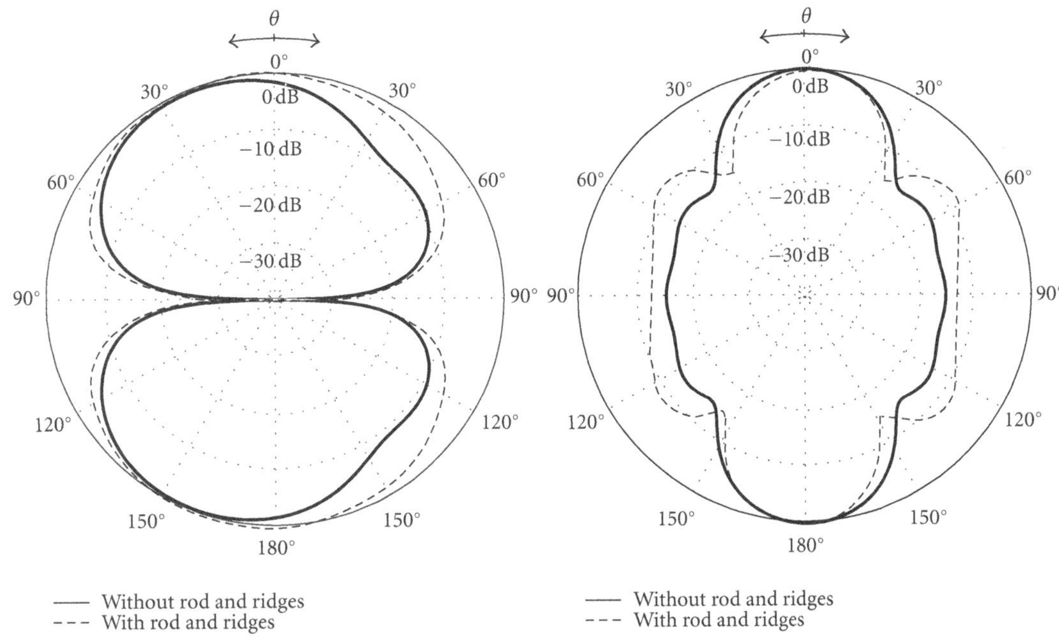

Figure 19: Comparison of radiation pattern with and without rod and double ridges in *yz* and *xz* planes at 10.60 GHz.

Table 1: Suitable antenna parameters.

Parameter	Physical size	Electrical size (λ)
a	4.0 cm	0.400
b	2.8 cm	0.280
c	1.5 cm	0.150
s_l	1.1 cm	0.110
s_u	4.0 mm	0.040
r_c	7.7 mm	0.077
l_p	6.0 mm	0.060
h_u	3.0 mm	0.030
h_l	3.0 mm	0.030

Figure 20: Photograph of the prototype antenna.

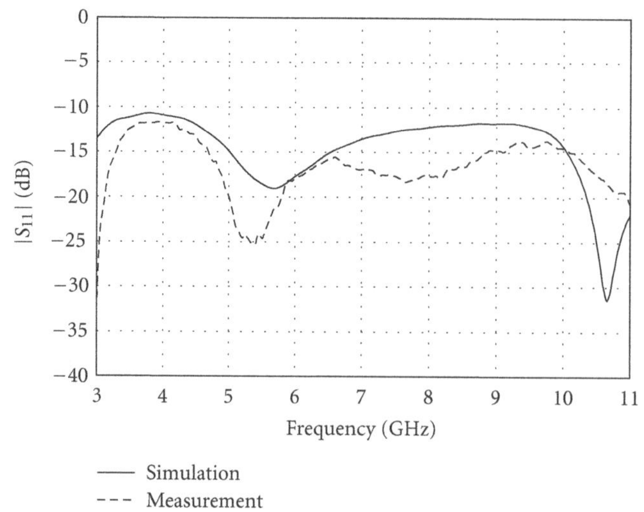

FIGURE 21: Compared $|S_{11}|$ between simulation and measurement.

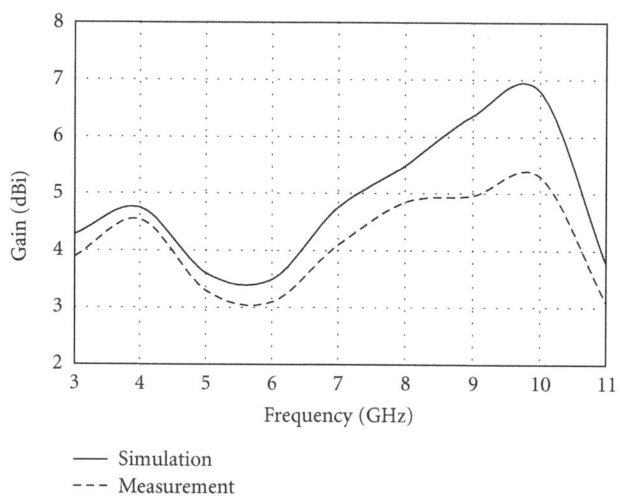

FIGURE 22: Simulated and measured gain.

range of 3.1–10.6 GHz. The minimum and maximum gains from the simulation and measurement are 3.5–6.9 dBi and 3.1–5.3 dBi, respectively.

4.2. Radiation Pattern. The simulated and measured radiation patterns in the *yz*- and *xz*-planes at 3.1, 6.8, and 10.6 GHz are shown in Figure 23. It is obvious that the beam peak directs to the $+z$ and $-z$ directions. Furthermore, it is found that the radiation pattern in the *yz*-plane is slightly tilted from z axis at the higher frequency because of the present CDM located at the lower wall of rectangular ring making unsymmetrical structure the upper and lower parts of the ring. Due to the symmetrical structure along left and right part of the ring, the radiation pattern is symmetry in *xz*-plane. The radiation patterns from simulation and measurement in the *yz*- and *xz*-planes are in good agreement.

5. Conclusions

The rectangular ring antenna fed by CDM with conducting rod and double ridges is proposed for an antenna at the base station of ultra-wideband applications. This antenna radiates bidirectional pattern with constant beam direction over the UWB frequency. The effect of antenna parameters to $|S_{11}|$ characteristics and the radiation patterns is investigated by parametric study. This antenna possesses the maximum gain from the measurement of 5.3 dBi at 9.8 GHz and higher than 3 dBi for the frequency range from 3.1 to 10.6 GHz. The impedance bandwidth of $|S_{11}| < -10$ dB can cover the frequency range of UWB. Since the proposed antenna possesses the compact size and good radiation performance, it can be a potential candidate for ultra-wideband applications.

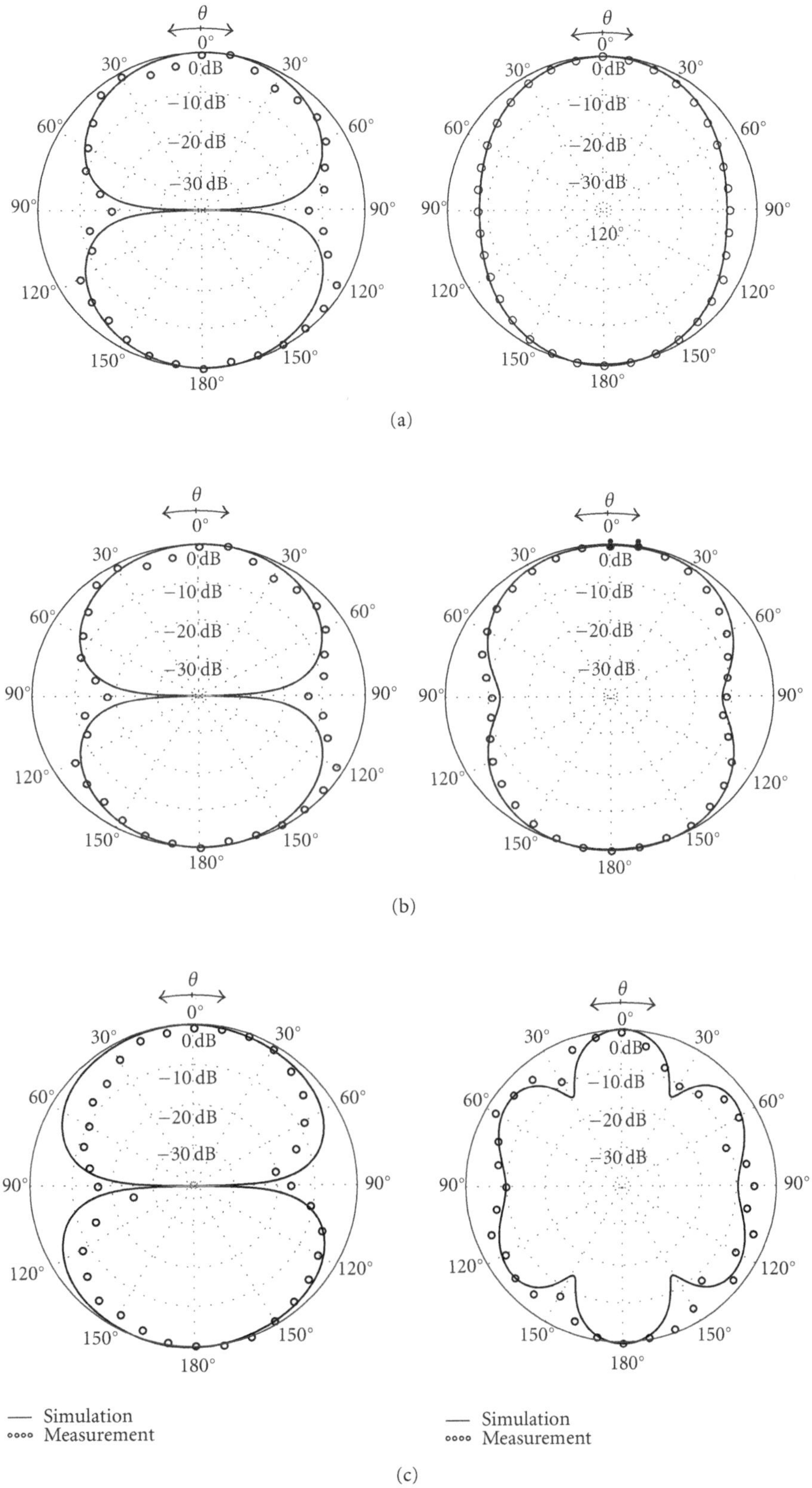

FIGURE 23: The comparison between the simulated and measured radiation pattern in *yz* and *xz* planes.

Acknowledgment

This work is supported by AUN/Seed-Net Program.

References

[1] Federal Communications Commission, "First order and report: revision of part 15 of the commission's rules regarding UWB transmission systems," April 2002.

[2] M. Hammoud, P. Poey, and F. Colombel, "Matching the input impedance of a broadband disc monopole," *Electronics Letters*, vol. 29, no. 4, pp. 406–407, 1993.

[3] C. Phongcharoenpanich, S. Lamultree, S. Kosulvit, and M. Krairiksh, "Theory and experiment of an antenna using a probe excited rectangular ring," in *Proceedings of the IEEE International Symposium on Antennas and Propagation and USNC/CNC/URSI North American Radio Science Meeting*, vol. 3, pp. 737–740, June 2003.

[4] K. Chawanonphithak, C. Phongcharoenpanich, S. Kosulvit, and M. Krairiksh, "Characteristics of an elliptical ring antenna excited by a linear electric probe," *International Journal of Electronics*, vol. 94, no. 10, pp. 973–984, 2007.

[5] S. Kosulvit, C. Phongcharoenpanich, M. Krairiksh, and T. Wakabayashi, "Radiation characteristics of a bidirectional antenna using a probe excited circular ring," in *Proceedings of the IEEE International Symposium. On Intelligent Signal Processing and Communication Systems*, pp. 713–716, December 1999.

[6] S. Lamultree and C. Phongcharoenpanich, "Full-wave investigations of a probe-excited rectangular ring antenna by method of moments with RWG basis functions," *PIERC*, vol. 8, pp. 161–177, 2009.

[7] L. Wei, Q. Jinghui, and S. Ying, "Design and simulation of novel ultra wideband planar reflector antenna," in *Proceedings of the International Conference on Microwave and Millimeter Wave Technology (ICMMT '07)*, pp. 1–4, April 2007.

[8] C. Phongcharoenpanich, S. Lamultree, S. Kosulvit, M. Krairiksh, and J. Takada, "Analysis of a dihedral corner reflector antenna excited by a probe inside rectangular ring," in *Proceedings of the International Symposium on Antennas and Propagation*, pp. 225–228, August 2004.

[9] J. D. Kraus, "The corner-reflector antenna," *Proceedings of the IRE*, vol. 28, pp. 513–519, 1940.

[10] P. T. Lam, S. W. Lee, K. C. Lang, and D. C. D. Chang, "Sidelobe reduction of a parabolic reflector with auxiliary reflectors," *IEEE Transactions on Antennas and Propagation*, vol. AP-35, no. 12, pp. 1367–1374, 1987.

[11] G. Ketwan, C. Phongcharoenpanich, and S. Kosulvit, "A wideband unidirectional antenna using conical reflector fed by circular ring," in *Proceedings of the Asia-Pacific Microwave Conference*, pp. 2939–2942, December 2005.

[12] N. P. Agrawall, G. Kumar, and K. P. Ray, "Wide-band planar monopole antennas," *IEEE Transactions on Antennas and Propagation*, vol. 46, no. 2, pp. 294–295, 1998.

[13] Z. N. Chen, M. J. Ammann, M. Y. W. Chia, and T. S. P. See, "Annular planar monopole antennas," *IEE Proceedings: Microwaves, Antennas and Propagation*, vol. 149, no. 4, pp. 200–203, 2002.

[14] M. J. Ammann and Z. N. Chen, "Wideband monopole antennas for multi-band wireless systems," *IEEE Antennas and Propagation Magazine*, vol. 45, no. 2, pp. 146–150, 2003.

[15] J. Liang, L. Guo, C. C. Chiau, X. Chen, and C. G. Parini, "Study of CPW-fed circular disc monopole antenna for ultra wideband applications," *IEE Proceedings Microwaves, Antennas and Propagation*, vol. 152, no. 6, pp. 505–508, 2005.

[16] CST-Microwave Studio, User's Manual, 2006.

24

A Systemic Approach to the Compensation of Rain Attenuation in Ka-Band Communication Satellites

Piero Angeletti[1] and Marco Lisi[2]

[1] *Directorate of Technical & Quality Management, European Space Agency, Keplerlaan 1, 2200 AG Noordwijk, The Netherlands*
[2] *Directorate of Galileo Programme & Navigation Related Activities, European Space Agency, Keplerlaan 1, 2200 AG Noordwijk, The Netherlands*

Correspondence should be addressed to Marco Lisi, marco.lisi@esa.int

Academic Editor: Borja Vidal

Rain attenuation at Ka-band is a severe phenomenon that drastically impairs satellite communications at these frequencies. Several adaptive compensation techniques have been elaborated to counteract its effects and most often applied one at a time. The present paper proposes the contemporary exploitation of different techniques in a combined approach. Such an integrated approach is thoroughly analyzed in a simplified scenario and will be shown to achieve a very effective solution, making the Ka-band spectrum fully available for broadband satellite applications and network-centric systems.

1. Introduction

The advent of Ka-band (The frequency bands 17.7–21.2 GHz (space to Earth) and 27.5–31 GHz (Earth to space) have been allocated by the International Telecommunication Union (ITU) for the use of Fixed Satellite Services (FSS)) satellite communications is driven by the diffusion of consumer applications (such as High Definition Television broadcasting) and by the increased demand for broadband networks in network-centric systems.

The network-centric architecture, originally conceived for military applications, is becoming a paradigm for large integrated systems, or "systems of systems", able to collect and to convey data seamlessly throughout a number of different possible media and to deliver useful information after a data fusion process.

Ka-band space segment and ground segment technologies, after a pioneering phase (Italsat [1, 2], ACTS [3]) and a successive consolidation, are now mature for cost-effective and reliable commercial missions.

One the major drawbacks still affecting Ka-band satellite communications is that related to rain attenuation.

Rain attenuation in satellite communication systems operating at Ka-band frequencies is much more severe than that usually experienced at lower frequency bands: rain attenuation at 20 GHz (Ka-band down-link), for instance, is almost three times that at 11 GHz (Ku-band down-link) [4, page 125]. This fact makes rain attenuation one of the most important limiting factors to be taken into account in the design of a 20/30 GHz satellite communication system.

A number of mitigation techniques has been envisioned and experimented over the years, in the attempt to overcome the problem and to make Ka-band satellite applications as commercially viable as those at Ku band [5, Chapter 8].

One first classification of such techniques is between ground-based and space-based solutions.

Ground-based techniques are those basically operated on ground (either in open- or closed-loop configurations) and only indirectly affecting the satellite payload. Among them are Up-Link Power Control (both for transparent and regenerative payloads), End-to-End Power Control (for transparent payloads only), Adaptive Coding and/or Modulation, Site Diversity, and Terrestrial Back-Ups.

Space-based techniques involve, in one way or another, an adaptive use of the payload resources to counteract rain attenuation.

Down-Link Power Control is a technique suitable for both "bent pipe" and regenerative payloads. Adaptive Coding and Adaptive Modulation techniques can be used, in regenerative payload, to adapt the down-link signal characteristics to rain attenuation, independently from the up-link signal.

Adaptive down-link antennas allow a reconfiguration of the gain distribution over the coverage to compensate for higher rain attenuations in specific areas. Contoured beam antennas (in array-fed reflector configurations) can achieve such capability by incorporating variable components (such as variable power dividers) into their beam forming networks.

Active phased-array antennas (either in "array-fed reflector" or in "direct radiating array" configurations) offer an even better control of the down-link EIRP. In an active downlink antenna each radiating element (feed or subarray) is individually fed by a transmit active module (solid-state power amplifier). These antennas not only allow a fast and effective reconfiguration of the gain distribution over the coverage, but also make possible, in case of multiple-beam configurations, a full reallocation of the RF power among the radiated beams. In an active phased-array antenna all active modules (power amplifiers in this case) contribute to the formation of any single beam; this implies that the overall RF power can be dynamically shared among the beams (Power Reconfigurability). On the opposite end, in conventional focused reflector antennas each feed (and its corresponding power amplifier) is virtually associated to only one beam, hence traffic reconfigurability is impossible.

An alternative adaptive antenna configuration is the Multibeam Adaptive Antenna (MAA) concept, which allows a flexible sharing of the total power from the power amplifier modules among the beams by means of a Multiport Power Amplifier (MPA). The application of this antenna configuration to counteract rain attenuation at Ka band was proposed for future generations of Italsat satellites as early as in 1989 [6, 7], and it is still subject of intense R and D [8–11]. While the above summarized adaptive compensation techniques are most often applied one at the time, the development of the technology and the highly integrated nature of today's satellite communication systems suggest combined approaches, implementing several techniques in parallel in a dynamic configuration.

In particular in the following paragraphs we will focus our attention on the downlink of a multibeam Ka-band system. After a general discussion about individual and joint benefits of Adaptive Coding and Modulation (ACM) and Down-Link Power Control, the proposed integrated approach will be thoroughly analyzed in a simplified scenario.

2. Overview of Adaptive Coding and Modulation Techniques

Adaptive Coding and Modulation (ACM) techniques, applied in interactive and point-to-point applications, provide efficient dynamic link adaptation to propagation conditions, targeting each individual earth station.

The adaptation is obtained by varying the air interface characteristics in terms of modulation constellation and coding scheme. The aim is to allow each user to have a modulation and coding such that at the considered BER the associated required Signal-to-Noise-plus-Interference Ratio (SNIR) is compatible with the experienced atmospheric losses and cochannel interference. It is worth noting that for its efficacy ACM has been selected as baseline for the next generation of the Digital Video Broadcasting via Satellite standard (DVB-S2) [12].

Link adaptation is achieved informing the satellite uplink station of the channel condition (i.e., SNIR) of each earth station via the satellite return link. Each link operates dynamically exploiting the SNIR to the maximum spectral efficiency. The set of standardized DVB-S2 modes (combination of symbol constellation and coding rates) hinges on the revitalized use of Low Density Parity Check (LDPC) error correction codes initially introduced by Gallager in the 60's [13].

The net result is the availability of a fine granularity in the Spectral Efficiency versus SNIR curve (refer to Figure 1), resembling the Shannon's bound,

$$\frac{R_b}{B} \leq \log_2\left(1 + \frac{E_S}{N_0 + I_0}\right) \text{[bits/Hz]}, \quad (1)$$

where R_s is the symbol rate, B is the bandwidth, and $E_S/(N_0 + I_0)$ is the SNIR in terms of symbol energy E_s, noise spectral density N_0, and interference spectral density density I_0.

Satellite throughput gains of ACM systems have been estimated to reach a capacity increase up to 100%–200% with respect to constant coding and modulation systems (e.g., DVB-S) [12]. In addition, service availability can be extended.

3. An Integrated Approach

Adaptive coding and modulation techniques allow to effectively comply with different link conditions providing to the users the maximum data-rate permitted with the experienced Signal-to-Noise-plus-Interference Ratio (SNIR). This approach maximizes the individual link exploitation (in terms of spectral efficiency) and the overall system capacity evaluated as sum over all the coverage of active users spectral efficiencies. The availability of a large number of coding and modulation combinations allows to close the link budget in severe interference and fading conditions. The result is an increased link availability at low data rates and a high overall system capacity on a best-effort base. When minimum guaranteed data-rates with certain availabilities are required, the system dimensioning must still rely on power margins on the worst case link budget condition. These power margins, later on, can be fully exploited at operational level to increase the overall capacity on a best effort base.

On the other hand, adaptive power allocation techniques aim at reducing individual power margins by means of an adaptive redistribution of the available RF power based on the real-time needs. The achievement of this objective is strictly dependent on the architecture of the payload high-power section. The idea is to have a common power pool from which a channel can adaptively draw the power it requires. Two candidate solutions are nowadays debated.

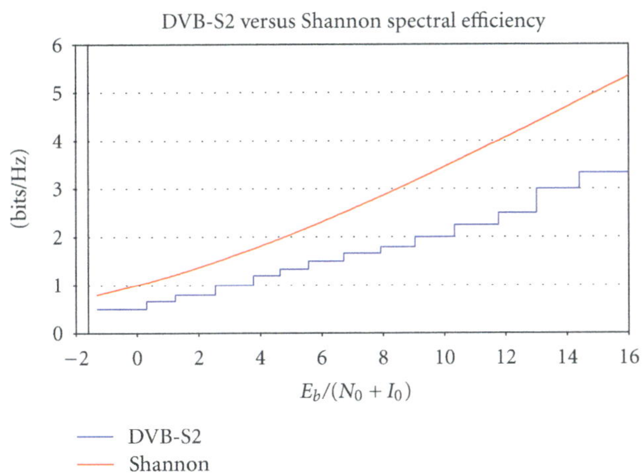

FIGURE 1: Spectral efficiency—DVB-S2 versus Shannon's bound.

(i) Flex TWTAs: the power pool is realized at the level of the satellite platform DC power capacity. Each individual TWTA can arbitrarily change (within a range that in current developments is about 3-4 dBs) its saturated RF output power while maintaining a constant efficiency (hence, a DC power consumption proportional to the RF output power) [14]. In nominal conditions, the TWTAs are operated at their minimum saturated RF output power, while the DC power subsystem is over dimensioned, in order to have headroom in case of rain fading. It is worth noting that a Flex TWTA does not allow a fast variation of its bias conditions.

(ii) Multiport amplifiers: the power pool is realized at RF power level. The sharing flexibility is achieved through the parallel amplification of all signals by a stack of Power Amplifiers [15–22].

A joint Adaptive Coding and Modulation (ACM) and Down-Link Power-Control approach can be proven to be advantageous in case a minimum data-rate and an associated availability are required and can be traded off with the overall best-effort system capacity. The approach consists in operating the payload in an RF power-pooled configuration [1, 2, 9], so that faded beams can be adaptively boosted to counteract rain attenuation. The system can be optimized to redistribute the power among beams in fading and clear sky conditions, guaranteeing the achievement of all the required data rates and of the availability performance.

To better understand the close relationship between the rain countermeasure on a faded earth station and the remaining population in clear sky we developed a simplified two-cochannel beams model that is reported in the following subparagraphs.

3.1. Rain Attenuation Impairments on the Signal-to-Noise Ratio. The receiver system temperature can be evaluated as:

$$T_S = T_A + \left(L_{\text{preRX}} - 1\right) \cdot T_{\text{Amb}} + L_{\text{preRX}}\, T_{\text{eqRX}}, \quad (2)$$

where T_A is the antenna temperature, L_{preRX} is the passive section loss (from the antenna to the receiver), T_{Amb} is the ambient temperature (working temperature of the passive section), and T_{eqRX} is the effective input noise temperature of the receiver. In clear sky, the antenna temperature (apex C) can be decomposed in sky ($T_{A,\text{Sky}}$) and the ground ($T_{A,\text{Gnd}}$) temperatures as seen by the antenna:

$$T_A^C = T_{A,\text{Sky}} + T_{A,\text{Gnd}}, \quad (3)$$

while in rain conditions (apex R), the antenna temperature (T_A^R) becomes:

$$T_A^R = \frac{T_{A,\text{Sky}}}{L_R} + \left(\frac{L_R - 1}{L_R}\right) T_R + T_{A,\text{Gnd}}, \quad (4)$$

where T_R is the rain temperature, and L_R is the rain attenuation. In summary we can write:

$$T_S^C = T_{A,\text{Sky}} + T_{A,\text{Gnd}} + \left(L_{\text{preRX}} - 1\right) \cdot T_{\text{Amb}} + L_{\text{preRX}} T_{\text{eqRX}}, \quad (5)$$

$$\begin{aligned} T_S^R &= \frac{T_{A,\text{Sky}}}{L_R} + \left(\frac{L_R - 1}{L_R}\right) T_R + T_{A,\text{Gnd}} \\ &\quad + \left(L_{\text{preRX}} - 1\right) \cdot T_{\text{Amb}} + L_{\text{preRX}} T_{\text{eqRX}} \\ &= T_S^C \left[1 + \left(\frac{L_R - 1}{L_R}\right)\left(\frac{T_R - T_{A,\text{Sky}}}{T_S^C}\right)\right]. \end{aligned} \quad (6)$$

The closure of the link budget requires a minimum Signal-to-Noise value that in the following we will consider in terms of symbol energy over noise spectral density:

$$\left(\frac{E_s}{N_0}\right) \geq \left(\frac{E_s}{N_0}\right)_{\text{req}} \equiv \rho_{\text{req}}. \quad (7)$$

In clear sky conditions we can explicit E_s^C in terms of the satellite transmitted power in the beam/channel (P^{Sat}), the satellite antenna gain (G^{Sat}), the free space loss (L), and the Earth Station antenna gain (G_{ES}):

$$E_s^C = \frac{1}{R_s} \cdot \frac{G^{\text{Sat}} P^{\text{Sat}}}{L} \cdot G_{\text{ES}} \quad (8)$$

and for the noise spectral density, K being the Boltzman constant:

$$N_0^C = K T_S^C. \quad (9)$$

In summary, in clear sky we have

$$\left(\frac{E_s}{N_0}\right)^C = \frac{G^{\text{Sat}} P^{\text{Sat}}}{R_s L G_{\text{ES}} K T_S^C}. \quad (10)$$

Similarly, substituting L with LL_R and introducing (6) in (10), we can explicit this ratio in rain conditions:

$$\left(\frac{E_s}{N_0}\right)^R = \left[L_R + (L_R - 1)\left(\frac{T_R - T_{A,\text{Sky}}}{T_S^C}\right)\right]^{-1} \left(\frac{E_s}{N_0}\right)^C. \quad (11)$$

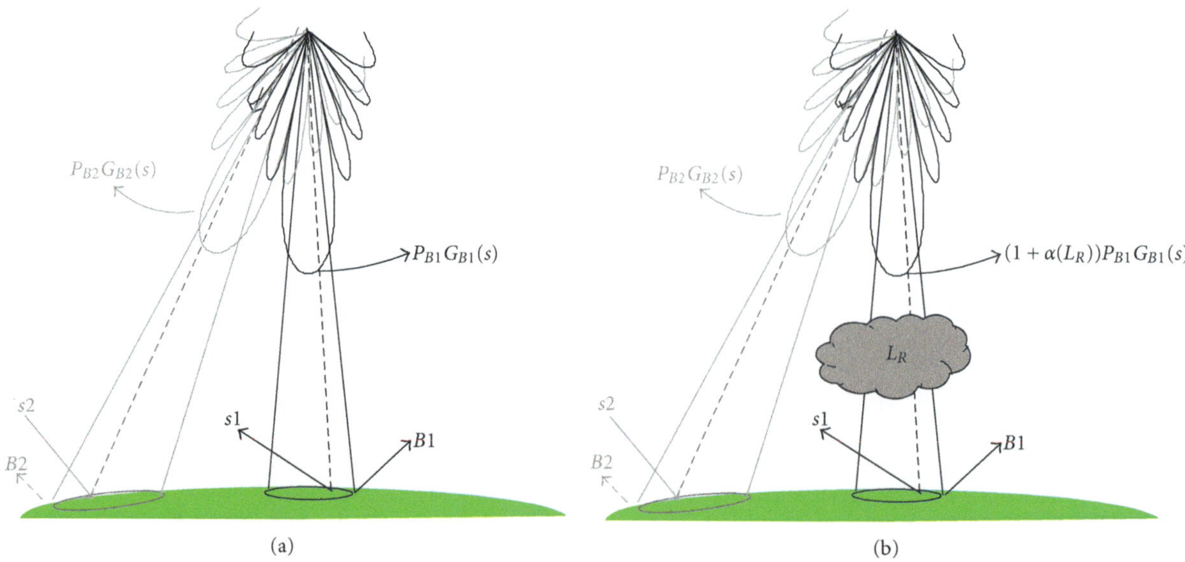

FIGURE 2: Multibeam antenna geometry ((a) clear sky, (b) fading in Beam-1).

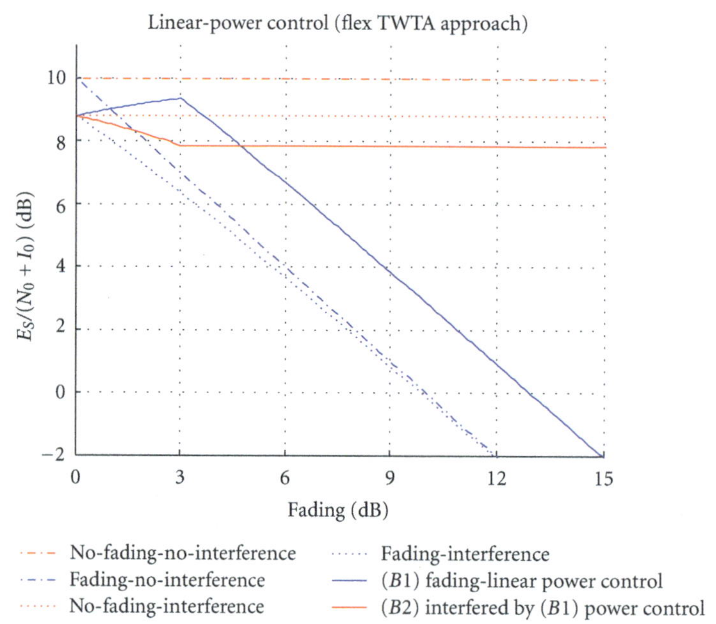

FIGURE 3: Bidimensional link budget analysis (flex TWTAs with linear power control).

3.2. Rain Attenuation Impairments on the Signal-to-Noise-Plus-Interference Ratio. In a multibeam system with frequency reuse, several beams make use of the same frequency causing cochannel interference that must be duly considered in the link closure condition (7), that becomes:

$$\left(\frac{E_s}{N_0+I_0}\right) \geq \left(\frac{E_S}{N_0+I_0}\right)_{\text{req}} \equiv \rho_{\text{req}}. \tag{12}$$

A pictorial representation of a two cochannel beams system is shown in Figure 2. The two beams are indicated as Beam-1 ($B1$) and Beam-2 ($B2$); $s1$ and $s2$ refer to two ground stations served by Beam-1 and Beam-2, respectively. The statistical independence of thermal noise (N_0) and cochannel interference (I_0) allows the separation of the two different contributions:

$$\begin{aligned}\left(\frac{E_s}{N_0+I_0}\right) &= \left(\left(\frac{E_s}{N_0}\right)^{-1}+\left(\frac{E_s}{I_0}\right)^{-1}\right)^{-1} \\ &= \left(\left(\frac{N_0}{E_s}\right)+\left(\frac{I_0}{E_s}\right)\right)^{-1}.\end{aligned} \tag{13}$$

A link equation can be written for each user/beam. In clear sky conditions we have (14) (reversing symbol energy and noise/interference eases the understanding of the complex coupled behavior); while in case of rain fading over $s1$ in Beam-1 we can write (15).

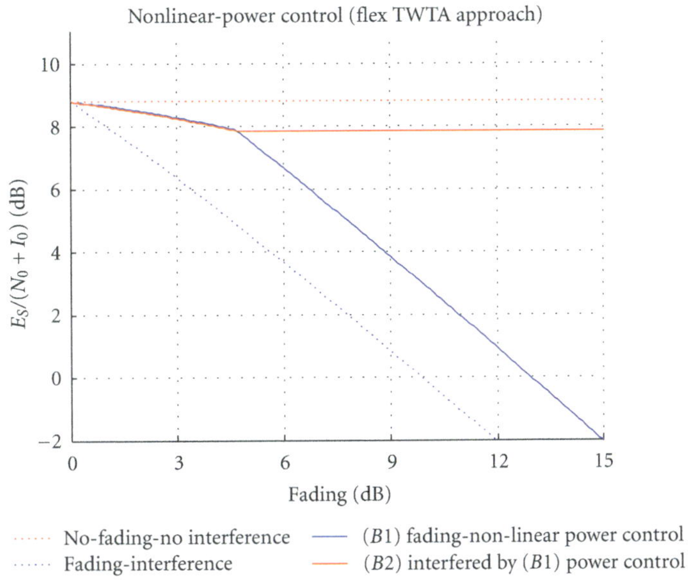

FIGURE 4: Bidimensional link budget analysis (flex TWTAs with nonlinear power control).

To counteract the fading of $s1$ in Beam-1, the satellite can increase the power of Beam-1 as a generic function of the rain attenuation ($\alpha(L_R)$). Two possible strategies can be foreseen based on the different high-power sections described above.

3.2.1. Power Control with Flex TWTAs. The faded beam output power is increased without changing the power level of the other beams. This can be implemented, for example, increasing the saturation power of a flex TWTA; however, the maximum delta-power is limited by the tube technology. The power increase has two beneficial effects on the faded beam: a compensation of the rain losses and an advantage in terms of Signal-to-Interference. However, an increase of the interference is experienced by the other beams, (refer to (16) for the two-beam model).

3.2.2. Power Control with Multiport Amplifiers. The overall RF power constitutes a unique pool. An increase of the power dedicated to one beam of an amount equal to $\alpha(L_R)$ means an equivalent decrease of the power available to the other beams. In this case the delta-power could reach the overall RF Power-Pool amount (e.g., the sum of the power of all the TWTAs constituting an MPA). The reduction in interfering power sums up to the already referred beneficial effects on the faded beam. On the other hand, nonfaded beams suffer, together with a higher interference level, a decreased power. The simplified equation for a two-beam system is (17).

A fair comparison between the two power control approaches with and without power-pooling must take into account that both systems should be dimensioned for the same maximum RF power (or DC power). In this case the power-pooling approach can benefit from a higher clear-sky operating point, equal to the total available RF power divided by the number of active channels/beams (this advantage has been considered in the following graphs).

Clear Sky. Consider

$$\left(\frac{N_0+I_0}{E_s}\right)^C_{(s1)} = \left(\frac{N_0}{E_s}\right)^C_{(s1)} + \left(\frac{I_0}{E_s}\right)^C_{(s1)},$$
$$\left(\frac{N_0+I_0}{E_s}\right)^C_{(s2)} = \left(\frac{N_0}{E_s}\right)^C_{(s2)} + \left(\frac{I_0}{E_s}\right)^C_{(s2)}. \quad (14)$$

Rain Fading over Beam-1. Consider

$$\left(\frac{N_0+I_0}{E_s}\right)^R_{(s1)} = \left[L_R + (L_R-1)\left(\frac{T_R - T_{A,\mathrm{Sky}}}{T_S^C}\right)\right] \times \left(\frac{N_0}{E_s}\right)^C_{(s1)} + \left(\frac{I_0}{E_s}\right)^C_{(s1)}, \quad (15)$$
$$\left(\frac{N_0+I_0}{E_s}\right)^R_{(s2)} = \left(\frac{N_0}{E_s}\right)^C_{(s2)} + \left(\frac{I_0}{E_s}\right)^C_{(s2)}.$$

Rain Fading over Beam-1, Power Control with Flex TWTAs. Consider

$$\left(\frac{N_0+I_0}{E_s}\right)^R_{(s1)} = \frac{1}{1+\alpha(L_R)}\left[L_R + (L_R-1)\left(\frac{T_R - T_{A,\mathrm{Sky}}}{T_S^C}\right)\right] \times \left(\frac{N_0}{E_s}\right)^C_{(s1)} + \frac{1}{1+\alpha(L_R)}\left(\frac{I_0}{E_s}\right)^C_{(s1)},$$
$$\left(\frac{N_0+I_0}{E_s}\right)^R_{(s2)} = \left(\frac{N_0}{E_s}\right)^C_{(s2)} + [1+\alpha(L_R)]\left(\frac{I_0}{E_s}\right)^C_{(s2)}. \quad (16)$$

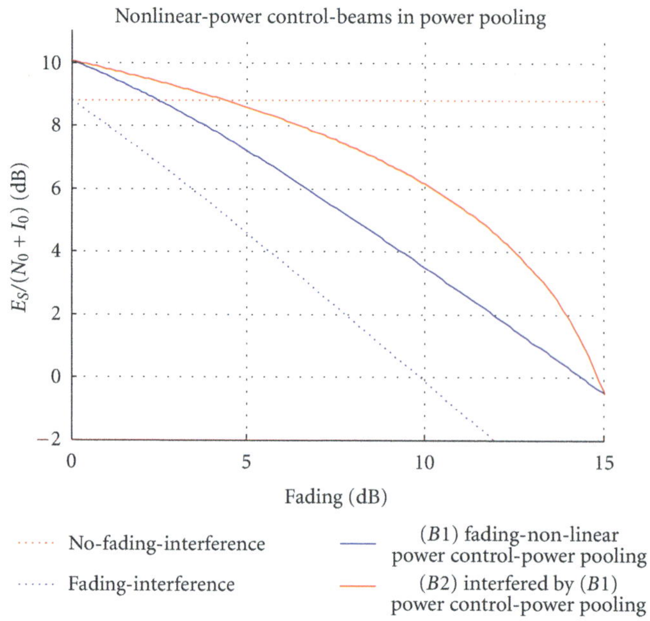

FIGURE 5: Bidimensional link budget analysis (multiport amplifier with nonlinear power control).

Rain Fading over Beam-1, Power Control with Multiport Amplifiers. Consider

$$\left(\frac{N_0+I_0}{E_s}\right)^R_{(s1)} = \frac{1}{1+\alpha(L_R)}\left[L_R+(L_R-1)\left(\frac{T_R-T_{A,\mathrm{Sky}}}{T_S^C}\right)\right] \times\left(\frac{N_0}{E_s}\right)^C_{(s1)} + \frac{1-\alpha(L_R)}{1+\alpha(L_R)}\left(\frac{I_0}{E_s}\right)^C_{(s1)},$$

$$\left(\frac{N_0+I_0}{E_s}\right)^R_{(s2)} = \frac{1}{1-\alpha(L_R)}\left(\frac{N_0}{E_s}\right)^C_{(s2)} + \frac{1+\alpha(L_R)}{1-\alpha(L_R)}\left(\frac{I_0}{E_s}\right)^C_{(s2)}. \quad (17)$$

3.3. Results. Results of the performed analysis are reported in the figures. SNR/SNIR referring to the faded beam are represented in blue, those referring to the clear-sky beam in red.

Examining Figure 3, dot-dash curves refer to an interference-free case: the faded beam SNR degrades linearly with rain attenuation while the clear-sky SNR stays constant (10 dB). Dot curves account for interference (15 dB in the examples) and no power control. Solid curves refer to power compensation proportional to the rain attenuation without power-pooling; it is clear that the overshooting of the blue curve is due to the advantage in Signal-to-Interference Ratio (SIR) that the faded beam gains with respect to the noncompensated beam. Once the maximum TWTA power is reached (twice the nominal level, in the example) further compensation is not possible and the decay follows the attenuation.

Figure 4 shows a similar analysis (power compensation with Flex TWTAs) with the compensation law optimized to maintain the system performance constant all over the coverage up to the maximum compensation range of a single TWTA (3 dB assumed as in Figure 3).

Finally, Figure 5 shows that the power pooling architecture based on multiport amplifiers (or active antennas as DRAs) with an optimized compensation law outperforms all the previous results in terms of achievable compensation depth (better SNIR at maximum attenuation depth), still maintaining a high overall capacity.

4. Conclusions

The present paper proposed and analyzed the contemporary exploitation of different rain fading countermeasures in a combined approach, based on advanced payload architectures exploiting power-pooling capabilities. Such an integrated approach is thoroughly analyzed in a simplified scenario and it is shown to achieve a very effective solution for the minimum data-rate/availability requirement; at the same time, it preserves the ACM characteristic feature of offering increased capacity on a best-effort basis. The proposed approach makes the Ka-band spectrum fully available for broadband satellite applications and network-centric systems.

References

[1] F. Marconicchio and A. Rodotà, "The ITALSAT program," *Rivista Tecnica Selenia*, vol. 2, no. 4, pp. 21–26, 1990.

[2] A. Paraboni, G. Masini, M. Mauri, R. Polonio, and C. Riva, "Summary of propagation results obtained in 5 years of ITALSAT experiment in Milano," *International Journal of Satellite Communications*, vol. 17, no. 2-3, pp. 169–175, 1999.

[3] R. T. Gedney, R. Schertler, and F. Gargione, *The Advanced Communications Technology Satellite*, SciTech Publishing, 2001.

[4] B. R. Elbert, *Introduction to Satellite Communication*, Artech House, 3rd edition, 2008.

[5] L. J. Ippolito, *Satellite Communications Systems Engineering*, John Wiley & Sons, 2008.

[6] M. Lisi and G. Perrotta, "An integrated 20/30 GHz multibeam/multiport antenna system for fixed satellite services," in *Proceedings of the ISAP*, Tokyo, Japan, 1989.

[7] M. Lisi and G. Panariello, "An adaptive multibeam antenna for italsat future generation satellites," in *Proceedings of the 15th AIAA International Communications Satellite Systems Conference (ICSSC '94)*, San Diego, Calif, USA, 1994.

[8] A. Paraboni, P. Gabellini, A. Martellucci, P. Rinous, C. Capsoni, and N. Gatti, "Proposed architectures for a reconfigurable Ka-Band antenna front-end for active rain fade compensation," in *Proceedings of the 28th ESA Antenna Workshop*, Noordwijk, The Netherlands, 2005.

[9] A. Paraboni, C. Capsoni, M. Buti et al., "Assessment of performance of a ka-band broadcasting reconfigurable satellite antenna with adaptive mitigation for atmospheric attenuation," in *Proceedings of the European Conference on Antennas and Propagation (EuCAP '06)*, Nice, France, November 2006.

[10] P. Angeletti, A. Bolea-Alamañac, M. Bousquet, and M. Lisi, "Joint exploitation of interference and fading mitigation techniques inmultibeam power-pooling satellite systems," in *Proceedings of the 25th AIAA International Communications Satellite Systems Conference*, Seoul, Korea, April 2007.

[11] P. Gabellini, C. Capsoni, A. Martellucci, N. Gatti, L. Resteghini, and P. Rinous, "A reconfigurable multi-matrix antenna front-end architecture for Ka-Band broadcasting satellite services," in *Proceedings of the 2nd ESA Workshop on Advanced Flexible Telecom Payloads*, Noordwijk, The Netherlands, April 2012.

[12] ETSI EN 302 307, Digital Video Broadcasting, "Second Generation framing structure, channel coding and modulation systems for Broadcasting, Interactive Services, News Gathering and other broadband satellite applications".

[13] R. G. Gallager, "Low density parity check codes," *IEEE Transactions on Information Theory*, vol. 8, no. 1, pp. 21–28, 1962.

[14] M. Khilla, W. Gross, H. Schreiber, and D. Leucht, "Flexible Ka-Band LCAMP for in-orbit output power adjustable MPM," in *Proceedings of the 23rd AIAA International Communications Satellite Systems Conference (ICSSC '05)*, Rome, Italy, 2005.

[15] W. A. Sandrin, "The butler matrix transponder," *COMSAT Technical Review*, vol. 4, no. 2, pp. 319–345, 1974.

[16] S. Egami and M. Kawai, "An adaptive multiple beam system concept," *IEEE Journal on Selected Areas in Communications*, vol. 5, no. 4, pp. 630–636, 1987.

[17] M. Lisi, "An introduction to multiport power amplifiers," Microwave Engineering Europe, pp. 43–48, 1992.

[18] M. Aloisio, P. Angeletti, and M. Lisi, "Multiport amplifiers for satellite applications: historical perspective and future directions," in *Proceedings of the IEEE International Vacuum Electronics Conference (IVEC '09)*, pp. 201–202, Rome, Italy, April 2009.

[19] S. D'Addio, M. Aloisio, E. Colzi, and P. Angeletti, "Performance analysis of satellite payload output sections based on multiport amplifiers," in *Proceedings of the IEEE International Vacuum Electronics Conference (IVEC '09)*, pp. 207–208, Rome, Italy, April 2009.

[20] P. Angeletti and M. Lisi, "Multiport pwer amplifiers for flexible satellite antennas and payloads," *Microwave Journal*, vol. 53, no. 5, 2010.

[21] M. Aloisio, P. Angeletti, and S. D'Addio, "Statistical characterization of isolation performance of multiport amplifiers: a critical discussion," in *Proceedings of the IEEE International Vacuum Electronics Conference (IVEC '11)*, pp. 83–84, Bangalore, India, February 2011.

[22] M. Aloisio, P. Angeletti, and S. D'Addio, "Accurate modeling and analysis of isolation performance in multiport amplifiers," *Active and Passive Electronic Components*, vol. 2012, Article ID 738367, 4 pages, 2012.

Location and Shape Reconstruction of 2D Dielectric Objects by Means of a Closed-Form Method: Preliminary Experimental Results

Gian Luigi Gragnani and Maurizio Diaz Mendez

Diten, University of Genoa, Via Opera Pia 11A, 16145 Genoa, Italy

Correspondence should be addressed to Gian Luigi Gragnani, gianluigi.gragnani@unige.it

Academic Editor: Andrea Randazzo

An analytical approach to location and shape reconstruction of dielectric scatterers, that was recently proposed, is tested against experimental data. Since the cross-sections of the scatterers do not depend on the z coordinate, a 2D problem can be formulated. A closed-form singular value decomposition of the scattering integral operator is derived and is used to determine the radiating components of the equivalent source density. This is a preliminary step toward a more complete solution, which will take into account the incident field inside the investigation domain in order to provide the dielectric features of the scatterer and also the nonradiating sources. Reconstructions of the equivalent sources, performed on some scattering data belonging to the Fresnel database, show the capabilities of the method and, thanks to the closed-form solution, results are obtained in a very short computation time.

1. Introduction

In the last decades, inverse electromagnetic scattering and near field imaging have been widely studied research topics [1]. Actually, electromagnetic imaging [2] is a very promising technique in many practical application fields. The more consolidated applications perhaps are those related to the use of ground penetrating radar, but a lot more are at present used or in advanced state of development. For example, it has been suggested that microwave imaging could be successfully used as a diagnostic technique in several areas, including civil and industrial engineering [3–5], nondestructive testing and evaluation [6–9], geophysical prospecting [10], and biomedical engineering [11–14]. Microwave imaging has also been proposed as a useful tool in wood industry [15]. Also, sophisticated techniques for plasma diagnostic can be based on microwave imaging [16]. Furthermore, inverse electromagnetic scattering can provide insights for approaching other problems of great interest nowadays, for example, plasmonic cloaking and the synthesis of metamaterials [17].

One of the aim of an inverse electromagnetic-scattering imaging system is to find the actual position of a dielectric object inside a bounded space region, as well as its shape. The techniques applied are based on the use of a known incident field illuminating the space region containing the object. By suitably measuring the scattered field, a solution to the problem can be derived.

The main difficulties with solving an inverse electromagnetic-scattering problem result from its nonlinearity and instability. To overcome these negative features, in the scientific literature, several regularization techniques have been proposed. Furthermore, the a priori knowledge usually available could be included in the formulation of the resolving algorithm. The problem could be recast as a global nonlinear optimization problem, and stochastic, as well as deterministic approaches can be used to achieve a solution. The interested reader is referred, for example, to [18–22].

Another common technique to overcome nonlinearity is based on the transformation of the original problem into an inverse source one, by replacing the scatterer with an equivalent (induced) source. Although the equivalent source density linearizes the inverse scattering equation, the transformation of the inverse scattering problem into an inverse source one involves the solution of an integral equation of the first kind. Hence, the resulting operator is ill-posed, and its solution is usually unstable and nonunique [23]. In this particular case, nonuniqueness is due to the so-called nonradiating sources, that do not radiate outside their domain of definition and hence belong to the null space of the scattering operator [24].

In the present work, an approach to the inverse source problem that allows to clearly reformulate it in terms of both the radiating and nonradiating parts of the induced sources is used. Through a singular value decomposition (SVD) of the scattering operator, a closed-form solution to the radiating sources is reached. By using such a solution, an expression for the reconstructed scattered field inside the dielectric object can also be obtained in closed form.

While the theory that is at the base of this work was already presented and tested by numerical simulations [25, 26], the present paper is devoted to the reconstruction of the scattering properties from experimental data. Testing inversion algorithms against experimental data is a very important topic to deal with. Actually, since inverse scattering problems are ill-posed, their solutions are not as easy to be obtained as in the case of direct problems, and the risk in not validating an imaging algorithm with real world data is that the simulations could not take into account all the parameters involved in reality. Simulations can be made more and more complex, thanks to the availability of computers with continuously increasing power, however researchers must validate the model with experiments, refine it, and simulate and test again, and iteratively repeat the process until the model matches the physical data. Fortunately, in the last decade, some research groups have dealt with this lack of experimental data by performing measurements on a controlled environment and by making available the results to the inverse scattering community. The first available data were those of the so-called Ipswich database [27–30]. A larger collection of data, on both metallic and dielectric scatterers, is now available thanks to the efforts of the group at the Institut Fresnel, Marseille, France [31, 32]. In the present work, some of the data of the Fresnel database are used to test the inversion method. Since one of the interesting features of the Fresnel database is that data are provided over a wide band frequency range, also some comparisons about the behavior of the method at different frequencies have been carried out.

Although the present work is at an early stage and only some features of the theoretical model have been checked against the experimental data, the method seems to perform well, and its capabilities to reconstruct both the location and the shape of the induced sources are demonstrated. Furthermore, thanks to the closed-form formulation of the solution, results are available in a very short time that can allow for a quasi real-time imaging system.

2. Mathematical Background

In this section, the closed-form solution to the inverse scattering problem is briefly resumed. For more details the reader is referred to [25, 26].

2.1. The Electromagnetic Scattering. Let us consider an inhomogeneous object whose dielectric parameters are unknown, and which is irradiated by an electromagnetic field produced by a known source. The presence of the object leads to a total electromagnetic field that is different from the one that the same source would radiate in a free-space environment. In particular, the scattered field due to the object can be modeled by means of the volume equivalence theorem [33], that allows one to replace the dielectric object with an induced (or equivalent) source density, radiating in free space. So the object acts as a secondary source of radiated electromagnetic field. In three dimensions, the scattered field vector $\mathbf{E}_s$ can be expressed as

$$\begin{aligned}\mathbf{E}_s(\mathbf{r}) &= \mathbf{E}_{\text{tot}}(\mathbf{r}) - \mathbf{E}_{\text{inc}}(\mathbf{r}) \\ &= \omega^2\mu\varepsilon \iiint \mathbf{P}_{\text{eq}}^{\text{TOT}}(\mathbf{r}')\mathcal{G}(|\mathbf{r}-\mathbf{r}'|)d\mathbf{r}' \\ &\quad + \nabla\nabla\cdot \iiint \mathbf{P}_{\text{eq}}^{\text{TOT}}(\mathbf{r}')\mathcal{G}(|\mathbf{r}-\mathbf{r}'|)d\mathbf{r}',\end{aligned} \tag{1}$$

where

(i) $\mathbf{E}_{\text{tot}}$ is the total electrical field vector;

(ii) $\mathbf{E}_{\text{inc}}$ is the incident field vector, that is the known field that would be radiated without the object;

(iii) $\mathbf{r} = \mathbf{r}(\rho, \vartheta, z)$ is the position vector of the observer;

(iv) $\mathbf{r}' = \mathbf{r}'(\rho', \vartheta', z')$ is the position vector of any point of the scattering domain;

(v) $\mathcal{G}(|\mathbf{r}-\mathbf{r}'|)$ is the Green's function for the electromagnetic radiation in free space;

(vi) ε is the dielectric permittivity of the propagation medium;

(vii) μ is the magnetic permeability of the propagation medium;

(viii) $\mathbf{P}_{\text{eq}}^{\text{TOT}}$ is a polarization field that acts as an equivalent source for the scattered field.

Within the assumption of dealing with nonmagnetic media (i.e., $\mu = \mu_0$ both inside and outside the scatterers, μ_0 being the permeability in vacuo), we have [34]

$$\mathbf{P}_{\text{eq}}^{\text{TOT}}(\mathbf{r}) = \tau(\mathbf{r})\mathbf{E}_{\text{tot}}(\mathbf{r}), \tag{2}$$

where

(i) $\tau(\mathbf{r}) = (\varepsilon_r(\mathbf{r}) - 1)$ is called *scattering potential* and $\varepsilon_r(\mathbf{r})$ is the complex relative dielectric permittivity of the scatterer, with respect to the background permittivity ε.

2.2. 2D TM Model

The problem is simplified into a two-dimensional and scalar one, when

(1) the scatterers are infinitely extended along z;

(2) the scattering potential does not depend on z;

(3) the incident electric field does not depend on z, is z directed, and propagates in a transverse direction (TM illumination).

In this case also the scattered field is directed along z and only the cross section with respect to z must be considered. The integral equation modeling the inverse scattering problem becomes [35]

$$E_s(\boldsymbol{\rho}) = E_{\text{tot}}(\boldsymbol{\rho}) - E_{\text{inc}}(\boldsymbol{\rho}) = -j\frac{k^2}{4}\iint_{D_{\text{inv}}} \tau(\boldsymbol{\rho}')E_{\text{tot}}(\boldsymbol{\rho}')H_0^{(2)}(k|\boldsymbol{\rho}-\boldsymbol{\rho}'|)d\boldsymbol{\rho}', \tag{3}$$

where, in a generic cross section of the object:

(i) E_{tot} is the z component of the total electrical field;

(ii) E_{inc} is the z component of the incident field;

(iii) $\boldsymbol{\rho} = \boldsymbol{\rho}(\rho, \vartheta)$ is the position vector of the observer;

(iv) $\boldsymbol{\rho}' = \boldsymbol{\rho}'(\rho', \vartheta')$ is the position vector of any point of the investigation domain D_{inv};

(v) $H_0^{(2)}(k|\boldsymbol{\rho}-\boldsymbol{\rho}'|)$ is the second-kind and zeroth-order Hankel function;

(vi) $k = \omega\sqrt{\mu\varepsilon}$ is the wave number.

In the paper, it is assumed that scatterers are placed inside a circular region of radius R, along its circumference the scattered electromagnetic field is measured. The circular region defines the geometry of the problem and is called the domain of investigation D_{inv}, while the circle of sensors is called the domain of observation D_{obs}.

As it was already mentioned, the inverse problem given by (3) is nonlinear because such an equation contains the product of the total field $E_{\text{tot}}(\boldsymbol{\rho}')$ and the scattering potential $\tau(\boldsymbol{\rho}')$, both of which are unknown.

The problem can be linearized by using the equivalent source $P_{\text{eq}}^{\text{TOT}}$. It must be noticed that since in this way an equivalent inverse source problem is faced instead of an inverse scattering one, both the radiating and nonradiating contributions, P_{eq} and $P_{\text{eq}}^{\text{NR}}$, of the source should be accounted for. Hence $P_{\text{eq}}^{\text{TOT}}$ can be rewritten as

$$P_{\text{eq}}^{\text{TOT}}(\boldsymbol{\rho}') = \tau(\boldsymbol{\rho}')E_{\text{tot}}(\boldsymbol{\rho}') = P_{\text{eq}}(\boldsymbol{\rho}') + P_{\text{eq}}^{\text{NR}}(\boldsymbol{\rho}'), \quad \boldsymbol{\rho}' \in D_{\text{inv}} \tag{4}$$

and the inverse problem can be formulated as

$$E_s(\boldsymbol{\rho}) = -j\frac{k^2}{4}\iint_{D_{\text{inv}}} P_{\text{eq}}^{\text{TOT}}(\boldsymbol{\rho}')H_0^{(2)}(k|\boldsymbol{\rho}-\boldsymbol{\rho}'|)dr'. \tag{5}$$

As compared to other nonlinear methods (see, e.g., [12, 13]) where there is no separation between radiating and nonradiating sources, in the present approach it is clearly identified which part of the reconstruction can be obtained by measurements. Further improvement can then only be obtained by means of the nonradiating sources and by using the constraint that (3) must hold also inside the investigation domain.

2.3. Radiating Source. The radiating part P_{eq} of the induced source can be computed after the derivation of the singular value decomposition of (5) and the definition of its singular system $\{w_\mu, \varphi_\mu; \sigma_\mu\}$. Detailed mathematical derivations can be found in [26]. In the present section, only the main results of the method are recalled.

For a lossless propagation medium, the singular system $\{w_\mu, \varphi_\mu; \sigma_\mu\}$ of (5) is

$$w_\mu(\vartheta) = \frac{1}{\sqrt{2\pi R}}e^{j\mu\vartheta},$$

$$\varphi_\mu(\rho,\vartheta) = j\frac{1}{\sqrt{2\pi}}\frac{H_\mu^{(2)}(kR)J_\mu(k\rho)}{\left|H_\mu^{(2)}(kR)\right|\sqrt{(R^2/2)\left[J_\mu^2(kR) - J_{\mu+1}(kR)J_{\mu-1}(kR)\right]}} \times e^{j\mu\vartheta},$$

$$\sigma_\mu = \frac{k^2\pi\sqrt{R}}{2}\left|H_\mu^{(2)}(kR)\right| \times \sqrt{\frac{R^2}{2}\left[J_\mu^2(kR) - J_{\mu+1}(kR)J_{\mu-1}(kR)\right]}, \tag{6}$$

where

(i) the functions $w_\mu(\vartheta)$ are the left singular functions of the decomposition;

(ii) the functions $\varphi_\mu(\rho,\vartheta)$ are the right singular functions of the decomposition;

(iii) the numbers σ_μ are the singular values,

and where $H_\mu^{(2)}$ and J_μ are the second-kind Hankel function of order μ and the first-kind Bessel function of order μ, respectively.

A solution to the radiating source contribution P_{eq} in term of singular system can be written as [36]

$$P_{\text{eq}}(\boldsymbol{\rho}) = \sum_{\mu=1}^{\infty}\frac{1}{\sigma_\mu}\varphi_\mu\int_0^{2\pi} E_{\text{scatt}}w_\mu^* R d\vartheta, \tag{7}$$

where the asterisk means complex conjugate. When the values given by (6) are substituted into (7), the following expression for the equivalent source is obtained:

$$P_{\text{eq}}(\rho, \vartheta) = \frac{2j}{k^2\pi} \times \sum_{\mu=-\infty}^{+\infty} c_\mu \frac{H_\mu^{(1)}(kR)}{\left|H_\mu^{(2)}(kR)\right|^2 (R^2/2)\left[J_\mu^2(kR) - J_{\mu+1}(kR)J_{\mu-1}(kR)\right]} \times J_\mu(k\rho)e^{j\mu\vartheta}, \tag{8}$$

where c_μ stand for the Fourier coefficients of $E_s(R, \vartheta)$ with respect to ϑ [26].

Since P_{eq} depends on the singular values of the scattering operator, care must be exercised in deriving an expression for the reconstructed radiating source. Actually, the nature of the involved integral equation is such that, in every case, after a certain index, the singular values sharply decrease and tend to zero. These small singular values increase the instability of the solution. In order to minimize such instability, only the singular values above a suitable threshold should be taken into account in the solution.

As for the nonradiating sources $P_{\text{eq}}^{\text{NR}}$, since they are not measurable, they are in general difficult to be reconstructed. In principle, their contributions could be found out by taking into account that they must be orthogonal to the radiating ones and do not radiate outside the domain of investigation, and by using some priors about the scatters. Some, works trying to deal in a formal way with $P_{\text{eq}}^{\text{NR}}$ (see, e.g., [37, 38]) and, in particular, the method presented in [39] fits very well with the present closed-form approach. All these approaches, however, are computationally very expensive. In the most of the imaging systems, nonradiating currents are put to zero. This is a common assumption in the solution of this kind of inverse problems, and this is the line followed in the present paper, in order to have results in short time. Since the nonradiating sources contribute to the higher spatial frequencies, the price to pay for assuming $P_{\text{eq}}^{\text{NR}}$ negligible is usually a smoothing effect in the reconstruction, but the solution is anyway good enough for many purposes in electromagnetic imaging.

2.4. Scattered Field and Scattering Potential. By using the expression given by (8), approximate scattered fields both inside and outside the scatterer can be analytically obtained. Again, the approximation in deriving the scattered fields stands in using only radiating contribution of the sources, that is:

$$E_s(\boldsymbol{\rho}) \approx -j\frac{k^2}{4}\int_{D_{\text{inv}}} P_{\text{eq}}(\boldsymbol{\rho}')H_0^{(2)}(k|\boldsymbol{\rho} - \boldsymbol{\rho}'|)d\boldsymbol{\rho}'. \tag{9}$$

And the scattering potential becomes

$$\tau(\boldsymbol{\rho}) \approx \frac{P_{\text{eq}}(\boldsymbol{\rho})}{E_{\text{tot}}(\boldsymbol{\rho})} \approx \frac{P_{\text{eq}}(\boldsymbol{\rho})}{E_{\text{inc}}(\boldsymbol{\rho}) - j(k^2/4)\int_{D_{\text{inv}}} P_{\text{eq}}(\boldsymbol{\rho}')H_0^{(2)}(k|\boldsymbol{\rho} - \boldsymbol{\rho}'|)d\boldsymbol{\rho}'}. \tag{10}$$

Also for these results, details about the closed-form derivation are given in [25, 26]. It should be however anticipated that, since the Fresnel database does not provide the values of the incident field inside the investigation domain, a direct implementation of (10) by using the experimental data is not possible. In some cases, the incident field inside D_{inv} has been inferred from the values in the measurement domain. However, in the present work, we have preferred to focus the attention on the reconstruction of P_{eq}, that is, anyway the first step of any inverse scattering process. It should be nevertheless noticed that reconstructing the scattering potential, not only can provide quantitative information, but also the qualitative information about the location and the shape of the scatterers is usually better that the one provided by the radiating equivalent source. Hence, attempting to reconstruct τ, by using the full procedures proposed in [26, 39], will be the core of future works involving experimental data.

3. Experimental Results

The method was tested against three sets of experimental data involving dielectric scatterers. In particular, the dielectric targets of the first Fresnel dataset [31] were considered, as well as the "FoamDielExt" of the second dataset [40]. Furthermore, although the method was not originally conceived to deal with conducting objects, a test was also performed by using the data of the U-shaped metallic scatterer present in the first dataset.

It should be pointed out that the present theory assumes that data are available over the full circumference. Instead, in any dataset, scattering field measures were taken over an arc of fixed radius with an angular span of only 240 degrees, whereas no data were available in the remaining 120 degrees. Hence, in order to compute the coefficients c_μ appearing in (8), scattering data were continued with zeros. Although this approach is not the best choice and could lead to some overshoot in the Fourier coefficients, it is the simplest one and proved sufficient to achieve reliable results in any testing case.

The tests were focused on the reconstruction of the radiating equivalent source, and, in particular, three parameters involved in the process were considered: the number of singular values used in computing P_{eq}, the operating frequency f, and the direction of arrival of the incident field ϑ_v. Furthermore, the capabilities of the method of dealing with multiview and multifrequency data were checked against some different strategies of combining the results.

Results are shown by means of normalized maps of the reconstructed domain. On the same maps also the shape and the location of the scattering object are shown, in order to give an immediate feeling about the fairness of the result.

Although both [31, 40] suggest that the investigation domains can be restricted to a small square around a scatterer, this information is not used in the present work, and the whole circle bounded by the measurement circle is considered to belong to the investigaed domain. However, to increase the legibility of the maps, only the square area of $30\times30\ \text{cm}^2$ around the scatter is shown. Anyway, it should be stressed that the value of the reconstructed equivalent source outside this area is negligible in any case.

Thanks to the closed-form formulation of the solution, all the results were obtained in a very short computation time. In particular, on a single-core class PC, a single view reconstruction takes about $2s$. Since no optimization of the code was performed, it is expected that this figure can become of magnitude smaller on a dedicated hardware with optimized code, hence allowing for a quasi real-time imaging system.

Since reconstructions are evaluated from the data collected by a single experiment, only a partial information is available at each view. Furthermore, inspections at different frequencies also provide different information. To enhance the overall result, three different possibilities were considered: summing the partial results over the different views at a fixed frequency, summing over the frequencies at a fixed view, and summing over both frequencies and views. These approaches are summarized by the expressions in (11), where the quantities P_F, P_V, and $\mathcal{P}_{VF}$ are defined as

$$P_F = \sum_{v=1}^{N_v} \left| \widetilde{P}_{\text{eq}}(f, \vartheta_v) \right|, \qquad P_V = \sum_{f=1}^{N_f} \left| \widetilde{P}_{\text{eq}}(f, \vartheta_v) \right|,$$
$$\mathcal{P}_{VF} = \sum_{f=1}^{N_f} P_f = \sum_{v=1}^{N_v} P_v, \tag{11}$$

and where the quantity $|\widetilde{P}_{\text{eq}}(f, \vartheta_v)|$ is the map of the amplitude of the reconstructed equivalent source at frequency f and at view ϑ_v.

3.1. Single Circular Cylinder. The first results are about the inversion of the scattering data produced by a singular dielectric cylinder with a circular cross section of radius $a = 15$ mm. According to [31], the relative dielectric constant is $\varepsilon_r = 3 \pm 0.3$ and the center of the object is shifted, in one direction, of about 30 mm, with respect to the origin of the axes. Data are collected over a circle of radius $R = 0.76$ m and 48 samples per frequency are available.

A single-view, single-frequency reconstruction is shown in Figure 1. It was obtained at $\vartheta_v = 0$ and $f = 8$ GHz, and using 8 singular values. The map clearly shows a radiating source in proximity of the true-scattering object, whose maximum value is centered at the side which gave the most significant contribution to the scattered field. Reconstructions obtained by using multiview and multiview-multifrequency gave encouraging results. In particular, in Figure 2 the quantity P_V computed at $f = 8$ GHz is shown. It can be seen that the circular shape of the object is clearly recognized, and also the maximum values of the reconstruction are contained inside the boundary of the true object.

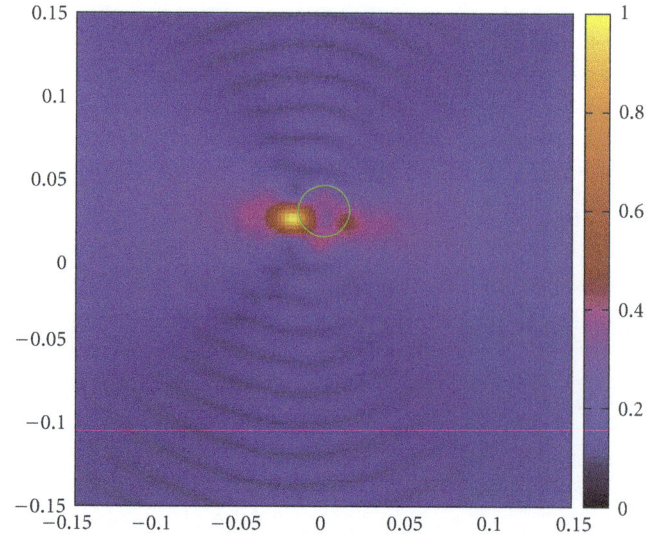

FIGURE 1: Reconstruction of the radiating source for a single dielectric cylinder. Single view $\vartheta_v = 0$ and single frequency $f = 8$ GHz. 8 singular values were used.

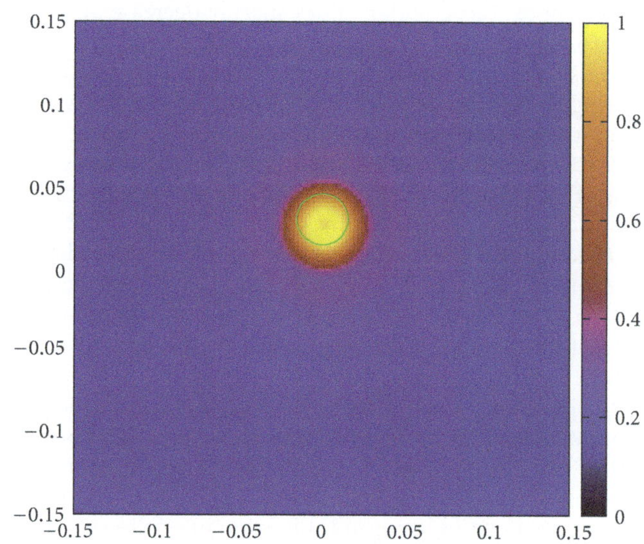

FIGURE 2: Reconstruction of a single dielectric cylinder. Multiview; map of the P_F values computed at $f = 8$ GHz.

The reconstruction using the multifrequency-multiview approach provided the result shown in Figure 3. Also in this case, the object is clearly reconstructed and, in comparison with Figure 2, the background is much more flat.

3.2. Pair of Circular Cylinders. This configuration was made by a couple of twin dielectric cylinders whose centers are spaced by 90 mm [31]. The dimensions of the cross-section and the dielectric constant of both cylinders were the same

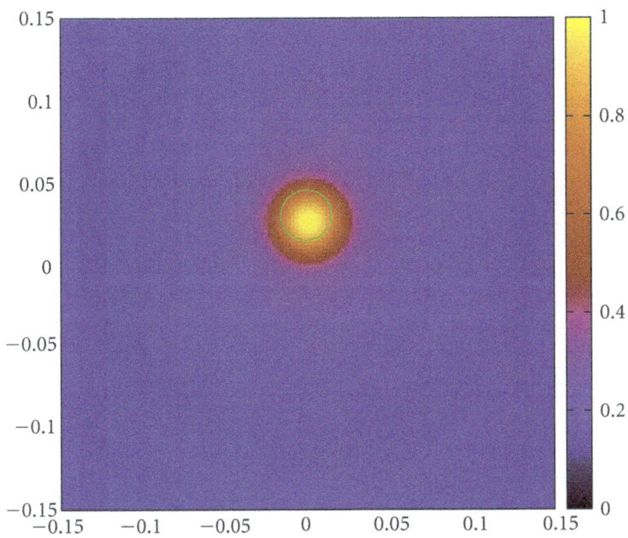

FIGURE 3: Reconstruction of a single dielectric cylinder. Multiview-multifrequency. Map of the $\mathcal{P}_{VF}$ values.

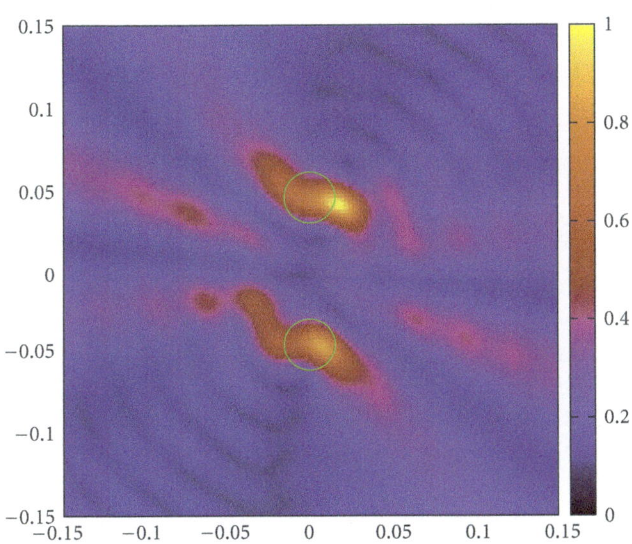

FIGURE 4: Reconstruction of the radiating source for a couple of twin dielectric cylinders. Single view $\vartheta_v = 0$ and single frequency $f = 8$ GHz. 8 singular values were used.

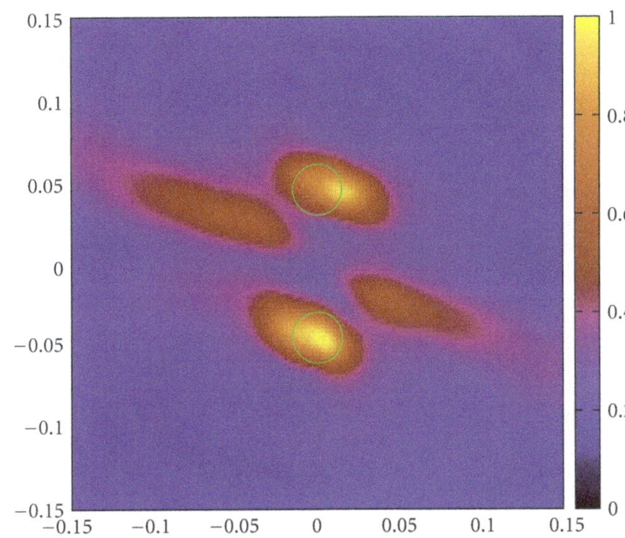

FIGURE 5: Reconstruction of a couple of twin dielectric cylinders. Multifrequency. Map of the P_V values computed at $\vartheta_v = 0$.

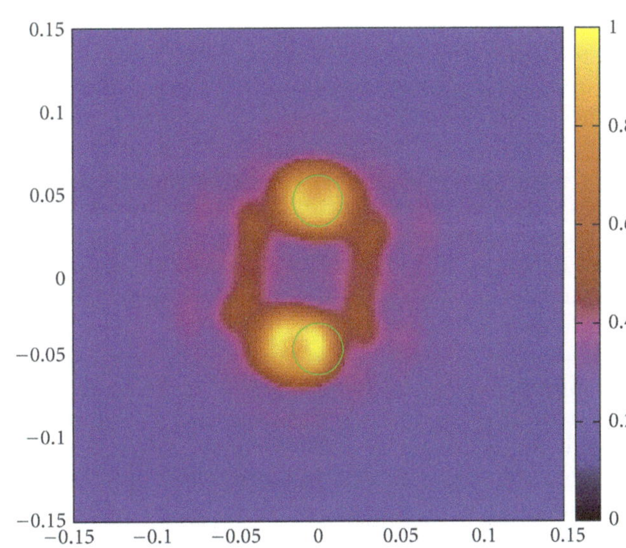

FIGURE 6: Reconstruction of a couple of twin dielectric cylinders. Multiview; map of the P_F values computed at $f = 8$ GHz.

of the cylinder of the previous case, as well as they were the radius of the measurement region and the number of samples.

Single view reconstructions, at each frequency, clearly show the presence of two distinct objects. As an example, in Figure 4 the reconstruction at $\vartheta_v = 0$ and $f = 8$ GHz is shown. As in the previous example, 8 singular values were used. In Figure 5, instead, the multifrequency result, for the same illuminating angle $\vartheta_v = 0$, is shown. With respect to Figure 4, the object is better located and the background appears much more regular. However, also two spurious "scattering queues" are evident.

Multiview reconstructions can provide a much more regular result, with a limited background noise. The case of $f = 8$ GHz is shown in Figure 6.

Finally, in Figure 7, the result given by the multiview-multifrequency method is shown. As it was in the case of a single cylinder, once again the combination of the results in frequency can provide a much better background.

3.3. Foam and Plastic Object. The third considered scatterer was the "FoamDielExt" object, included in the second Fresnel dataset [40]. The scatterer was made by two tangent cylinders with circular cross sections; one cylinder (radius $r_1 = 15.5$ mm) was plastic, with $\varepsilon_{r,1} = 3 \pm 0.3$, while the

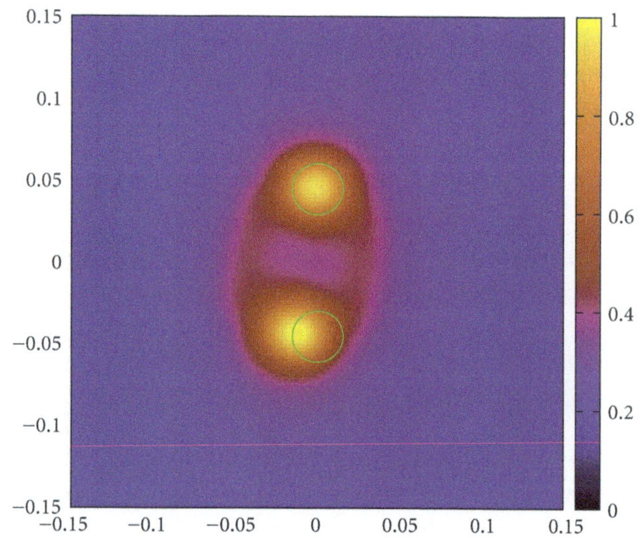

FIGURE 7: Reconstruction of a couple of twin dielectric cylinders. Multiview-multifrequency. Map of the $\mathcal{P}_{VF}$ values.

other one (radius $r_2 = 40$ mm) was made of a foam having $\varepsilon_{r,2} = 1.45 \pm 0.15$. The radius of the observation domain was $R = 1.67$ m and 240 samples were available at each frequency.

For this case, an accurate investigation about the right number of singular values was needed to achieve a good reconstruction. As an example, the limiting values of 8 and 210 singular values are considered, respectively, in Figures 8(a) and 8(b). It can be seen that, by using a small number of singular values, the scattering from the foam cylinder is much more clear and sharp than in the case of 210 singular values. However, when only 8 singular values are used, information about the dielectric cylinder gets mostly lost and hence a trade-off is in order. Actually, depending on the frequency, a variable number of singular values, ranging from 35 to 350 were used for achieving the best results.

Another example of reconstruction is given in Figure 9, where the multiview map at $f = 4$ GHz is shown. In this case, 140 singular values were used to achieve the better result. It can be seen that the different scattering given by the foam and by the plastic parts of the object can be clearly appreciated.

A further enhancement, with a pretty nice contrast between the two scattering cylinders, is achieved when results are combined also in frequency. The related map is shown in Figure 10.

3.4. Metallic Object. The metallic object was an U-shaped cylinder, having a cross-section with dimensions 80 × 50 mm^2, and the thickness of the metal was 5 mm [31]. Also in this case the analytical method is capable of recovering the presence of the target in the investigation domain, and a rectangular shape can be clearly recognized, when the various views are combined, especially at the higher frequencies.

Information provided by multifrequency can give a sharp reconstruction of a partial object, as it is shown in Figures 11 and 12, but fail to give a reconstruction of the whole object.

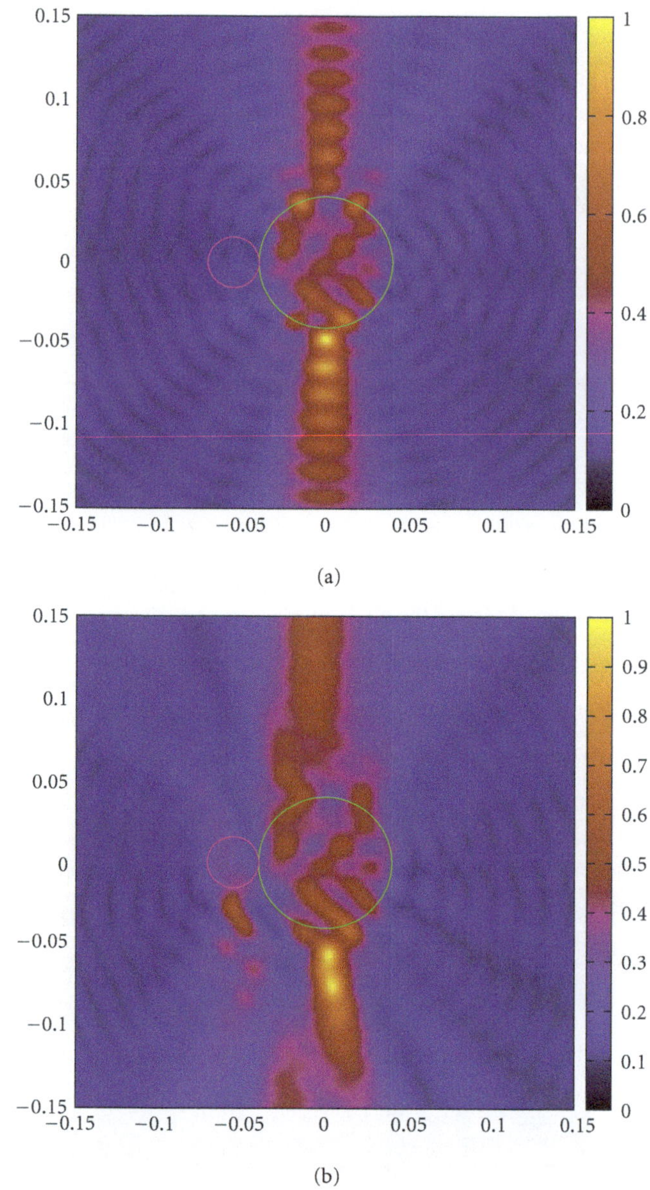

FIGURE 8: Reconstructions of the radiating source for the FoamDielExt object. Single view $\vartheta_v = 90°$ and single frequency $f = 10$ GHz. (a) 8 singular values were used. (b) 210 singular values were used.

As already mentioned, instead, multiview gives a much more better overall result. In particular, in Figure 13, the map of the multiview result, computed at $f = 16$ GHz, is shown. Although some resonances seem to appear inside the U, it can be seen that the object can be recognized without any doubt, and it is worth noting that the open side is very different from the other sides.

As it was in the cases of dielectric objects, the multiview-multifrequency approach can smooth the noise in the background. However, in this case, the results is not as good as it was for dielectric objects and the true object is smoothed, too. This result is shown in Figure 14.

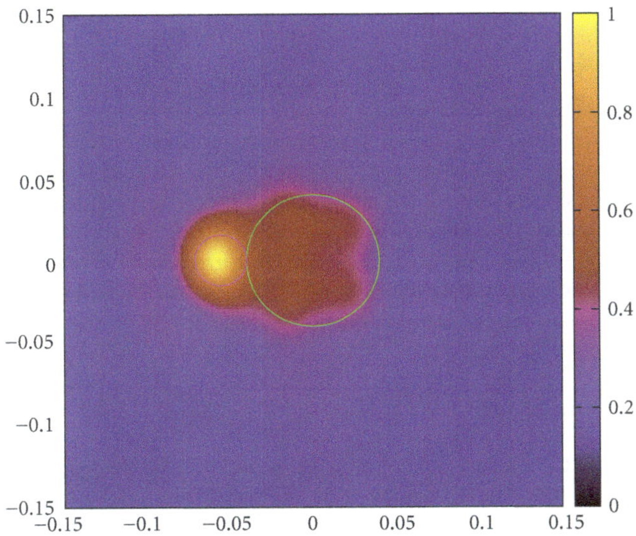

FIGURE 9: Reconstruction of a couple of the FoamDielExt object. Multiview; map of the P_F values computed at $f = 4\,\text{GHz}$. At this frequency 140 singular values were used.

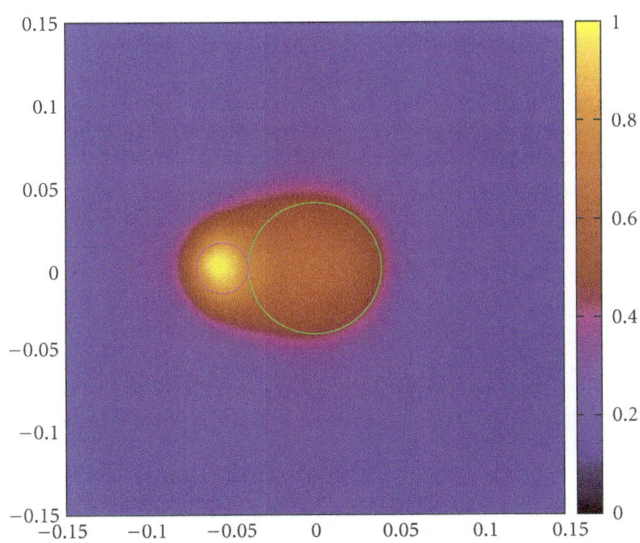

FIGURE 10: Reconstruction of a couple of twin dielectric cylinders. Multiview-multifrequency. Map of the $\mathcal{P}_{VF}$ values.

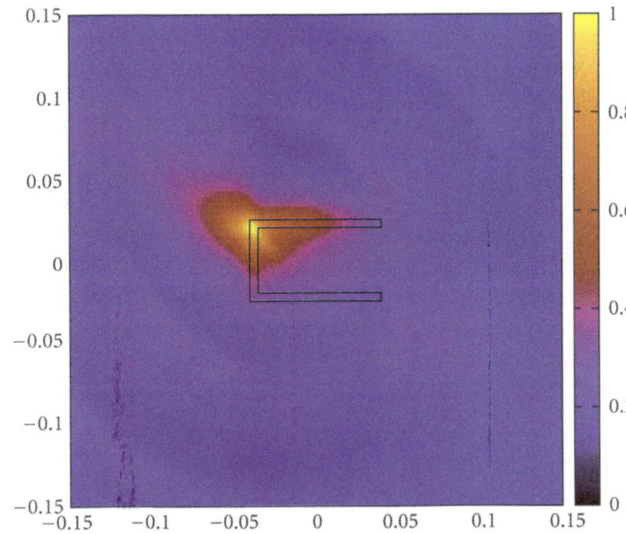

FIGURE 11: Reconstruction of a metallic U-shaped cylinder. Multi-frequency; map of the P_V values computed at $\vartheta_v = 160^\circ$.

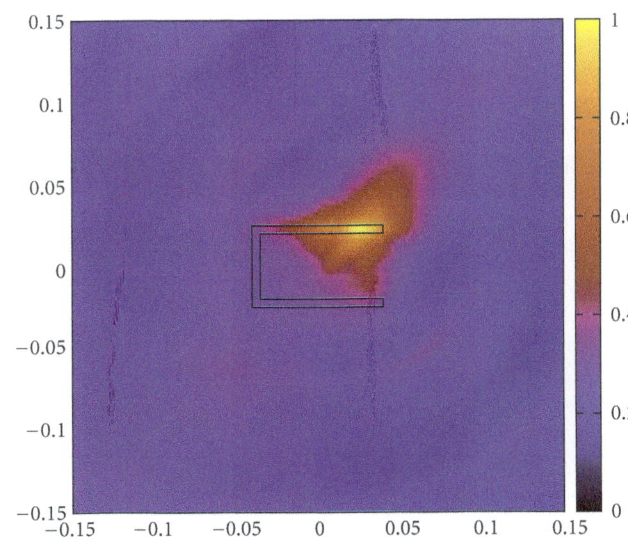

FIGURE 12: Reconstruction of a metallic U-shaped cylinder. Multi-frequency; map of the P_V values computed at $\vartheta_v = 60^\circ$.

4. Conclusions

In this work, an analytical method for inverse scattering problems was used to successfully reconstruct both homogeneous and inhomogeneous targets from multifrequency multistatic experimental data. The proposed method uses the singular values decomposition of the scattering operator to achieve a closed solution to the integral equation of the electromagnetic scattering. The derivation of the radiating source density inside the domain of investigation is the first step of the work, as well as the preliminary result useful to locate a scattering object inside such a region.

The dependence of the reconstructions on various parameters was investigated. In particular, results have shown that a correct choice of the significant number of singular values in the solution can lead to an optimal reconstruction, however; the method has proved robust enough to provide good reconstructions for a wide range of parameters.

Dielectric objects as well as objects made of conductive material were successfully tested.

As a general comment it should be stressed that the algorithm is very robust, and that, even for the very large domains used and with a suboptimal number of singular values, the object location was always clearly recovered.

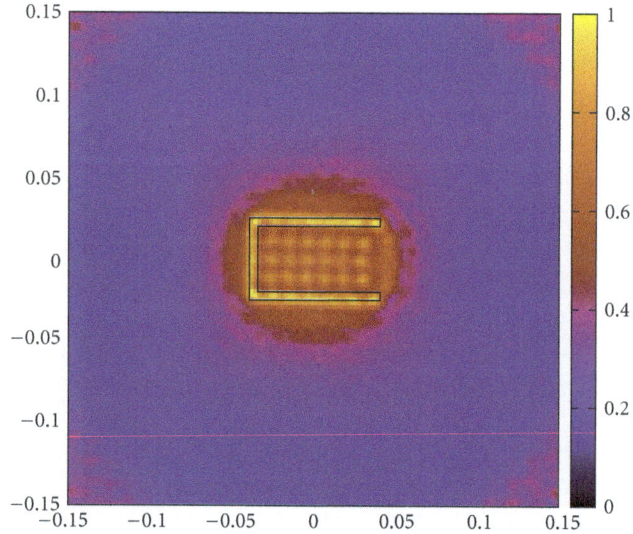

FIGURE 13: Reconstruction of a metallic U-shaped cylinder. Multiview; map of the P_F values computed at $f = 16\,\text{GHz}$.

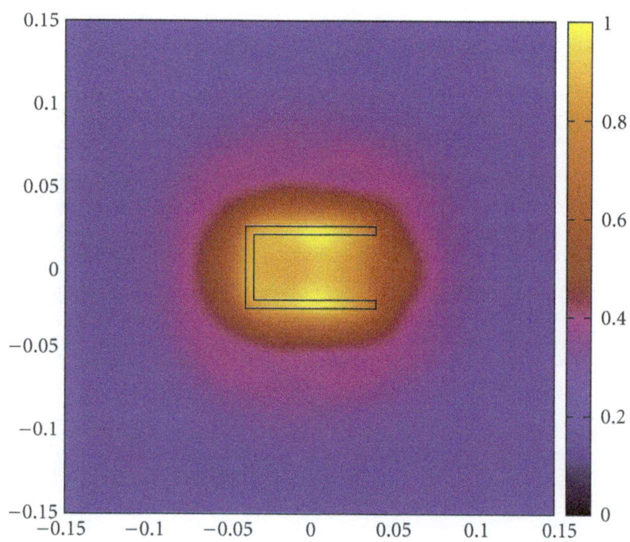

FIGURE 14: Reconstruction of the shape of a metallic U-shaped cylinder. Multiview-multifrequency. Map of the $\mathcal{P}_{VF}$ values.

Furthermore, thanks to the closed-form formulation of the solution, all the results were obtained in a very short computation time. It must also be pointed out, in particular, that, once the operating frequency and the dimensions of the investigation domain have been chosen, the singular system of the problem can be computed "offline." For the same reason also some other operations, for example, the inspection of the singular values, can be performed before the reconstruction process. As for the "online" part of the algorithm, also in this case things go very fast. As an extra bonus, since the closed-form virtually provides the solution at any point inside the investigation domain, high resolution maps of the reconstructed area can be obtained, without the constraints on grid dimensions and on the number of pixels usually imposed by algorithms based on numerical discretization of the problem.

Future work will be focused on the reconstruction of the scattering potential of the objects under test and on the possible use of a set of nonradiating sources to refine the results. It is expected that these steps can provide a notable enhancement in the results, as it was already proved by numerical simulations.

A further goal will be also testing the method on other experimental datasets, for example, on the series provided by the University of Manitoba [41].

References

[1] J. C. Bolomey and L. Jofre, "Three decades of active microwave imaging achievements, difficulties and future challenges," in *Proceedings of the IEEE International Conference on Wireless Information Technology and Systems (ICWITS '10)*, pp. 1–4, Honolulu, Hawaii, USA, September 2010.

[2] M. Pastorino, *Microwave Imaging*, John Wiley & Sons, New York, NY, USA, 2010.

[3] Y. J. Kim, L. Jofre, F. De Flaviis, and M. Q. Feng, "Microwave reflection tomographic array for damage detection of civil structures," *IEEE Transactions on Antennas and Propagation*, vol. 51, no. 11, pp. 3022–3032, 2003.

[4] K. J. Langenberg, K. Mayer, and R. Marklein, "Nondestructive testing of concrete with electromagnetic and elastic waves: modeling and imaging," *Cement and Concrete Composites*, vol. 28, no. 4, pp. 370–383, 2006.

[5] M. Benedetti, M. Donelli, A. Martini, M. Pastorino, A. Rosani, and A. Massa, "An innovative microwave-imaging technique for nondestructive evaluation: applications to civil structures monitoring and biological bodies inspection," *IEEE Transactions on Instrumentation and Measurement*, vol. 55, no. 6, pp. 1878–1884, 2006.

[6] N. Qaddoumi, R. Zoughi, and G. W. Carriveau, "Microwave detection and depth determination of disbonds in low-permittivity and low-loss thick sandwich composites," *Re-earch in Nondestructive Evaluation*, vol. 8, no. 1, pp. 51–63, 1996.

[7] S. Kharkovsky and R. Zoughi, "Microwave and millimeter wave nondestructive testing and evaluation," *IEEE Instrumentation and Measurement Magazine*, vol. 10, no. 2, pp. 26–38, 2007.

[8] R. Zoughi and S. Kharkovsky, "Microwave and millimetre wave sensors for crack detection," *Fatigue and Fracture of Engineering Materials and Structures*, vol. 31, no. 8, pp. 695–713, 2008.

[9] Y. Deng and X. Liu, "Electromagnetic imaging methods for nondestructive evaluation applications," *Sensors*, vol. 11, pp. 11774–11808, 2011.

[10] M. Pastorino, "Recent inversion procedures for microwave imaging in biomedical, subsurface detection and nondestructive evaluation applications," *Measurement*, vol. 36, no. 3-4, pp. 257–269, 2004.

[11] M. Bertero and M. Piana, "Inverse problems in biomedical imaging: modeling and methods of solution," in *Complex Systems in Biomedicine*, A. Quarteroni, L. Formaggia, and A. Veneziani, Eds., pp. 1–33, Springer, Milan, Italy, 2006.

[12] S. C. Hagness, "Microwave imaging in medicine: promises and future challenges," in *Proceedings of the URSI General Assembly*, pp. 7–16, Chicago, Ill, USA, August 2008.

[13] S. Semenov, "Microwave tomography: review of the progress towards clinical applications," *Philosophical Transactions of the Royal Society A*, vol. 367, no. 1900, pp. 3021–3042, 2009.

[14] P. M. Meaney and K. D. Paulsen, "Challenges on microwave imaging supported by clinical results," in *Proceedings of the International Workshop on Biological Effects of Electromagnetic Fields*, pp. 10–14, Bodrum, Turkey, October 2010.

[15] A. Salvadè, M. Pastorino, R. Monleone et al., "Microwave imaging of foreign bodies inside wood trunks," in *Proceedings of the IEEE Workshop on Imaging Systems and Techniques (IST '08)*, pp. 88–93, Chania, Greece, September 2008.

[16] H. Park, C. C. Chang, B. Deng et al., "Recent advancements in microwave imaging plasma diagnostics," *Review of Scientific Instruments*, vol. 74, pp. 4239–4262, 2003.

[17] A. Alu and N. Engheta, "Plasmonic and metamaterial cloaking: physical mechanisms and potentials," *Journal of Optics A*, vol. 10, no. 9, Article ID 093002, 2008.

[18] S. Caorsi, A. Costa, and M. Pastorino, "Microwave imaging within the second-order born approximation: stochastic optimization by a genetic algorithm," *IEEE Transactions on Antennas and Propagation*, vol. 49, no. 1, pp. 22–31, 2001.

[19] M. Benedetti, M. Donelli, G. Franceschini, M. Pastorino, and A. Massa, "Effective exploitation of the a priori information through a microwave imaging procedure based on the SMW for NDE/NDT applications," *IEEE Transactions on Geoscience and Remote Sensing*, vol. 43, no. 11, pp. 2584–2592, 2005.

[20] G. Bozza, C. Estatico, M. Pastorino, and A. Randazzo, "Application of an inexact-Newton method within the second-order Born approximation to buried objects," *IEEE Geoscience and Remote Sensing Letters*, vol. 4, no. 1, pp. 51–55, 2007.

[21] G. Bozza, C. Estatico, A. Massa, M. Pastorino, and A. Randazzo, "Short-range image-based method for the inspection of strong scatterers using microwaves," *IEEE Transactions on Instrumentation and Measurement*, vol. 56, no. 4, pp. 1181–1188, 2007.

[22] A. Randazzoa and C. Estatico, "A regularisation scheme for electromagnetic inverse problems: application to crack detection in civil structures," *Nondestructive Testing and Evaluation*. In press.

[23] B. Hofmann, *On Ill-Posedness and Local Ill-Posedness of Operator Equations in Hilbert Spaces*, Chemnitz University of Technology, 1997.

[24] A. J. Devaney and G. C. Sherman, "Nonuniqueness in inverse source and scattering problems," *IEEE Transactions on Antennas and Propagation*, vol. 30, pp. 1034–1042, 1982.

[25] M. Chiappe and G. L. Gragnani, "An analytical approach to the reconstruction of the radiating currents in inverse electromagnetic scattering," *Microwave and Optical Technology Letters*, vol. 49, no. 2, pp. 354–360, 2007.

[26] G. L. Gragnani, "Shape reconstruction of 2-D dielectric objects by an analytical method," *International Journal of Signal and Imaging Systems Engineering*, vol. 3, no. 2, pp. 81–92, 2010.

[27] R. V. McGahan and R. E. Kleinman, "Special session on image reconstruction using real data," *IEEE Antennas and Propagation Magazine*, vol. 38, no. 3, pp. 39–40, 1996.

[28] R. V. McGahan and R. E. Kleinman, "Second annual special session on image reconstruction using real data," *IEEE Antennas and Propagation Magazine*, vol. 39, no. 2, pp. 7–9, 1997.

[29] R. V. McGahan and R. E. Kleinman, "The third annual special session on image reconstruction using real data, Part 1," *IEEE Antennas and Propagation Magazine*, vol. 41, no. 1, pp. 34–51, 1999.

[30] R. V. McGahan and R. E. Kleinman, "The third annual special session on image reconstruction using real data Part 2," *IEEE Antennas and Propagation Magazine*, vol. 41, no. 1, pp. 20–40, 1999.

[31] K. Belkebir and M. Saillard, "Special section: testing inversion algorithms against experimental data," *Inverse Problems*, vol. 17, no. 6, pp. 1565–1571, 2001.

[32] K. Belkebir and M. Saillard, "Testing inversion algorithms against experimental data: inhomogeneous targets," *Inverse Problems*, vol. 21, pp. S1–S3, 2005.

[33] C. A. Balanis, *Advanced Engineering Electromagnetics*, John Wiley & Sons, New York, NY, USA, 1989.

[34] D. E. Livesay and K. M. Chen, "Electromagnetic field induced inside arbitrarily shaped biological bodies," *IEEE Transactions on Microwave Theory and Techniques*, vol. 22, no. 12, pp. 1273–80, 1974.

[35] S. Caorsi, A. Frattoni, G. L. Gragnani, E. Nortino, and M. Pastorino, "Numerical algorithm for dielectric-permittivity microwave imaging of inhomogeneous biological bodies," *Medical and Biological Engineering and Computing*, vol. 29, no. 6, pp. NS37–NS38, 1991.

[36] D. G. Dudley, T. M. Habashy, and E. Wolf, "Linear inverse problems in wave motion: nonsymmetric first-kind integral equations," *IEEE Transactions on Antennas and Propagation*, vol. 48, no. 10, pp. 1607–1617, 2000.

[37] A. Gamliel, K. Kim, A. I. Nachman, and E. Wolf, "A new method for specifying nonradiating, monochromatic, scalar sources and their fields," *Journal of the Optical Society of America*, vol. 6, no. 9, pp. 1388–1393, 1989.

[38] E. Wallacher and A. K. Louis, "Complete sets of radiating and nonradiating parts of a source and their fields with applications in inverse scattering limited-angle problems," *International Journal of Biomedical Imaging*, vol. 2006, Article ID 93074, 2006.

[39] G. L. Gragnani and M. DiazMendez, "Improved electromagnetic inverse scattering procedure using non-radiating sources and scattering support reconstruction," *IET Microwaves Antennas & Propagation*, vol. 5, no. 15, pp. 1822–1829, 2011.

[40] J.-M. Geffrin, P. Sabouroux, and C. Eyraud, "Free space experimental scattering database continuation: experimental set-up and measurement precision," *Inverse Problems*, vol. 21, no. 6, pp. S117–S130, 2005.

[41] C. Gilmore, A. Zakaria, P. Mojabi, M. Ostadrahimi, S. Pistorius, and J. LoVetri, "The University of ManitobaMicrowave Imaging Repository: a two-dimensionalmicrowave scattering database for testing inversion and calibration algorithms," *IEEE Antennas and Propagation Magazine*, vol. 53, no. 5, pp. 126–133, 2011.

26

A Novel Internal Dual-Polarized EBG Antenna for Indoor Reception of UHF Terrestrial Digital TV Broadcasting

Mohamed Sanad[1] and Noha Hassan[2]

[1] *Amant Antennas, Giza 12451, Egypt*
[2] *Department of Biomedical Engineering and Systems, Cairo University, Giza 12613, Egypt*

Correspondence should be addressed to Mohamed Sanad, msanad@hotmail.com

Academic Editor: Safieddin Safavi-Naeini

A novel internal antenna has been developed for indoor reception of UHF terrestrial digital TV broadcasting. The overall size of some configurations of the new antenna is less than 2 cm^3, and its weight is less than 1 gm. It is made of a flexible material that can be bent or folded and shaped in any form. It is an unbalanced resonant antenna that does not need a matching circuit. The new antenna can be fully embedded inside TV sets or portable computers. It has a bandwidth of about 68%. Thus, it can also cover the bands of GSM and CDMA, which is advantageous in case of portable computers. The new antenna is linearly polarized. It can be easily modified to be dual polarized by combining two orthogonal antennas with one or two feed points. The overall efficiency of some configurations of the new indoor digital TV receiving antenna is more than 80%, and its peak gain is about 2 dBi over the whole UHF band. The peak gain can be increased to more than 5 dBi by adding EBG (electromagnetic bandgap) structures. The EBG structure also increases the efficiency to around 90%.

1. Introduction

The start of terrestrial digital TV broadcasting may considerably reduce the need for on-roof directional receiving antennas and increase the use of indoor antennas. Most terrestrial digital TV broadcasting channels are in the UHF band. Of course, analogue TV indoor receiving antennas can still be used with digital TV indoor reception. The most common UHF indoor digital TV receiving antennas are loop antennas and triangular dipoles [1]. However, they have large sizes and they are not rigid on top of TV sets. Also, they are balanced antennas and, hence, they need baluns. Furthermore, they are sensitive to only one polarization and, therefore, they have a poor indoor performance. Moreover, they are pure electric field antennas and, hence, their performance is significantly deteriorated in vicinity of conductive objects such as concrete walls. In addition, conventional indoor TV receiving antennas cannot be used in multi-input (MI) configurations and techniques for space and/or polarization diversity. In order to overcome some of the above problems, other indoor TV receiving antennas have been developed and published as in [2–4]. However, they are also external antennas that have to be mounted outside TV sets and they are still large in size.

On the other hand, introduction of digital TV, built into portable computers, is expected in the near future. Since portable computers are getting smaller and smaller in size, they add volume limitations on digital TV antennas. Hence, it is very challenging to cover the whole UHF band of digital TV with a single small-size resonant antenna that can be integrated into portable computers or handsets. Therefore, matching circuits have been usually used to tune mobile TV antennas [5–8]. Matching circuits increase the complexity of the antenna and reduce the efficiency. In the mean time, cellular network radios have been already integrated into some modern laptop, notebook, and palmtop computers in order to give the user access to the Internet in areas not covered by WLANs. The problem is that the frequency bands of GSM (824–894 MHz) and UHF digital TV (470–862 MHz) are overlapping. As a result, there will be a severe interference between their antennas.

In this research, a novel internal antenna that can solve all the above problems is developed for indoor reception of UHF terrestrial digital TV broadcasting. The new antenna covers a bandwidth of around 68%. It resonates from 470 MHz to more than 960 MHz. Hence, a single antenna can cover all the bands of UHF digital TV (470–862 MHz), 700 MHz WiMax, CDMA/GSM800 (824–894 MHz), and E-GSM900 (880–960 MHz). The new antenna is resonant antenna that does not need matching circuits. Furthermore, it does not need an additional extended ground plane or any other components. Thus, it can be mounted anywhere inside or outside any TV set or any portable computer because the antenna does not use a part of the equipment as an extended ground plane as usually happens with internal antennas. Moreover, the new antenna is made of a flexible material and it can be bent and/or folded in different forms. The new antenna can be fully embedded inside TV sets and portable computers. Actually, it can be used as internal, external or partially internal, and partially external antenna. In order to increase the peak gain of the antenna, an EBG (electromagnetic bandgap) structure has been used [9]. The new antenna is linearly polarized. It can also be modified in order to make it sensitive to two orthogonal polarizations, which is a very important factor in all indoor applications. Dual-polarized antennas can have a single feed point or two different feed points in order to be used in multi-input, multioutput (MIMO) configurations.

2. Geometry of the New Antenna

Figure 1 shows the geometry of the new developed UHF digital TV antenna. The new antenna consists of two narrow printed metallic arms connected together by a shorting metallic strip. The two arms may be parallel to each other or may have any angle between them. The length of the short arm is L_1, and its width is W_1 while the length of the long arm is L_2, and its width is W_2. The thickness of the antenna is T, and it is fed with a coaxial feed at a distance F from the shorted edge. The two arms of the antenna can have equal or unequal widths W_1 and W_2. Furthermore, the two arms can be shaped in different ways in order to optimize the antenna performance. As shown in Figure 1, each arm has a set of slots having different configurations. These slots can be circular, rectangular, square, or have other shapes. The arm lengths of the new antenna, especially the length of the short arm, are the main parameters that determine the operating frequency of the antenna. The feed location is adjusted in each configuration in order to improve the return loss as much as possible. The bandwidth, the peak gain, and the efficiency of the antenna are mainly determined by the widths of the two arms, the angle between them, the thickness of the antenna and the configurations of the slots which are all optimized together in order to enhance the antenna performance, especially the bandwidth. The basic concept of using such slots to increase the antenna bandwidth was initially used in E-shaped antennas as explained in [10], where only 2 of such slots were utilized. Thus, this new antenna is an extension to the concept of E-shaped antennas, which were originally developed by the authors of this paper [11].

Different prototypes of the new digital TV antenna have been designed, manufactured, and tested. The results of a selected sample antenna configuration will be presented. The sample is made of a flexible printed material "PET" with a dielectric constant $\varepsilon_r = 3.5$ and a tangent loss $\delta = 0.015$. The geometry of the selected antenna configuration is shown in Figure 2. The two arms of the selected sample antenna are parallel to each other. The length L_1 of the short arm is 11.5 cm while the length L_2 of the long arm is 25 cm. The width W_1 of the short arm is 2.6 mm while the width W_2 of the long arm is 3.5 mm, and the antenna thickness T is 2 mm. The overall size of the antenna is $25 \times 0.35 \times 0.2 = 1.75\,\text{cm}^3$, and its weight is less than 1 gm. It should be noted that this is the overall volume of the antenna because it does not require an additional ground plane, a matching circuit or any other components. It should also be noted that the length of the antenna can be reduced from 25 cm to 16 cm without any significant effect on the performance by folding the two ends of the antenna [12–14].

The slots in both arms are selected to be rectangular in shape. The length of each slot is 5 mm and its width is 2 mm. The distance between the shorted edge and the first slot is D_1. The distances between the successive slots are $D_2, D_3, \ldots,$ and D_8, respectively. The locations of the first five slots are similar in both arms. This means that the first five slots in the short arm are located exactly above the first five slots in the long arm. However, since the long arm is wider than the short arm, the first five slots are positioned close to the middle of the long arm forming rings while they are located at the edge of the short arm as shown in Figure 2(b). The values of $D_1, D_2, \ldots,$ and D_8 are shown in Table 1.

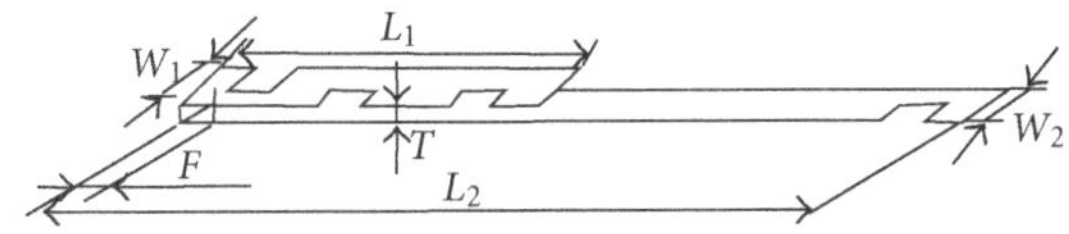

Figure 1: Geometry of the new wideband resonant antenna.

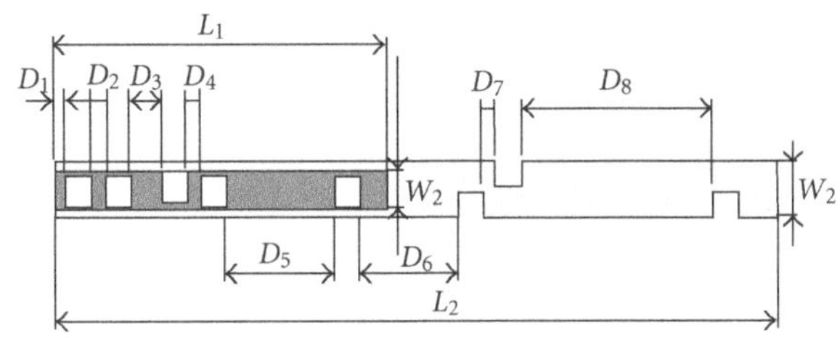

(a) Top view

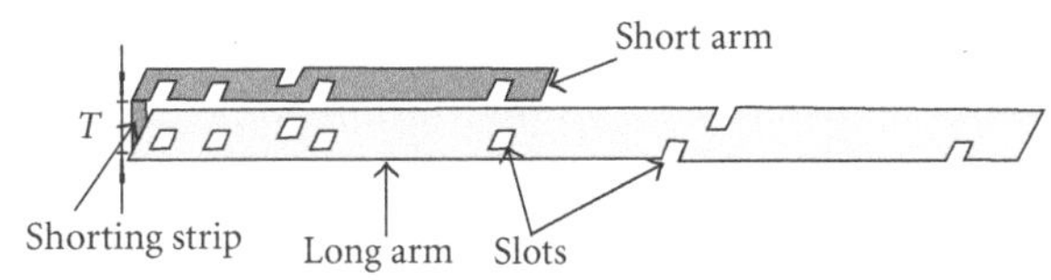

(b) 3D view

Figure 2: The selected sample antenna configuration.

Table 1

D_1	D_2	D_3	D_4	D_5	D_6	D_7	D_8
5 mm	10 mm	15 mm	5 mm	45 mm	35 mm	5 mm	75 mm

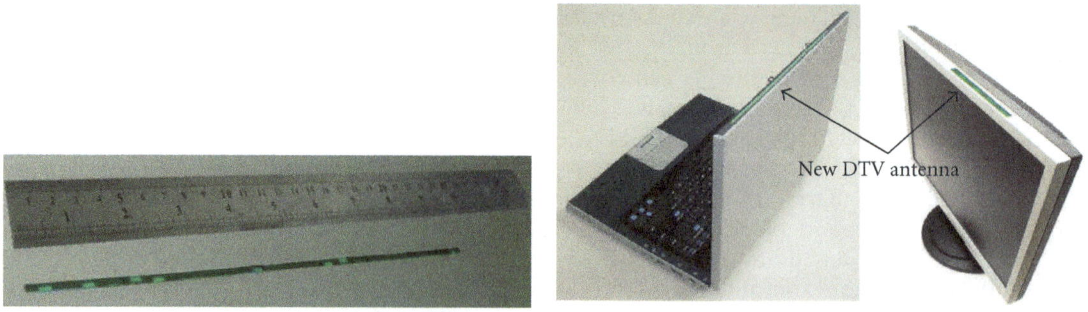

(a) (b)

FIGURE 3: The new digital TV antenna prototype on the display rim of a portable computer and a digital TV set.

Figure 3 shows the first manufactured prototype in free space and while it is mounted on the display rim of a portable computer and a TV set. In this configuration, the new antenna has a minimum blockage by the TV set and the portable computer housing. This unique configuration is feasible with all portable computers and TV sets because the widths of the new antenna can always be made narrower than the width of the display rim of any portable computer and any TV set. Furthermore, the new antenna can be mounted anywhere because it does not use a part of the TV set or the portable computer as an extended ground plane as usually happens with internal antennas. Moreover, the new antenna does not need to be customized for any portable computer or any TV set because the effect of the housing of these sets on the antenna was found to be negligible [15].

The return loss and the radiation patterns of the new digital TV receiving antenna were numerically calculated using Zeland (http://www.zeland.com/) software package that utilizes the moment method. The above selected configuration and some other prototypes were also measured at IMST antenna labs in Germany [16] in order to verify the numerical simulations. Both numerical and experimental results of the above selected configuration will be presented in order to show the agreement between them. For the following configurations, only measured return losses and simulated patterns and efficiencies will be presented.

Figure 4 shows the calculated and the measured return loss of the new antenna. The measured return loss is less than −6 dB from about 470 MHz to 960 MHz which is more than 68% bandwidth. This band can cover digital TV, GSM, and CDMA. Figure 5 shows the calculated and the measured peak gain of the antenna. The measured peak gain is higher than 0 dBi over most of the band. This peak gain is much higher than MBRAI specifications of the UHF DVB-H mobile TV [17]. The calculated and the measured efficiency of the antenna is shown in Figure 6. The average efficiency is more than 45% over the whole band. The calculated radiation patterns at 600 MHz, as a sample frequency, are shown in Figure 7. The corresponding measured patterns are shown in Figure 8.

From all the above results, it can be seen that there are some considerable differences between the calculated and measured results. This is because the used version of the software package assumes an infinite size substrate.

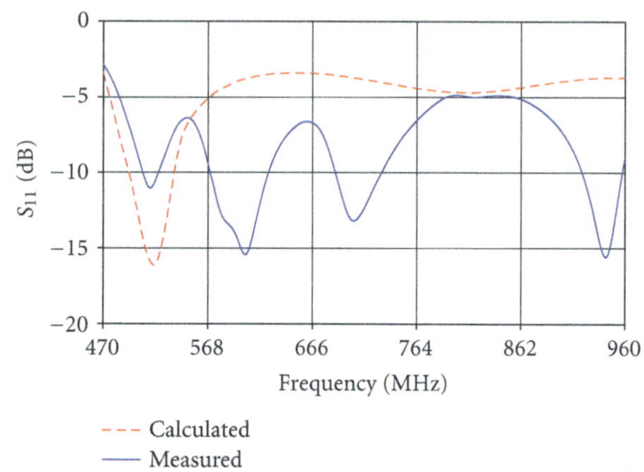

FIGURE 4: Calculated and measured return loss of the new antenna from 470 MHz to 960 MHz.

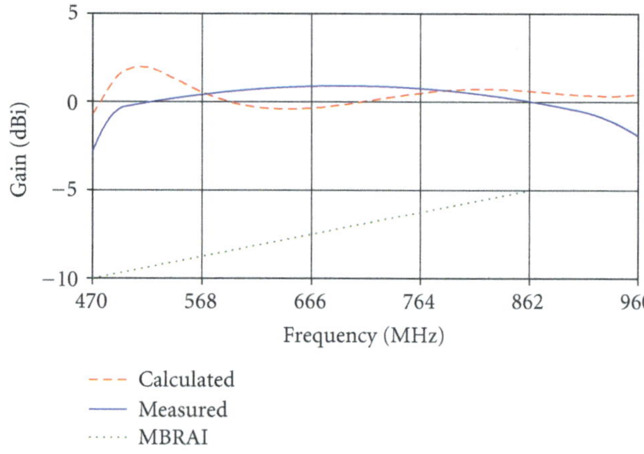

FIGURE 5: Calculated and measured peak gain of the new antenna from 470 MHz to 960 MHz.

This assumption affects the accuracy of calculations when the width of the antenna substrate is very narrow as in the above antenna configurations. The difference between calculations and measurements is significant in efficiency curves because the efficiency depends on radiation patterns in all directions and in all planes and, hence, there is an accumulated reduction in the accuracy of calculations. This

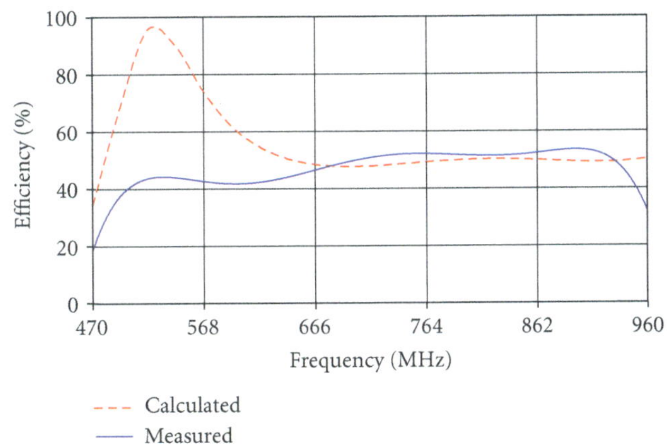

FIGURE 6: Calculated and measured efficiency of the new antenna from 470 MHz to 960 MHz.

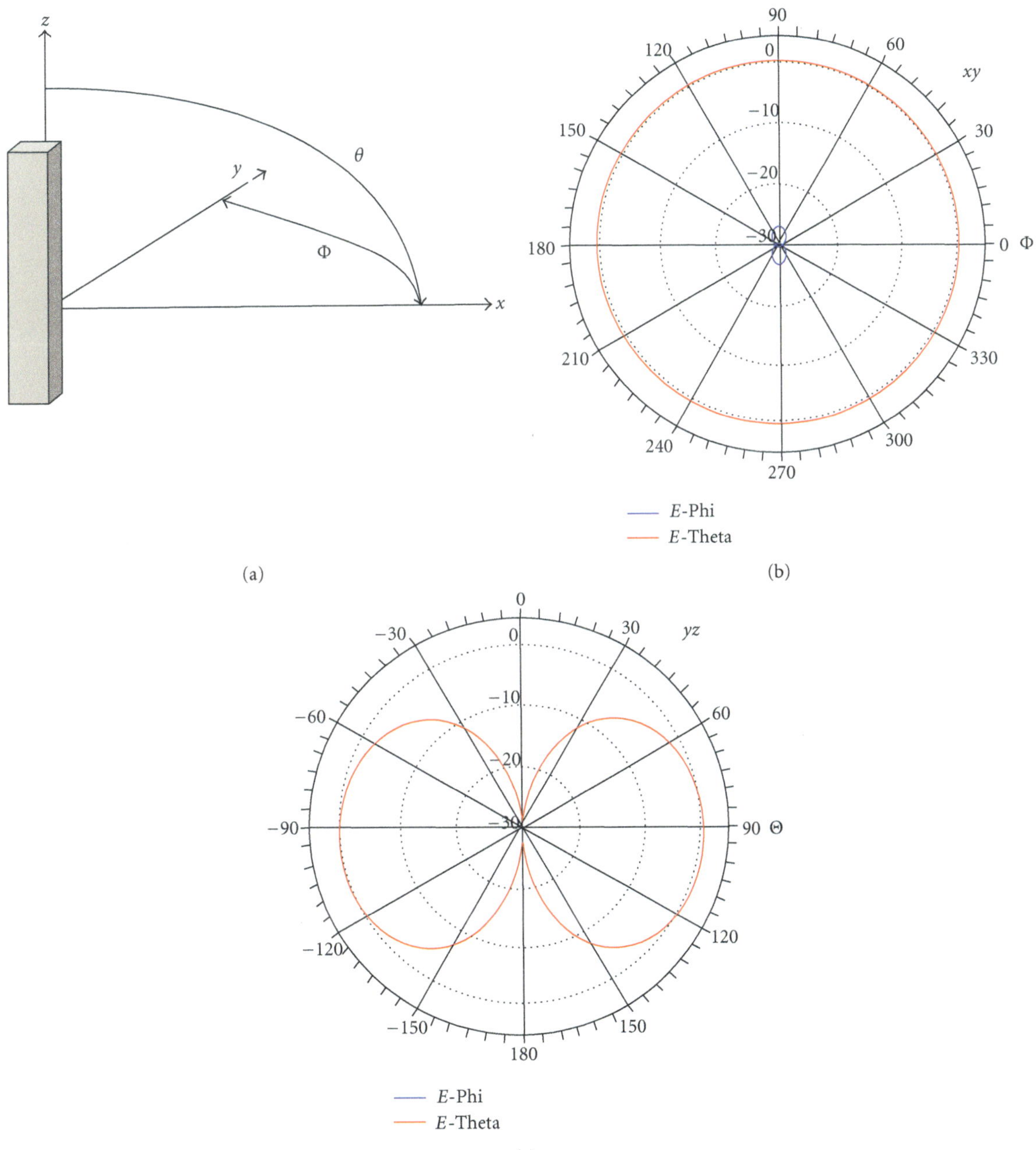

FIGURE 7: Calculated radiation patterns of the new antenna at 600 MHz.

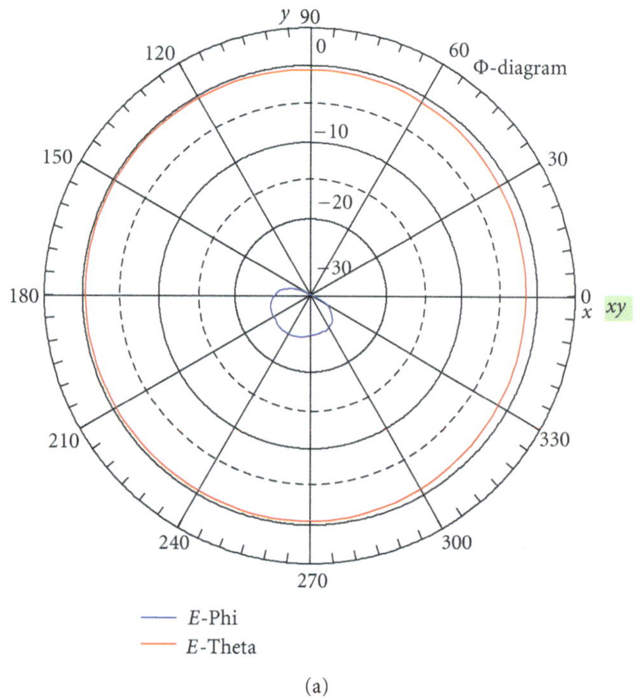

Figure 8: Measured radiation patterns of the new antenna at 600 MHz.

was more clear at the lower part of the frequency range where the antenna width is very narrow in terms of wavelengths.

3. Increasing the Gain and the Efficiency

As shown above, the new antenna could cover the whole frequency band of UHF digital TV, and its performance was much better than MBRAI specifications of the DVB-H mobile TV. However, for DVB-T (digital video broadcasting-terrestrial TV), it will be desirable to further increase the gain and the efficiency of the new antenna. This can be easily achieved by increasing the width, and the thickness of the antenna as much as possible according to the available space in the portable computer or the TV set. A prototype of the antenna has been manufactured with an increased width and thickness. The dimensions of the increased-size antenna are: $L_1 = 11.5$ cm, $L_2 = 25$ cm, $W_1 = 8$ mm, $W_2 = 12$ mm, and $T = 4$ mm. The overall size of the increased-size antenna is $25 \times 1.2 \times 0.4 = 12\,\text{cm}^3$. Figure 9 shows the return loss of the increased-size antenna, which is around −8 dB over the whole band. The overall efficiency of the increased-size antenna is shown in Figure 10. The average efficiency was more than 80% over the whole bandwidth. Figure 11 shows the peak gain of the new low-band antenna which is about 2 dBi over the whole bandwidth. Figure 12 shows the radiation patterns of the new antenna at 600 MHz which are omnidirectional with about 2 dBi peak gain.

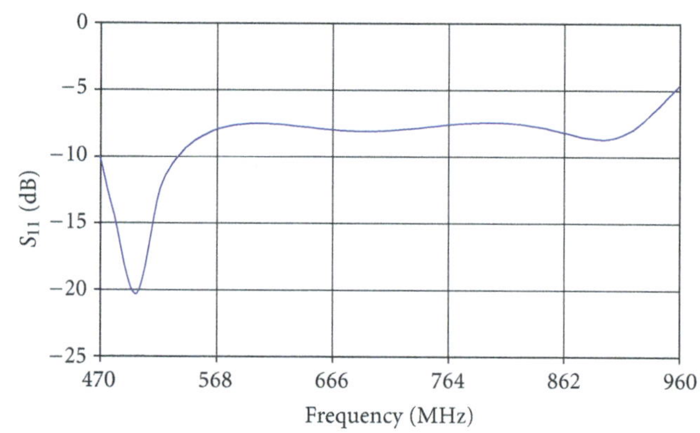

Figure 9: Return loss of the increased-size antenna.

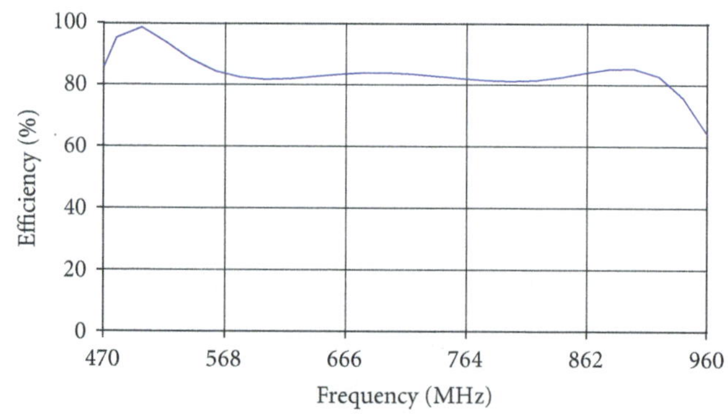

Figure 10: Efficiency of the increased-size antenna.

4. Dual Polarized Antennas

From Figure 12, it can be seen that the new antenna is sensitive to only one polarization while the orthogonal

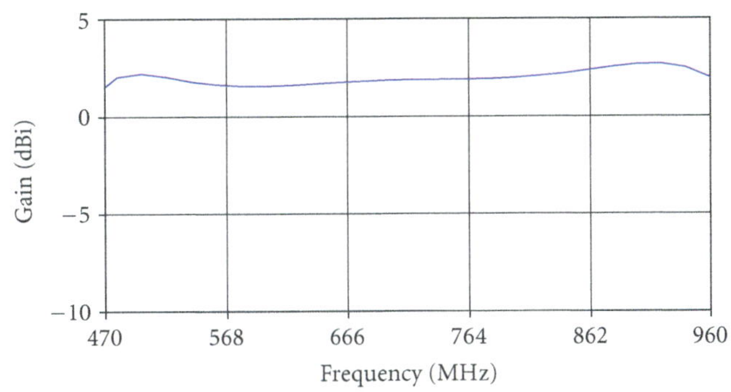

FIGURE 11: Peak gain of the increased-size antenna.

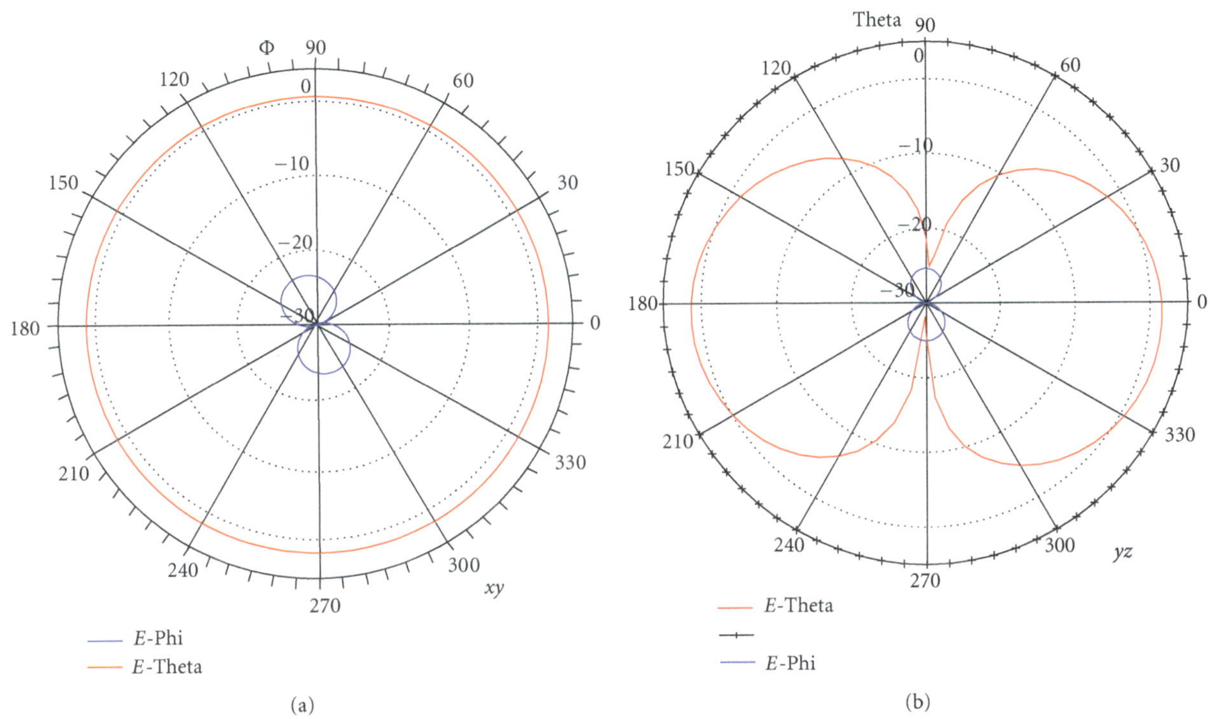

FIGURE 12: Radiation patterns of the increased-size antenna at 600 MHz.

polarization is negligible. However, the new antenna can be modified in order to make it sensitive to two perpendicular polarizations, which is a very important factor in all indoor applications. Dual-polarization can be accomplished by combining two orthogonal antennas with a common feed point as shown in Figure 13. The unique advantage of this dual-polarized antenna configuration is that it can still be mounted on the display rim of portable computers and TV sets and, thus, both polarizations of the antenna have a minimum blockage by the TV set and the portable computer housing. As mentioned above, this unique advantage is feasible with all portable computers and TV sets because the widths of both antenna polarizations can always be made narrower than the width of the display rim of any portable computer and any TV set.

The return loss of the dual-polarized antenna is shown in Figure 14 while the efficiency is shown in Figure 15. The efficiency is still higher than 80% over most of the band. The peak gain of the dual-polarized antenna is shown in Figure 16, and it is still higher than 2 dBi over most of the band. Figure 17 shows the radiation patterns of the dual-polarized antenna at 600 MHz. It is clear from the figure that the antenna is equally sensitive to two orthogonal polarizations.

Dual-polarization can also be achieved by using two orthogonal antennas with two different feed points as shown in Figure 18. Again, this dual-polarized configuration has the unique advantage of being mounted on the display rim of portable computers and TV sets. This dual-feed configuration is important for multi-input, multioutput (MIMO) techniques. Since the interference between the orthogonal MIMO antennas is negligible, the return loss and the efficiency of the orthogonal antennas are not affected by the MIMO configuration, and they are almost the same as

FIGURE 13: Two orthogonal antennas on a TV set with a common feed point.

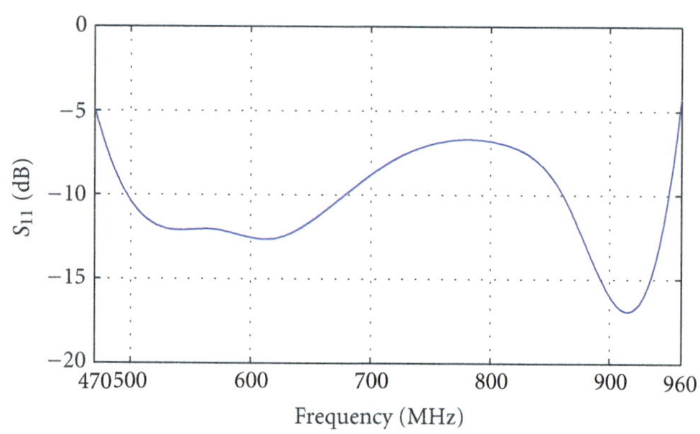

FIGURE 14: Return loss of the dual-polarized antenna.

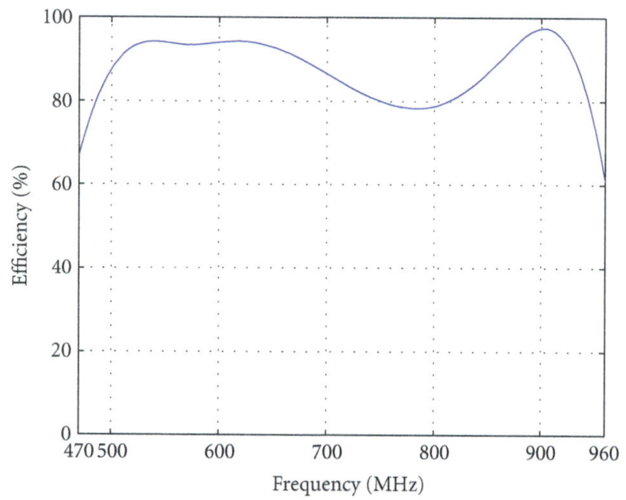

FIGURE 15: Efficiency of the dual-polarized antenna.

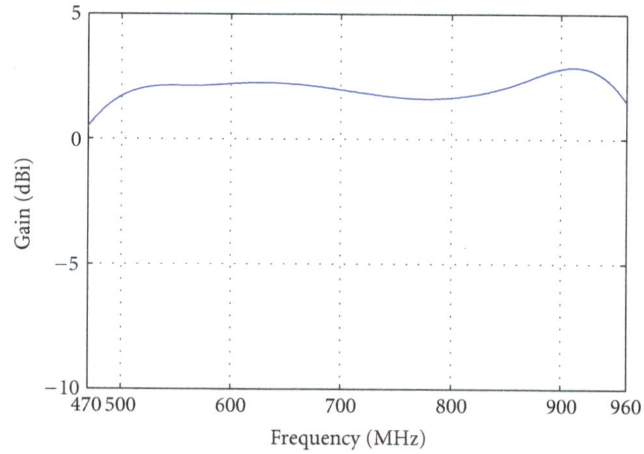

FIGURE 16: Peak gain of the dual-polarized antenna.

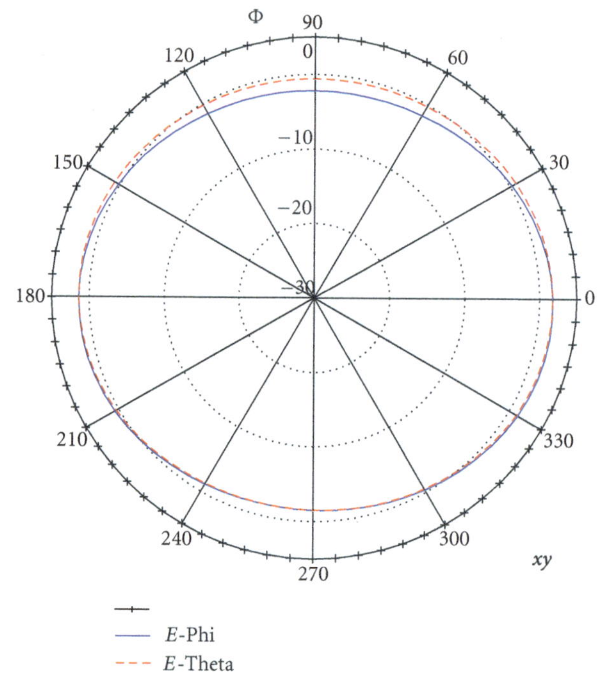

FIGURE 17: Copolar and cross-polar radiation patterns of the dual-polarized antenna at 600 MHz.

the above presented results of single antennas. However, as expected, the radiation patterns of the orthogonal antennas are significantly affected by MIMO configurations. The overall radiation patterns of MIMO antennas are the summations of the radiation patterns of the orthogonal antennas. Figure 19 shows the copolar and cross-polar components of the radiation patterns of two orthogonal antennas at 600 MHz in the plane of the antennas. The orthogonal antennas are 1 cm apart from each other. The copolar and cross-polar components are exactly the same.

5. Digital TV Antennas with EBG Structures

In order to further increase the peak gain of the new digital TV antenna, a rectangular EBG structure was added to the

FIGURE 18: Two orthogonal antennas on a TV set with two different feed points for MIMO configurations.

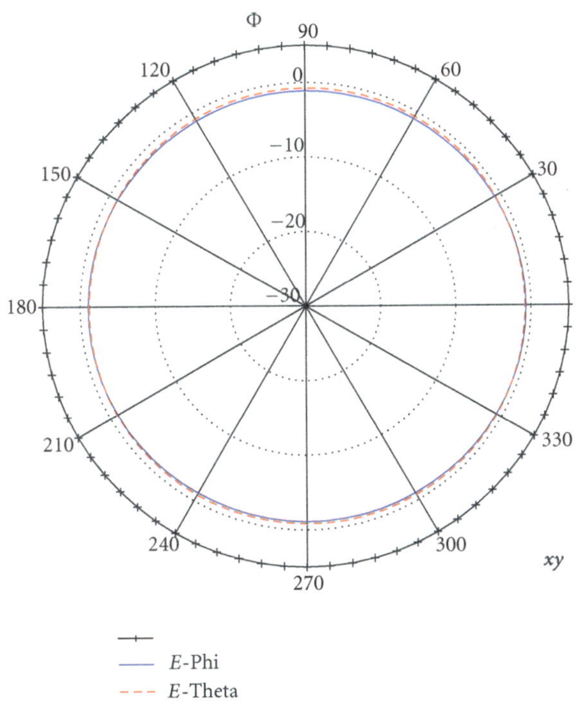

FIGURE 19: Radiation patterns of two orthogonal antennas at 600 MHz.

antenna. The EBG consists of 3 × 11 elements as shown in Figure 20. The dimensions of each element are 24 × 24 mm, and the gap between each two elements is 4 mm. The dimensions of the EBG ground plane are 10 × 32 cm, and it is located at 12 mm under the EBG elements. The antenna is located at 9 mm above the EBG elements. Figure 21 shows the return loss of the new digital TV antenna with the EBG structure. The return loss is better than −10 dB over most of the band. Figure 22 shows the efficiency of the EBG digital

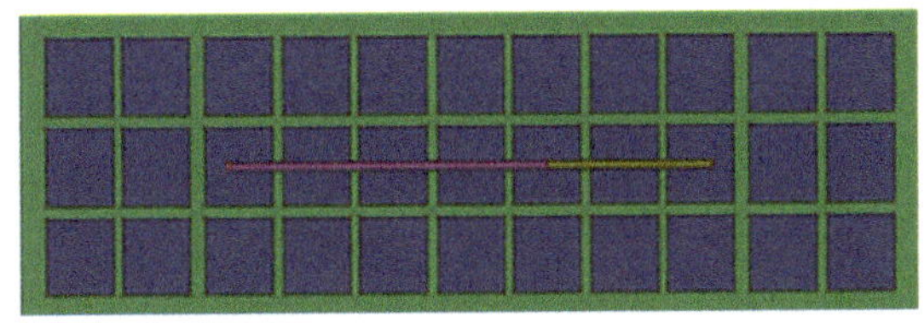

FIGURE 20: The new antenna on the EBG structure.

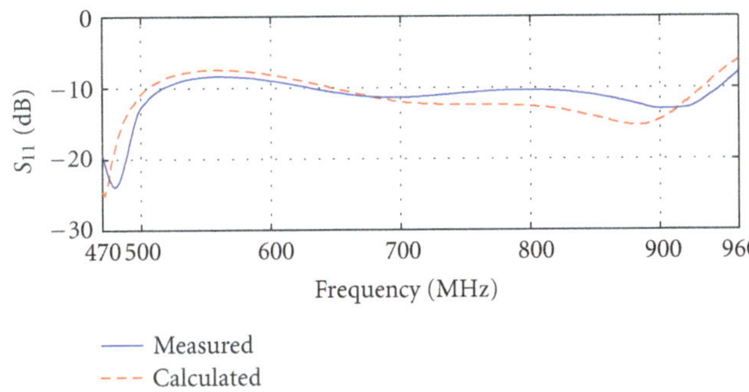

FIGURE 21: Return loss of the EBG digital TV antenna.

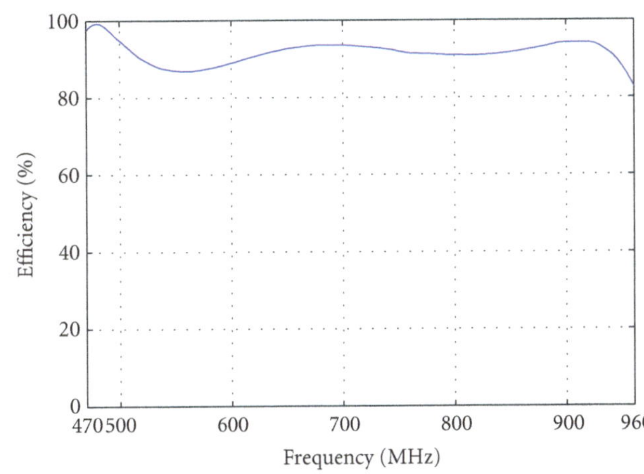

FIGURE 22: Efficiency of the EBG digital TV antenna.

TV antenna. The antenna efficiency exceeds 90% over most of the band. The peak gain of the antenna is shown in Figure 23, which is more than 5 dBi over most of the band.

6. Conclusions

A novel internal antenna was developed for indoor reception of UHF terrestrial digital TV broadcasting. The overall size of the antenna was 1.75 cm^3 and its weight was less than 1 gm. This was the overall volume of the antenna because it did not require an additional ground plane, a matching circuit, or any other components. The new digital TV receiving antenna was made of a flexible material, and it could be bent or folded and shaped in any form. It could be fully embedded inside TV sets and portable computers. The new antenna had a bandwidth of more than 68%. It could also cover the bands of GSM and CDMA, which is advantageous in case of

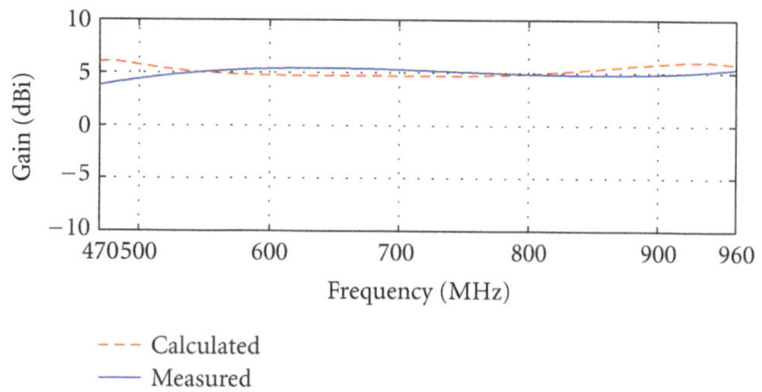

Figure 23: Peak gain of the EBG digital TV antenna.

portable computers. The peak gain of the new internal digital TV antenna was about 0 dBi over the whole UHF band, and the average efficiency was more than 45%. The peak gain was increased to about 2 dBi by increasing the width and the thickness of the antenna. Increasing the antenna size also increased the overall efficiency to more than 80% over most of the UHF band. The peak gain could be further increased to more than 5 dBi by adding EBG (electromagnetic bandgap) structures to the new digital TV antenna. The EBG structure consisted of 3 × 11 elements. The EBG also improved the return loss of the antenna and increased its efficiency to around 90%.

The new internal digital TV antenna was linearly polarized. It could be easily modified in order to make it sensitive to two perpendicular polarizations, which is a very important factor in all indoor applications. Dual-polarization could be accomplished by combining two orthogonal antennas with a common feed point. Dual-polarization could also be achieved by two orthogonal antennas with two different feed points for multi-input, multioutput (MIMO) techniques.

Acknowledgment

The authors would like to thank Mohamed Osama, Lamiaa Mamdouh, and Mahetab Salama for their great effort in preparing this paper.

References

[1] R. C. Johnson and H. Jasik, *Antenna Engineering Handbook*, McGraw-Hill Book Company, New York, NY, USA, 1984.

[2] S. Kashihara and F. Kuroki, "J-shaped monopole antenna array as an antenna for terrestrial digital broadcasting at UHF band," in *Proceedings of the IEEE MTT-S International Microwave Symposium (IMS '09)*, pp. 293–296, Boston, Mass, USA, June 2009.

[3] F. Kuroki and H. Ohta, "L-shaped monopole array for terrestrial broadcasting reception in the UHF band," in *Proceedings of the 37th European Microwave Conference (EUMC '07)*, pp. 1070–1073, Munich, Germany, October 2007.

[4] F. Kuroki, H. Ohta, M. Yamaguchi, and E. Suematsu, "Wall-hanging type of self-complementary spiral patch antenna for indoor reception of digital terrestrial broadcasting," in *Proceedings of the IEEE MTT-S International Microwave Symposium Digest*, pp. 194–197, San Francisco, Calif, USA, June 2006.

[5] J. Kim, G. Kim, W. Seong, and J. Choi, "A tunable internal antenna with an epsilon negative zeroth order resonator for DVB-H service," *IEEE Transactions on Antennas and Propagation*, vol. 57, no. 12, pp. 4014–4017, 2009.

[6] L. Huang and P. Russer, "Tunable antenna design procedure and harmonics suppression methods of the tunable DVB-H antenna for mobile applications," in *Proceedings of the 37th European Microwave Conference (EUMC '07)*, pp. 1062–1065, Munich, Germany, October 2007.

[7] L. Huang and P. Russer, "Electrically tunable antenna design procedure for mobile applications," *IEEE Transactions on Microwave Theory and Techniques*, vol. 56, no. 12, pp. 2789–2797, 2008.

[8] C. M. Chiang, A. Yang, and C. C. Chien, "Active Digital TV Antenna," US Patent No. 0066116, 2008.

[9] J. Liang and H. Y. D. Yang, "Radiation characteristics of a microstrip patch over an electromagnetic bandgap surface," *IEEE Transactions on Antennas and Propagation*, vol. 55, no. 6, pp. 1691–1697, 2007.

[10] F. Yang, X. X. Zhang, X. Ye, and Y. Rahmat-Samii, "Wide-band E-shaped patch antennas for wireless communications," *IEEE Transactions on Antennas and Propagation*, vol. 49, no. 7, pp. 1094–1100, 2001.

[11] M. Sanad, F. Yang, Y. Rahmat-Samii, and X. X. Zhang, "Comments on wide-band E-shaped patch antennas for wireless communications," *IEEE Transactions on Antennas and Propagation*, vol. 51, no. 9, pp. 2541–2542, 2003.

[12] M. Sanad and N. Hassan, "A resonant handset antenna that can cover all bands of UHF mobile TV, GSM and CDMA without using matching circuits," in *Proceedings of the IEEE International Symposium on Antennas and Propagation (APSURSI '10)*, pp. 1–4, July 2010.

[13] M. Sanad and N. Hassan, "An internal EBG antenna for indoor reception of UHF terrestrial digital TV broadcasting," in *Proceedings of the 10th Mediterranean Microwave Symposium (MMS '10)*, pp. 178–181, August 2010.

[14] M. Sanad and N. Hassan, "A 470 to 960 MHz resonant antenna: covering uhf mobile TV and CDMA/GSM without tuning circuits," *Microwave Journal*, vol. 53, no. 11, pp. 56–72, 2010.

[15] M. Sanad and N. Hassan, "A novel dual resonant antenna configuration for mobile laptop, notebook and palmtop computers," in *Proceedings of the IEEE International Symposium on Antennas and Propagation*, pp. 1–4, Toront, canada, July 2010.

[16] http://www.imst.com/en/home.php.

[17] http://www.eicta.org/index.php?id=33&id_article=152.

Permissions

The contributors of this book come from diverse backgrounds, making this book a truly international effort. This book will bring forth new frontiers with its revolutionizing research information and detailed analysis of the nascent developments around the world.

We would like to thank all the contributing authors for lending their expertise to make the book truly unique. They have played a crucial role in the development of this book. Without their invaluable contributions this book wouldn't have been possible. They have made vital efforts to compile up to date information on the varied aspects of this subject to make this book a valuable addition to the collection of many professionals and students.

This book was conceptualized with the vision of imparting up-to-date information and advanced data in this field. To ensure the same, a matchless editorial board was set up. Every individual on the board went through rigorous rounds of assessment to prove their worth. After which they invested a large part of their time researching and compiling the most relevant data for our readers. Conferences and sessions were held from time to time between the editorial board and the contributing authors to present the data in the most comprehensible form. The editorial team has worked tirelessly to provide valuable and valid information to help people across the globe.

Every chapter published in this book has been scrutinized by our experts. Their significance has been extensively debated. The topics covered herein carry significant findings which will fuel the growth of the discipline. They may even be implemented as practical applications or may be referred to as a beginning point for another development. Chapters in this book were first published by Hindawi Publishing Corporation; hereby published with permission under the Creative Commons Attribution License or equivalent.

The editorial board has been involved in producing this book since its inception. They have spent rigorous hours researching and exploring the diverse topics which have resulted in the successful publishing of this book. They have passed on their knowledge of decades through this book. To expedite this challenging task, the publisher supported the team at every step. A small team of assistant editors was also appointed to further simplify the editing procedure and attain best results for the readers.

Our editorial team has been hand-picked from every corner of the world. Their multi-ethnicity adds dynamic inputs to the discussions which result in innovative outcomes. These outcomes are then further discussed with the researchers and contributors who give their valuable feedback and opinion regarding the same. The feedback is then collaborated with the researches and they are edited in a comprehensive manner to aid the understanding of the subject.

Apart from the editorial board, the designing team has also invested a significant amount of their time in understanding the subject and creating the most relevant covers. They scrutinized every image to scout for the most suitable representation of the subject and create an appropriate cover for the book.

The publishing team has been involved in this book since its early stages. They were actively engaged in every process, be it collecting the data, connecting with the contributors or procuring relevant information. The team has been an ardent support to the editorial, designing and production team. Their endless efforts to recruit the best for this project, has resulted in the accomplishment of this book. They are a veteran in the field of academics and their pool of knowledge is as vast as their experience in printing. Their expertise and guidance has proved useful at every step. Their uncompromising quality standards have made this book an exceptional effort. Their encouragement from time to time has been an inspiration for everyone.

The publisher and the editorial board hope that this book will prove to be a valuable piece of knowledge for researchers, students, practitioners and scholars across the globe.

List of Contributors

H. Ku
Faculty of Engineering and Surveying, University of Southern Queensland, Toowoomba, QLD 4350, Australia
Centre of Excellence in Engineered Fibre Composites, University of Southern Queensland Toowoomba, QLD 4350, Australia

M. Prajapati and M. Trada
Faculty of Engineering and Surveying, University of Southern Queensland, Toowoomba, QLD 4350, Australia

Kazuyoshi Sakamoto and Yasushi Itoh
Graduate School of Electrical and Information Engineering, Shonan Institute of Technology, 1-1-25 Tsujido-Nishikaigan, Kanagawa, Fujisawa 251-8511, Japan

Arun Kumar
AITTM, Amity University, NOIDA 201303, India

Vipin Kumar and Kh. S. Singh
Department of Physics, Digamber Jain (P.G.) College, Baraut 250611, India

B. Suthar
Department of Physics, Govt. College of Engineering & Technology, Bikaner 334004, India

A. Bhargava
Nanophysics Laboratory, Department of Physics, Govt. Dungar College, Bikaner 334001, India

S. P. Ojha
Director General, IIMT Group of Colleges, Noida 201303, India

Khair Al Shamaileh
Waseela for Integrated Telecommunication Solutions, P.O. Box 962487, Amman 11196, Jordan

Abdullah Qaroot
Electrical Engineering Department, King Faisal University, P.O. Box 400, Al-Ahsa 31982, Saudi Arabia

Nihad Dib
Electrical Engineering Department, Jordan University of Science and Technology, P.O. Box 3030, Irbid 22110, Jordan

Abdelfattah Sheta and Majeed A. Alkanhal
Electrical Engineering Department, King Saud University, P.O. Box 800, Riyadh 11421, Saudi Arabia

Derek Wray and Hosahalli S. Ramaswamy
Department of Food Science and Agricultural Chemistry, McGill University, Macdonald Campus, 21,111 Lakeshore Road, Ste. Anne de Bellevue, Montreal, QC, Canada H9X 3V9

Dalia Elsheakh and Esmat Abdallah
Microstrip Department, Electronics Research Institute, ELtahrir Street, Dokki, Giza 12622, Egypt

Mohammad Asif Zaman and Md. Abdul Matin
Department of Electrical and Electronic Engineering, Bangladesh University of Engineering and Technology, Dhaka 1000, Bangladesh

Awinash Anand, Nischal Koirala, Haruichi Kanaya and Keiji Yoshida
Graduate School of Information Science and Electrical Engineering, Kyushu University, 744 Motooka, Fukuoka 819-0315, Japan

Ramesh K. Pokharel
E-JUST Center Kyushu University, 744 Motooka, Fukuoka 819-0315, Japan

Ehsan Adabi
Mobile and Wireless Group, Broadcom Corporation, Irvine, CA 92617, USA

Ali M. Niknejad
Berkeley Wireless Research Center (BWRC) and EECS Department, University of California, Berkeley, CA 94720-1770, USA

Marius Voicu and Domenico Zito
Tyndall National Institute, Lee Maltings, Dyke Parade, Cork, Ireland
Department of Electrical and Electronic Engineering, University College Cork, Cork, Ireland

Domenico Pepe
Tyndall National Institute, Lee Maltings, Dyke Parade, Cork, Ireland

K. Yousef, A. Allam and M. Ragab
Electronics and Communications Engineering Department, Egypt-Japan University of Science and Technology, New Borg Al-Arab, 21934 Alexandria, Egypt

H. Jia
E-JUST Center, Kyushu University, Nishi-ku, Fukuoka 819-0395, Japan

R. Pokharel, H. Kanaya and K. Yoshida
Graduate School of ISSE, Kyushu University, Nishi-ku, Fukuoka 819-0395, Japan

Sergio E. Melais
Tampa Microwave, Tampa, FL 33626, USA

David Cure and Thomas M. Weller
Department of Electrical Engineering, University of South Florida, Tampa, FL 33620, USA

B. Kapilevich, B. Litvak, M. Anisimov, D. Hardon and Y. Pinhasi
Department Electrical and Electronics Engineering, Ariel University Center of Samaria, Ariel 40700, Israel

E. V. Anichenko, V. S. Arykov and E. V. Erofeev
Microelectronics Department, Research and Production Company "Micran", 47 Vershinina Street, Tomsk 634045, Russia

V. A. Kagadei
RF Electronics Department, Submicron Technologies, 8 Akademicheskiy Avenue, Tomsk 634021, Russia

Thomas Baron, Eric Lebrasseur, Florent Bassignot, Haixia Wang and Sylvain Ballandras
FEM TO-ST UMR 6174 CNRS-UFC-ENSMM-UTBM, ENSMM, 26 Chemin de l Epitaphe, 25030 Besanc on Cedex, France

Ji Fan, Lise Catherinot and Matthieu Chatras
Xlim, UMR 6172 CNRS, Universite de Limoges, 87060 Limoges, France

Philippe Monfraix
Thales Alenia Space, CCEL/LPH/EA, Toulouse, France

Laetitia Estagerie
CNES, 18 Avenu Edouard Belin, 31401 Toulouse Cedex 9, France

Leung Chiu and Quan Xue
Department of Electronic Engineering, City University of Hong Kong, Hong Kong

Andrea Randazzo
Department of Naval, Electrical, Electronic, and Telecommunication Engineering, University of Genoa, Via Opera Pia 11A, 16145 Genova, Italy

Sang-yeop Lee, Hiroyuki Ito, Shuhei Amakawa, Noboru Ishihara and Kazuya Masu
Solutions Research Laboratory, Tokyo Institute of Technology, 4259-S2-14 Nagatsuta, Midori-ku, Yokohama 226-8503, Japan

Diego Palombini, Maciej Jankowski and Ernesto Limiti
Department of Electronic Engineering, University of Roma Tor Vergata, Roma 00133, Italy

O. Abedi and M. C. E. Yagoub
School of Electrical Engineering and Computer Science, University of Ottawa, Ottawa, ON, Canada

L. C. Costa
I3N and Physics Department, University of Aveiro, 3810-193 Aveiro, Portugal

F. Henry
LRP, CNRS, 2 rue Henry Dunand, 94320 Thiais, France

Matteo Pastorino and Andrea Randazzo
Department of Naval, Electrical, Electronic, and Telecommunication Engineering, University of Genoa, Via Opera Pia 11A, 16145 Genova, Italy

S. Vongsack
International College, King Mongkut's Institute of Technology Ladkrabang, Bangkok 10520, Thailand

S. Lamultree
Faculty of Engineering and Technology, Asian University, Chonburi 20260, Thailand

P. Osklang, C. Phongcharoenpanich and S. Kosulvit
Faculty of Engineering, King Mongkut's Institute of Technology Ladkrabang, Bangkok 10520, Thailand

K. Hamamoto
School of Information and Telecommunication Engineering, Tokai University, Kanagawa 259-1292, Japan

T. Wakabayashi
Malaysia-Japan International Institute of Technology (MJIIT), Universiti Technologi Malaysia, Kuala Lumpur 54100, Malaysia

Piero Angeletti
Directorate of Technical & Quality Management, European Space Agency, Keplerlaan 1, 2200 AG Noordwijk, The Netherlands

Marco Lisi
Directorate of Galileo Programme & Navigation Related Activities, European Space Agency, Keplerlaan 1, 2200 AG Noordwijk, The Netherlands

Gian Luigi Gragnani and Maurizio Diaz Mendez
Diten, University of Genoa, Via Opera Pia 11A, 16145 Genoa, Italy

Mohamed Sanad
Amant Antennas, Giza 12451, Egypt

Noha Hassan
Department of Biomedical Engineering and Systems, Cairo University, Giza 12613, Egypt